# Microbial Metabolism in the Digestive Tract

Editor

**Michael J. Hill, Ph.D.**
Director
Bacterial Metabolism Research Laboratory
PHLS Centre for Applied Microbiology and Research
Salisbury, England
and
Consultant to Research Department
St. Mark's Hospital
London, England

CRC Press
Taylor & Francis Group
Boca Raton London New York

CRC Press is an imprint of the
Taylor & Francis Group, an **informa** business

First published 1986 by CRC Press
Taylor & Francis Group
6000 Broken Sound Parkway NW, Suite 300
Boca Raton, FL 33487-2742

Reissued 2018 by CRC Press



**Library of Congress Cataloging-in-Publication Data**
Main entry under title:

Microbial metabolism in the digestive tract.

Includes bibliographies and index.
1. Gastrointestinal system — Microbiology.
2. Microbial metabolism. I. Hill, M. J.
[DNLM: 1. Bacteria — metabolism. 2. Digestive System — microbiology. 3. Digestive System Diseases — microbiology. WI 100 M626]
QR171,I6M55 1986 616.3'3014 85-19563
ISBN 0-8493-5936-8

A Library of Congress record exists under LC control number: 85019563

Publisher's Note
The publisher has gone to great lengths to ensure the quality of this reprint but points out that some imperfections in the original copies may be apparent.

Disclaimer
The publisher has made every effort to trace copyright holders and welcomes correspondence from those they have been unable to contact.

ISBN 13: 978-1-315-89544-4 (hbk)
ISBN 13: 978-1-351-07454-4 (ebk)

Visit the Taylor & Francis Web site at http://www.taylorandfrancis.com and the
CRC Press Web site at http://www.crcpress.com

# PREFACE

In this book an attempt has been made to give an update on the flora of the human digestive tract and its role in disease. This is a subject that has implications in many disciplines; the problem was to decide whether to write for specialists or for the nonspecialist with an interest in the subject. It was decided to opt for the latter course and the book is therefore aimed not only at microbiologists but also at clinicians, dentists, medical researchers, biochemists, and toxicologists who have a background knowledge of bacteriology but are not necessarily directly involved in research into the metabolic actions of gut bacteria. The authors have, however, generously referenced their chapters to indicate where the detailed source material is to be found.

I am very grateful to the contributors not only for producing excellent chapters but also for producing them on time and with a minimum of fuss. The book would have been impossible without the excellent secretarial help of Valerie Parry and Joanna Duthoit and I am lost in admiration for their stoicism during the many changes and rewrites.

Finally, I should acknowledge the invaluable support given by my wife during the preparation of this book.

**Michael Hill**

## EDITOR

**Michael Hill** is the Director of the Bacterial Metabolism Research Laboratory (BMRL) at the Public Health Laboratory Service Centre for Applied Microbiology and Research (PHLS-CAMR), Salisbury, England, and a Consultant to the Research Department, St. Mark's Hospital, London.

He was born in November 1939 in Watford, England and graduated in Chemistry from University College London in 1960. He obtained his doctorate (University of London) in 1963 on the physical chemistry of bacterial surfaces, specifically *Streptococcus pyogenes*, then worked for 2 years at the University of Minnesota Medical School, Minneapolis, in the Pediatrics Department, on *Streptococcus pyogenes*. He returned to London (St. Mary's Hospital Medical School) in 1963 to work first on *Staphylococcus aureus* and then, in 1967, began his work on the role of gut bacteria in colorectal carcinogenesis. In 1973 he moved to the PHLS and became Director of BMRL in 1976. He is a member of a number of societies including the Royal Society of Chemistry, Royal College of Pathologists, Biochemical Society, British Association for Cancer Research, and British Society of Gastroenterology; in addition he is a member of the Scientific Advisory Committee of the European Group for Cancer Prevention (ECP) as well as being chairman of the Diet and Cancer section of ECP. He has published more than 200 papers in the field of bacterial metabolism in relation to disease.

# CONTRIBUTORS

**S. P. Borriello, Ph.D.**
Scientist
Division of Communicable Diseases
Clinical Research Centre
Harrow, England

**B. S. Drasar, Ph.D., D.Sc.**
Reader in Bacteriology
London School of Hygiene and Tropical Medicine
London, England

**Michael J. Hill, Ph.D.**
Director
Bacterial Metabolism Research Laboratory
PHLS Centre for Applied Microbiology and Research
Salisbury, England
and
Consultant to Research Department
St. Mark's Hospital
London, England

**C. W. Keevil**
Principal Grade Microbiologist
Bacterial Metabolism Research Laboratory
PHLS Centre for Applied Microbiology and Research
Salisbury, England

**P. D. Marsh**
Head of Oral Microbiology
Bacterial Metabolism Research Laboratory
PHLS Centre for Applied Microbiology and Research
Salisbury, England
and
Honorary Lecturer in Oral Microbiology
Eastman Dental Hospital
London, England

**Robert W. Owen**
Senior Bile Acid Chemist
Bacterial Metabolism Research Laboratory
PHLS Centre for Applied Microbiology and Research
Salisbury, England

**Andrew G. Renwick, Ph.D.**
Senior Lecturer
Department of Clinical Pharmacology
University of Southampton
Southampton, England

**W. E. W. Roediger, D. Phil., F.R.A.C.S.**
Senior Lecturer, Surgery
University of Adelaide
Woodville, South Australia
Austrialia

**M. H. Thompson**
Deputy Director
Bacterial Metabolism Research Laboratory
PHLS Centre for Applied Microbiology and Research
Salisbury, England

**Angela J. Vince, Ph.D.**
Senior Lecturer
Department of Medicine
University College Hospital Medical School
London, England

TABLE OF CONTENTS

Chapter 1

# MICROBIAL FLORA OF THE GASTROINTESTINAL TRACT

**S. P. Borriello**

## TABLE OF CONTENTS

# I. INTRODUCTION

The microbial flora of the gastrointestinal tract forms an extremely complex ecosystem. There are at least 17 families of bacteria yielding at least 50 different genera. There are then countless species, subspecies, and biotypes, with at least 400 to 500 different bacterial species thought to be present in the fecal flora of a single person.[1-3] Each gram of gut content contains up to $10^{12}$ organisms and these massive numbers of bacteria in feces would account for up to 40% of the total fecal mass. We each of us harbor more bacteria within our gastrointestinal tract than there are (or have been) people in the world. Obviously, the biochemical and disease causing potential of this complex ecosystem is immense. In many ways the gut flora could be viewed as an organ of the body in its own right.

Identifying the components of the gut flora and investigating their biochemical activities has occupied the mind of microbiologists from the very earliest days of this branch of science. What this chapter will try to do will be to give an overview of the components of the gut flora at different sites within the gastrointestinal tract in health and disease. This will be done after outlining the difficulties involved in studies of the gut flora to allow for better interpretation of the results presented.

# II. LIMITATIONS TO THE STUDY OF THE FLORA OF THE GASTROINTESTINAL TRACT

In order to interpret correctly the qualitative and quantitative data generated from studies on the flora of the gastrointestinal tract it is important to be aware of the limitations imposed on such studies. The major problems are those of the extreme complexity of the ecosystem under investigation and the inaccessibility of many parts of that ecosystem in a healthy subject. The difficulties that may influence the results fall into two major groups, sampling transport and storage, and cultivation, enumeration, and identification.

## A. Sampling, Transport, and Storage

One of the major problems encountered is that of obtaining specimens from any site other than the mouth and anus. However, even at these two sites problems may arise. For example, in analysis of fecal material in swine it has been shown that there are differences in the recoverable flora, dependent on where the stool specimen sample is taken from.[4] Also, in the mouth there are a number of sites where it is difficult to perform quantitative bacteriological investigations. For parts of the gastrointestinal tract other than these sites, invasive techniques are employed. This means either intubation or collection of material at operation. In both cases there will be associated factors that could be expected to influence the results. There may be contamination of the specimen due to bacteria from other sites. For example, salivary contamination of juice collected when intubating the jejunum. In most cases the subject was anesthetized and had fasted. These factors will obviously have an effect on the flora, especially the reduction of peristalsis induced by anesthetics which will have an effect on the flora of the small bowel. It must also be remembered that in many cases the subjects studied will have been abnormal and will have received antibiotics and/or bowel washout.

Having collected the specimen the nature of the collection vessel could influence the results. Obviously, the optimal collection vessel would be one in which an anaerobic atmosphere could be maintained. In addition, the length of time in which the specimen remains in the collection vessel during transport will also influence the results. Further details of sampling and transport techniques are given in the review by Borriello et al.[5] If the specimen is not to be analyzed immediately then the length of time that it is stored prior to examination and the method of storage will also greatly affect the results obtained. Even with a recommended storage method[6] it can be seen that there are qualitative and quantitative changes in the flora even after a relatively short storage time.[6] If frozen the rate of freezing and thawing of the specimen will also affect the bacteria present.[7]

## B. Cultivation, Enumeration, and Identification

One of the major problems in defining the flora of the gastrointestinal tract is that of being able to cultivate the microorganisms present. In many cases the cultivation media used may not be able to support the growth of certain bacteria. So although present and surviving the transport and storage systems used they will not form colonies visible to the naked eye. Some organisms may be able to grow under the conditions used, but will do so only very slowly, therefore the length of incubation is also an important variable. In addition to the type of media used the degree of anaerobiosis will also be an important factor influencing recovery. The major problem with respect to cultivation, enumeration, and identification is that of the complexity of the flora. If only nonselective media are employed then the less dominant bacteria will be masked by the dominant organisms that can grow under the in vitro conditions used. The relative numers of bacteria present in the samples from two different subjects may not only give different findings because of differences between the samples but may in themselves result in apparent differences. For example, we could envisage in one population $10^9$ per gram or milliliter of specimen of organism A and only $10^5$ of organism B, where organism A is the major component of the flora. The situation in the other population may be that there are $10^5$ of organism B but only $10^1$ of organism A and organism B is the major component. If no selective methods exist for these organisms then the findings would indicate that one population harbors only organism A and the other only organism B because the dominant organism outnumbers the other by $10^4$:1 and would be overgrown on agar media. In reality both groups carry both types of organism and more significantly both have the same number of organism B. The relative limits of detection of different organism imposed by their relative numbers are rarely stated or discussed. Although selective media or procedures can sometimes be employed to overcome this problem there are two major drawbacks to the use of selective media that must be borne in mind. First, few selective media are totally selective. Some unwanted organisms may be isolated by this method and some desired organisms will not be isolated and will fail to grow. For example, vancomycin, an antibiotic active against Gram-positive organisms, is frequently a component of selective media used for the isolation of *Bacteroides*. However, some species of *Bacteroides* are sensitive to this antibiotic.[8] Second, from a quantitative point of view selective media will give an underestimate of the true number. Despite the drawbacks selective media have been extremely useful in helping to define the flora of the gastrointestinal tract and in some cases have helped to isolate organisms for which the medium was not originally designed (e.g., the isolation of anaerobiospirillum using selective media and procedures designed for the isolation of *Campylobacter*).[9] The optimal methods for cultivating the bacteria present would be a combination of the Hungate roll-tube technique using a selection of selective and nonselective prereduced anaerobically sterilized media with added growth factors. However, many different methods and combination of methods have been used and these are outlined in Table 1 in increasing order of preference. Enumeration of bacteria is difficult to perform accurately. Counts can be obtained from roll-tubes, pour plates, spread plates, or dilution in broth to end-points to give most probable number determinations. Whether triplicate, duplicate, or single counts are performed varies from study to study. For fecal specimens some people record the results as number per gram dry weight and some as number per gram wet weight. For biopsy material when looking at the mucosally associated flora the efficiency of washing the specimen to remove nonadherent bacteria is not always stated. In addition the original size of the specimen becomes important here because in general they are very small, so the limit of detection of organisms expressed as per gram of tissue could vary greatly from study to study and from subject to subject within a study.

**Table 1**
**CONDITIONS FOR THE ISOLATION OF BACTERIA IN INCREASING ORDER OF EXCELLENCE**

| Technique | Media |
|---|---|
| Open bench and anaerobe jar | Routine |
| Anaerobic cabinet | Addition of reducing agents |
| Hungate roll-tube method[a] | Prereduced anaerobically sterilized (PRAS) |
| | PRAS with specialized nutritional requirements[a] |

[a] Ideal combination is Hungate roll-tube method (technique) with PRAS with specialized nutritional requirements (media).

Also there is as yet no satisfactory way of expressing the data. To express it as per gram of tissue introduces variables of tissue thickness etc., however it is not yet possible to express accurately the number of bacteria per unit of surface area which would be much more meaningful. Another problem associated with enumeration, in addition to the accuracy of the techniques, is that of attempting to quantify individual species within a specimen. Unless every isolated colony present is identified these figures will almost invariably be an estimate based on an estimated proportional representation as seen at a specific dilution. In addition to these problems of enumeration and interpretation of data, the concentration of the first dilution and the type of diluent used will have an effect on the results. Finally antagonism in vitro between different bacteria growing in or on the same medium will also influence the results. Having isolated your bacteria you then are left with three additional problems. What is a representative number of colonies to subculture for the purpose of identification? There is no good answer to this question other than as many as you can cope with. Once subcultured you must be able to retain viability especially during storage if identification cannot be done immediately. The last problem is that of identification itself and there are many problems in identifying bacteria from this complex ecosystem to species level, though fewer problems if only identifying to genus level. Obviously, accuracy of identification plays a major part in delineating the components of the flora of the gastrointestinal tract.

## III. MICROBIAL FLORA OF THE MOUTH

The oral cavity consists of several distinct habitats, each of which is associated with a characteristic microbial community. The various major habitats consists of lips, saliva, cheek, palate, tongue, gums, and teeth. Within these areas are additional habitats, such as the areas between adjacent teeth, gingival crevices, crevicular fluid, the buccal (cheek-facing), lingual (inside surface), and occlusal (biting) tooth surfaces, as well as pits and fissures in the teeth. In addition to the variety of habitats present the ecological conditions at these sites are generally unstable, so that the oral ecosystem is subject to frequent variation. Factors that can affect the ecology of the mouth are outlined in Table 2. In addition, many factors that in themselves may vary, help to control the microbial flora (Table 3). Some of these factors stimulate the growth of certain organisms and some of them inhibit the growth of bacteria and/or help to localize them.

The normal oral flora is extremely complex and consist of a wide range of species of yeasts, bacteria, protozoa, and viruses. The types of microorganisms most commonly encountered are listed in Table 4. In general it is the *Streptococci* that comprise a large

**Table 2**
**FACTORS AFFECTING THE ECOLOGY OF THE MOUTH**

| | |
|---|---|
| Dentition | Scaling |
| Change from primary to secondary dentition | Polishing |
| Tooth extraction | Oral hygiene |
| Disease | Type of food |
| Dentures | Saliva flow |
| Fillings | Antibiotic therapy |

**Table 3**
**FACTORS AFFECTING THE MICROBIAL FLORA OF THE MOUTH**

| | |
|---|---|
| Factors affecting oral ecology (see Table 2) | B and T lymphocytes |
| Lysozyme | Blast cells |
| Lactoferrin | Hemin (stimulates growth of *B. melaninogenicus*) |
| Lactoperoxidase system | |
| SIgA, IgG, IgM | α-2-globulin (stimulates growth of *T. denticola*) |
| Saliva flow rate | Wide range of nutrients available |
| pH | Anaerobiosis |
| Complement | Temperature |
| Neutrophils | Adherence |
| Leucocytes | Mucins |
| Monocytes | Microbial interactions |

proportion of the normal oral flora and they can be isolated from all sites in the mouth. It is interesting that *Staphylococci* and *Micrococci* are not commonly isolated in significant numbers, despite being common colonizers of the skin and mucous membranes of the nose. The only Gram-positive catalase-positive coccus that is considered as a normal component of the oral flora is *Micrococcus mucilagenosus (Staphylococcus salivarius)*. In contrast a variety of Gram-positive rods and filaments are commonly found. They are also the dominant component of the flora of dental plaque. It is of interest that *Bacterionema matruchotti*, which could be considered as the only true coryneform in the mouth, is not found elsewhere in the gastrointestinal tract, and that *Clostridia* which seem capable of colonizing most known ecosystems are rarely found in the mouth. A number of other bacteria not outlined in Table 4 can occasionally be found in the mouth and include *Bacilli* and *Mycobacteria*.

The majority of anaerobic Gram-negative rods in the mouth belong to the genus *Bacteroides*. The Gram-negative flora of the mouth is similar to that of the large bowel flora in this respect, however unlike the rest of the gastrointestinal tract the Gram-negative rods commonly associated with the gut, such as coliforms, are rarely found in the mouth.

As isolation methods and taxonomic tools improve, new bacterial genera are being described and isolated from this complex ecosystem and include the genera *Walinella*[10] and *Capnocytophaga*.[11]

## IV. MICROBIAL FLORA OF THE STOMACH

In general in the healthy individual the stomach can be considered sterile; though in disease states or after administration of antacids, a fairly rich flora may develop due to associated hypochlorhydria (Table 5). Although a number of factors may contribute to the overall control of the gastric flora, the most important factor is that of an acid pH. The stomach

## Table 4
## ORGANISMS COMMONLY ISOLATED FROM THE ORAL CAVITY

| Rods | Cocci |
|---|---|
| Gram-positive | |
| *Actinomyces* | *Streptococci* |
| *Arachnia* | Anaerobic streptococci |
| *Bifidobacteria* | *Micrococci* (only *M. mucilogenosus)* |
| *Lactobacilli* | *Enterococci* |
| *Bacterionema* | |
| *Eubacteria* | |
| *Propionibacteria* | |
| *Rothia* | |
| Gram-negative | |
| *Haemophilus* | *Neisseria* |
| Eikenella | *Branhamella* |
| *Bacteroides* | *Veillonella* |
| Walinella | |
| Capnocytophaga | |
| *Selenomonas* | |
| *Vibrios* | |
| *Fusobacteria* | |
| Treponemes | |
| *Simonsiella* | |
| *Leptotrichia* | |
| Others | |
| | *Mycoplasmas* |

## Table 5
## THE MICROBIAL FLORA OF THE STOMACH

| Disease | Health |
|---|---|
| All those found in health | *Streptococci* (aerobic) |
| *Bacteroides* | *Lactobacilli* |
| *Streptococci (anaerobic)* | Fungi |
| Diptheroids | *Micrococci* |
| *Klebsiella* | *Staphylococci* |
| *Campylobacter* | *Campylobacter* (?) |
| *Pseudomonas* | |
| *Neisseria* | |
| *Haemophilus* | |
| *Veillonella* | |
| *Bifidobacteria* | |
| *Eubacteria* | |
| *Proprionibacteria* | |
| Occasional *Fusobacteria* | |
| *Citrobacter* | |
| *Bacilli* | |
| *Clostridia.* | |

receives a fairly heavy microbial load from oral bacteria washed with saliva into the stomach and during feeding. These are rapidly killed by the gastric secretions. Obviously for a short period just post prandial a more complex flora will be present as the gastric pH increases due to the presence of ingested food.[12] In general the concentration of bacteria in the stomach is less than $10^3$/mℓ of content and consist primarily of *Streptococci, Lactobacilli*, and fungi.

**Table 6**
**FACTORS ASSOCIATED WITH COLONIZATION OF THE STOMACH**

| | |
|---|---|
| Pernicious anemia | Age |
| Hypogammaglobulinemia | Bleeding duodenal ulcer |
| Carcinoma | Obstructing duodenal ulcer |
| Gastrectomy | Gastric ulcer |
| Antacids | Hypochlorhydria |
| Redox potential | Microbial interactions |

Although present, one must wonder at what times the conditions in the stomach are favorable to the expression of the metabolic capability of these bacteria. In the colonized stomach the environment is much less acidic and the bacteria present are undoubtedly metabolically active. A number of factors can be associated with abnormal colonization of the stomach (Table 6). In general colonization is due to an associated increase in the pH of the stomach, with active colonization occurring at pH values of equal to or greater than four.

There has been a great deal of interest in abnormal gastric colonization due to the possible association with induction of gastric cancer. In its simplest form the theory is that the high numbers of bacteria that colonize the achlorhydric stomach reduce nitrate present to nitrite, which is then available for nitrosamine formation with dietary derived amines. The consequences of this microbial metabolism are dealt with in much more detail in Chapter 16.

Although the association with gastric cancer is still theoretical there is good evidence for an association between gastric colonization and postoperative sepsis.[13,14] In those individuals where gastric aspirates yielded $5 \times 10^6$ or more bacteria per milliliter, 93% of them developed wound sepsis, compared with 16% of individuals where the counts were less than this.[13] There are two interesting groups of patients, both of which have a gastric cancer risk greater than that of the normal population, namely pernicious anemia and hypogammaglobulinemia. Both may have achlorhydria and both become colonized in the achlorhydric state. The major difference in these two patient groups, with respect to factors controlling the stomach flora, is the presence or absence of an effective immune system, in particular SIgA. A comparison of the bacterial colonization in these two patient groups has shown that the degree and type of bacterial colonization was the same suggesting that antibodies do not play a significant role in determining the flora in the achlorhydric stomach from either a qualitative or quantitative point of view.[15]

Although gastric anacidity is rare in young people in Northern Europe and North America it is fairly common in other parts of the world, such as in South America.[16] It is possible that gastric colonization is the normal state of affairs when viewed on a worldwide basis.

Compared to the work done on gastric juice, there has been very little work done on the flora associated with the gastric mucosa. Recent work in this laboratory[17] has looked at the flora associated with gastric mucosa from the fundus, body, and antrum in patients with pernicious anemia, hypogammaglobulinemia, hypochlorhydria of unknown cause, and subjects with normal gastric pH. In general the flora was similar to that of the juice with *Streptococci, Staphylococci, Micrococci, Veillonella*, and *Bifidobacteria* predominating. Unlike the juice, *Klebsiella, Pseudomonas*, and *Proprionibacteria* were not found. There was little difference in mucosal flora between pernicious anemia and hypogammaglobulinemia patients, both of which had a richer flora than the hypochlorhydria subjects. Those subjects with a normal pH had a normal simple gastric flora of *Staphylococci, Streptococci, Micrococci*, and *Lactobacilli*. Bacteria were rarely isolated from the mucosa of these patients, but when present consisted mostly of *Streptococci* and *Staphylococci*. Interestingly, it was the mucosa from the body that was most commonly colonized in this group.

In the light of recent findings we may have to re-evaluate our ideas of the composition of the flora of the gastric mucosa in both health and disease with the report of the frequent

**Table 7**
**BACTERIA ISOLATED FROM THE BILIARY TRACT IN DISEASE**

| Anaerobes | Aerobes |
|---|---|
| *Veillonella* | *Escherichia* |
| *Peptostreptococci* | *Klebsiella* |
| *Peptococci* | *Proteus* |
| *Streptococci* | *Citrobacter* |
| *Bacteroides* | *Staphylococci* |
| *Fusobacteria* | Enterococci |
| *Propionibacteria* | *Aeromonas* |
| *Lactobacilli* | *Enterobacter* |
| *Bifidobacteria* | *Pseudomonas* |
| *Eubacteria* | *Streptococci* |
| *Actinomyces* | Candida |
| *Clostridia* | |

**Table 8**
**FACTORS CONTROLLING THE MICROBIAL FLORA OF THE SMALL BOWEL**

| | |
|---|---|
| Bile (?) | Seeding from the stomach |
| Peristalsis | Hypochlorhydria |
| Mucosal adhesion | Redox potential |
| SIgA (?) | Microbial interactions |

isolation of *Campylobacter pyrolidus* from gastric mucosa.[18,19] It is evident that there is still much to learn about the flora of the stomach.

## V. MICROBIAL FLORA OF THE BILIARY TRACT

A variety of bacteria, including both aerobes and anaerobes, can be isolated from the biliary tract (Table 7). Aerobic organisms are the most frequently encountered but anaerobes can be isolated from up to 50% of the culture positive samples. Although it has been known for some time that anaerobes can be found[20] it is only more recently that the variety of anaerobes that may be present has become evident (Table 7). In general terms up until the 1970s our understanding of the flora at this site was restricted to a variety of aerobes, most commonly *E. coli* (Table 7) and *Clostridium perfringens*. All workers up to that time agreed that *C. perfringens* was found in a surprisingly high number of cases and was the most common anaerobe isolated. However, with the advent of improved anaerobic techniques it has become obvious that many different anaerobes can be found.[21,22] Both of these recent studies again highlighted the frequency of isolation of *C. perfringens* but in addition it was shown that *Bacteroides fragilis* can also be frequently found.

It is likely that the organisms found at this site are metabolically active and that they may be involved in the pathogenesis of biliary tract inflammation due to their ability to deconjugate bile acids releasing the more toxic nonconjugated form.

## VI. MICROBIAL FLORA OF THE PROXIMAL SMALL BOWEL

The proximal small bowel consists of the duodenum, jejunum, and proximal ileum. In general this part of the gut, especially the duodenum and jejunum, is sparsely colonized. Any bacteria at this site are frequently exposed to high concentrations of bile acids and proteolytic enzymes as well as the normal range of antimicrobial substances present in natural secretions. An unknown antibacterial activity has been detected in deuodenal juice.[23] In addition there is rapid peristalsis continually washing the bacteria through towards the large bowel (Table 8). Peristalsis is probably the single most important factor that controls the microbial population at this site. Any procedure that inhibits peristalsis can lead to bacterial overgrowth in the small bowel. Some of the other factors that have been proposed as having a role in the control of the small bowel flora have been questioned. Although it is well known that unconjugated bile acids inhibit the growth of many bacteria in vitro investigators

**Table 9**
**MICROBIAL FLORA OF THE PROXIMAL SMALL BOWEL IN CAUCASIANS**

| Health | Disease |
|---|---|
| Anaerobic *Streptococci* | All those noted in health |
| Aerobic *Streptococci* | A fecal type flora consisting of as many as 20 different species existing in concentrations as high as $10^{11}$ colony forming units per milliliter of luminal content. |
| *Lactobacilli* | |
| *Staphylococci* | |
| *Micrococci* | |
| Also, less commonly found | |
| *Enterobacteria* | |
| Fungi | |
| *Bacteroides* | |
| *Bifidobacteria* | |
| Veillonella | |

have been unable to confirm any significant antibacterial activity in vivo.[24-26] In addition, the presence of unconjugated bile acids in the small intestine in patients with blind loops and bacterial overgrowth does not appear to inhibit enteric bacterial growth.[27,28] The role of antibodies, especially those of the SIgA class in controlling the flora of the small bowel has also been questioned. In general there is little evidence to show that immunodeficient patients have intestinal bacterial overgrowth.[29-32] In one of the better studies fluids were taken under nitrogen from several levels of the small bowel and great care was taken to protect these samples from exposure to air.[32] In this study the selectively IgA deficient subjects, all but one of which had undetectable levels of IgA in secretions, had microbial populations in their small bowel generally similar to those of control subjects. In contrast, greater numbers of anaerobes were cultured from the small bowel fluids of three patients with hypogammaglobulinemia. At least two of these patients were achlorhydric. These findings would therefore be in keeping with the generalized finding that a significant increase in the microbial population of the small bowel is present in most subjects with hypochlorhydria.[12,33,34] This is most likely due to the increased load of viable bacteria surviving passage through the stomach.

The bacterial flora of the proximal small bowel is predominantly Gram-positive in nature generally consisting of *Streptococci* and *Lactobacilli*, although *Enterobacteria* and *Bacteroides* can occur fairly frequently though in much lower numbers (Table 9). As in the stomach an increased number of bacteria can be isolated from jejunal samples for a period after a meal.[12] There is also evidence from studies in one subject that the flora of the proximal small bowel is stable, at least over a 2 week period.[35]

As for most gut flora studies these generalizations are based on observations of Western subjects. However, studies in South America and India indicate that normal people in these areas have a much more profuse proximal small bowel flora than that found in Western people.[36-40] In general the flora is predominantly composed of Gram-positive cocci such as *Peptostreptococci, Peptococci,* and *Streptococci*, and rarely consists of less than $10^4$ bacteria per milliliter of luminal content. The degree of colonization in western subjects in disease states (Table 10) can vary from moderate colonization to profuse fecal type flora ranging in counts from $10^5$ to $10^{10}$ organisms per milliliter of content.[41] Excessive proximal small bowel colonization can also be found in Western populations in those individuals who acquire tropical sprue.[42] The situation in countries where proximal small bowel colonization is more prevalent is much more difficult to interpret. Gorbach and co-workers[36,43] found that the flora of the small bowel in residents of Calcutta, India, who had no evidence of malnutrition, was similar to that of Westerners. However they did find colonization in malnourished

**Table 10**
**FACTORS ASSOCIATED WITH SMALL BOWEL OVERGROWTH**

| | |
|---|---|
| Hypochlorhydria | Disorders of intestinal motility, e.g., |
| Gastrectomy | idiopathic intestinal pseudo-ob- |
| Diverticula | struction, scleroderma, diabetic |
| Blind loops | autonomic neuropathy |
| Strictures | Cirrhosis |
| Radiation therapy | Malnutrition |
| Crohns disease | Extensive small bowel resection |
| Jejunocolic fistula | Age |
| Jejunoileal by-pass | Tropical sprue |

subjects, including low numbers of coliforms, whereas the colonization noted in individuals with tropical sprue consisted of high numbers of coliforms. In contrast to these findings Bhat et al.[38] could find no difference in the luminal or mucosal flora of control subjects compared to tropical sprue patients in India, a finding also noted by Jarumilinta and co-workers[40] for South American subjects.

There has been relatively little work published on the nature of the bacterial flora associated with the mucosa of the proximal small bowel in man, since the early observations of Gram-positive cocci and rods present in the jejunal mucosa.[44] Nelson and Mata[45] showed high numbers of bacteria, predominantly anaerobic *Streptococci*, present in jejunal biopsies from four of seven healthy subjects. The presence of *Streptococci* associated with the jejunal mucosa has also been shown in healthy Caucasians.[42] In keeping with the high degree of luminal colonization of the proximal small bowel noted by Jarumilinta and co-workers[40] and Bhat and co-workers[37,38] in South Americans and Indians, they have also noted a high degree of colonization of the jejunal mucosa. Jarumilinta et al.[40] found all 18 healthy control and 20 acute diarrhea jejunal biopsy samples studied to be colonized. In both groups the most common organisms found were Gram-positive cocci, *Peptostreptococci*, aerobic, and anaerobic *Streptococci*, *Peptococci*, and *Micrococci*. Interestingly, the number of bacteria associated with the jejunal mucosa in subjects with acute diarrhea was significantly higher than in control subjects. This pattern of Gram-positive cocci colonizing the mucosa was also found in three healthy children studied.[40] Bhat et al.[38] could find no significant difference between the luminal or mucosal flora of control subjects and tropical sprue patients. In contrast to the work of Jarumilinta et al.,[40] this group frequently found Veillonellae.

It is evident that there is colonization of the lumen of the proximal small bowel in the majority of people and that the flora at this site consists predominantly of Gram-positive cocci. It may also be the case that there is a relatively rich mucosal flora in the majority of healthy people. Many factors can predispose to small bowel overgrowth and some of the consequences of this as well as methods for detecting overgrowth at this site are dealt with in Chapter 12.

## VII. MICROBIAL FLORA OF THE DISTAL SMALL BOWEL

In both the distal and terminal ileum, which together constitute the lower small intestine, there is a much more profuse bacterial flora on both a qualitative and quantitative basis. In essence the flora of the lower ileum is qualitatively similar to that of feces. In contrast to the flora of the proximal small bowel and stomach, which is predominantly Gram-positive in nature and consists mostly of cocci, there is a reversal of this trend in the distal small bowel where Gram-negative bacteria begin to outnumber the Gram-positive ones (Table 11). In addition anaerobes are found in substantial concentrations (Table 11) which again contrasts with the findings in the upper small bowel and stomach. In keeping with the finding of

**Table 11**
**MICROBIAL FLORA OF THE DISTAL SMALL BOWEL IN CAUCASIANS**

| | |
|---|---|
| *Enterobacteria* | *Bacteroides* |
| *Proteus* | *Fusobacteria* |
| *Staphylococci* | *Bifidobacteria* |
| *Streptococci* | *Lactobacilli* |
| *Peptococci* | *Clostridia* |
| *Peptostreptococci* | *Eubacteria* |
| *Acidaminococci* | Fungi |
| Veillonella | |

**Table 12**
**FACTORS AFFECTING THE MICROBIAL FLORA OF THE LARGE BOWEL**

| | |
|---|---|
| Nutrient availability | Diarrhea |
| Diet | Bacterial antagonism |
| pH | Bacterial co-operation |
| Redox potential | Mucin |

anaerobes at this site is a redox potential of $-150$ mV which is not that dissimilar to the redox potential of $-200$ mV which is found in the cecum. All of the bacteria in Table 11 can be found in quite substantial concentrations.

The microbial flora of the lower ileum has also been studied in immuno-deficient patients.[46] Analysis of the microflora of the distal ileum of selectively IgA deficient patients failed to show any significant differences in the flora at that site compared to healthy age-matched controls. Interestingly, there were tendencies towards larger populations of *Bacteroides* and coliforms in control subjects than in the patient group. When looking at hypogammaglobulinemia patients analysis of the data implies that there may be higher concentrations of anaerobic *Streptococci, Lactobacilli*, and *Enterococci* in the patient group. However, only three hypogammaglobulinemia patients were studied.

## VIII. MICROBIAL FLORA OF THE LARGE BOWEL

The large bowel is the most heavily colonized part of the gastrointestinal tract and most of the different bacterial families and genera in the gut can be found at this site. There are at least 400 to 500 different bacterial species thought to be present in the fecal flora of a single person[1,3] and up to $10^{12}$ bacteria per gram of feces.[1] Most of the work on the bacterial flora of the large bowel has concentrated on the analysis of fecal specimens. There is much less information available on the flora of the cecum or that associated with the mucosa.

The factors that have been proposed as being involved in the control of the microbial flora of the large bowel are listed in Table 12. Of these factors microbial interference, both co-operative and antagonistic, probably plays a major role. This is best exemplified by the frequent occurrence of diarrhea after antibiotic treatment. In many of these cases the antibiotics disrupt the normal large bowel flora sufficiently to allow pathogens such as *C. difficile*[47] and enterotoxigenic *C. perfringens*[48] to establish. It is a well known phenomenon that a much lower infectivite dose of gut pathogens is required to induce disease in germfree or antibiotic pretreated animals.

The role of the other factors proposed as having a role in the control of the large bowel flora is much more difficult to discern. In most cases this is because attempts to monitor the effects of changes in these parameters have depended on microbial analysis of the fecal flora. It is obvious that this need not necessarily reflect changes occurring elsewhere in the large bowel. Of the factors that may be associated with determining the large bowel flora, diet has probably received the most attention. The reasons for this are twofold. First, it is the factor that is the most easily manipulated and second, there is a great deal of interest in the possible interrelationships between diet, gastrointestinal flora, and disease: especially large bowel cancer. In general it has proven difficult to show any significant changes in the composition of the fecal flora after feeding volunteers different diets. However, some changes have been noted (Table 13). Although dietary manipulation studies have rarely yielded

**Table 13**
**EFFECTS OF DIET ON THE FECAL FLORA**

| Diet | Effect |
|---|---|
| High bran, bagasse, or pectin supplement | No change |
| High carbohydrate | Increase in *Bifidobacteria* |
| High fat | Increase in *Bacteroides* |
| High protein | No change |
| Elemental | Decrease in *Enterococci* and *Lactobacilli* |
| Vegetarian | No change other than a possible increase in *Sarcina ventricula* |

**Table 14**
**EFFECT OF DIET ON MICROBIAL ACTIVITY**

| Diet | Effect |
|---|---|
| Elemental (man) | Decreases in metabolism of cholesterol and bile acid and production of phenols |
| High beef (rat) | Increases in β-glucuronidase, nitroreductase and azoreductase<br>Decrease in β-glucosidase |

striking changes in the fecal flora, more impressive results have been obtained when looking at the effects of changes in diet on the metabolic activities of the flora (Table 14).

Although bacteria can induce diarrhea in man, it is also probable that diarrhea itself affects the gut flora. The effect of cathartic-induced diarrhea on the fecal flora has been evaluated.[49-51] Induction of diarrhea by a peroral saline perfusion of the terminal ileum resulted in an increase in fecal coliforms,[50] whereas the use of lactulose[51] or cascara and milk of magnesia[49] resulted in an overall decrease in coliforms and *Enterococci*. In pathogen-induced diarrhea there are also major changes in the fecal flora, in general a marked reduction in the anaerobic component of the flora. This is exemplified by the finding of a $10^5$-fold reduction in *Bacteroides* in the feces of patients with cholera.[52]

## A. Microbial Flora of the Cecum and Appendix

Due to the problems involved in investigating these sites there have been relatively few studies on the microbial flora of the intestine other than fecal flora studies. In general, it would appear that the flora of the large intestine is qualitatively similar to that of feces (Table 15). However, evidence about quantitative aspects vary. Moore and colleagues[53] indicated that the numbers of bacteria in the colon was higher than in the feces, but at a later date Moore and Holdeman[2] stated that the fecal flora is representative of the flora of the colon. Analysis of the data of Gorbach and colleagues[35] in which samples were obtained by intubation of healthy volunteers shows the numbers to be lower in the cecum than in feces. In animal studies the cecal and fecal flora has been shown to be the same for the rat and mouse, but for the rabbit and guinea pig a number of differences are evident.[54] It would appear that the flora of the appendix may be different to that of other areas of the colon (Table 16). However, these possible differences must be treated with caution as the data for the flora of the appendix were derived from material from patients at operation and may not therefore be indicative of the flora at this site in health.

**Table 15**
**MICROBIAL FLORA OF THE CECUM[35]**

| Organism | $\log_{10}$ No./mℓ (range) |
|---|---|
| *Streptococci* | 2.2 (0 —3.6) |
| *Staphylococci* | 4.3 (2.6—6.9) |
| Coliforms | 6.2 (5.6—7.4) |
| Aerobic *Lactobacilli* | 1.8 (1.0—2.8) |
| Total aerobes | 6.8 (6.1—7.6) |
| Anaerobic *Lactobacilli* | 5.2 (2.3—7.6) |
| *Bacteroides* | 7.9 (6.4—9.1) |
| Total anaerobes | 8.0 (6.4—9.1) |
| Fungi | present |

**Table 16**
**MICROBIAL FLORA OF THE APPENDIX[67]**

| Organism | $\log_{10}$ No./mℓ (range) |
|---|---|
| Enterobacteriaceae | 6.9 (6.7—9.0) |
| *Bacteroides* | 7.1 (4.0—9.3) |
| *Streptococci* | 7.0 (5.0—9.7) |
| *Lactobacilli* | 6.4 (4.0—9.0) |
| Gram-positive nonsporing anaerobes | 5.6 (4.0—8.6) |

## B. Fecal Flora

By far the greatest amount of work on the flora of the gastrointestinal tract has been done on feces. As stated above, the flora at this site is extremely complex on both a qualitative and quantitative basis. All of the bacteria indicated in Table 17 have been isolated from this site. In general terms the fecal flora within an individual is stable[55-58] though there is a great deal of person to person variation. Although the flora within an individual is stable to genus level it is evident from the work of Borriello[58] that the apparent stability of the fecal flora may mask major changes occurring at the species and subspecies level. In this study all the major genera of bacteria, including *Clostridia*, were shown to be stable in terms of total numbers per gram of stool; however one of the components of the clostridial flora, *C. perfringens*, showed tremendous week to week variation in numbers, and in addition the major serotype of *C. perfringens* present also changed.

At one time only aerobes were known to exist and it was thought that the fecal flora was composed predominantly of coliforms. It became increasingly obvious as techniques improved that anaerobes were also present and that the fecal flora consisted mainly of nonsporing anaerobes. In one of the most thorough normal human fecal flora studies performed nonsporing Gram-negative and Gram-positive organisms such as species of *Bacteroides, Fusobacterium, Peptostreptococcus,* and *Eubacterium* were shown to be most dominant.[1] In total these workers detected 113 different kinds of organisms among the 1147 isolates examined obtained from the feces of 20 people. This type of work resulted in the description of 12 new species of bacteria, and one new genus, *Coprococcus*.[59] Since this work methanogens have been shown to be the most predominant organism present in some people, existing in concentrations of up to $10^{10}$/g dry wt. of stool.[60]

A number of interesting organisms are now being recovered from the feces of patients with disease. With improved microbiological techniques and awareness *Spirochaetes* are now being isolated from patients with rectal proctitis[61,62] and *Anaerobiospirillum* species from subjects with diarrhea.[9] It is still unknown to what degree these types of organisms may be isolaed from the feces of healthy people.

**Table 17**
**BACTERIAL FLORA OF FECES IN HEALTH**

| Family | Genus |
|---|---|
| Actinomycetaceae | *Bifidobacterium* |
| Bacillaceae | *Bacillus, Clostridium* |
| Bacteroidaceae | *Bacteroides, Fusobacterium, Butyrivibrio, Desulphomonas, Anaerobiospirillum* (?) |
| Budding bacteria | Gemmiger |
| Corynebacteriaceae | *Corynebacterium, Kurthia* |
| Enterobacteriaceae | *Escherichia, Klebsiella, Proteus, Enterobacter, Edwardsiella, Citrobacter, Morganella, Providencia, Hafnia, Moellerella, Salmonella* (asymptomatic carriage) |
| Lactobacillaceae | *Lactobacillus* |
| Methanobacteriaceae | *Methanobrevibacter*, others |
| Micrococcaceae | *Staphylococcus, Micrococcus* |
| Neisseriaceae | *Neisseria* |
| Peptococcaceae | *Peptococcus, Peptostreptococcus, Ruminococcus, Sarcina,* coprococcus |
| Propionibacteriaceae | *Propionibacterium, Eubacterium* |
| Pseudomonadaceae | *Pseudomonas* |
| Spirochaetaceae (Treponemataceae) | Treponema, Brachyspira (?) |
| Streptococcaceae | *Streptococcus* |
| Veillonellaceae | *Veillonella, Acidaminococcus, Megasphaera* |
| Vibrionaceae | *Vibrio* |

## C. Mucosal Flora

Most of the work on the composition of the microbial flora associated with the mucosa of the gastrointestinal tract has been with laboratory animals. Although there is little published work on this aspect of microbial ecology in the large bowel of man, the work done has looked at both normal and diseased sections of a number of different sites within the large bowel.[45,63-66] In general a specialized bacterial flora associated with the intestinal epithelium similar to that found in animals does not exist. The mucosal flora at this site in man is fairly complex and most of the representatives of the fecal/luminal flora can be found (Table 18). It is interesting to note that in two studies[64,66] no major differences in the flora could be detected between diseased (cancer, polyps, inflammatory bowel disease) or adjacent normal mucosa, although a slightly more profuse flora appeared to be associated with the mucosal lining of polyps.[66] When comparing the mucosal flora of the ascending, transverse, descending, and sigmoid colon the only apparent difference was the finding of a less complex flora in the sigmoid colon,[66] which was attributed to the preparations for colonoscopy. In a similar type of study in which the cecal and rectal mucosa were also studied there were no discernible differences in the numbers of *Escherichia coli* found in the 15 patients who were found to be grossly normal during the examination.[65] In general there is a lack of agreement between various groups with respect to the overall gross composition of the mucosal flora as determined by the anaerobe to aerobe ratio. Peach and colleagues[64] found this ratio to be about 1:1 and not exceeding 1.6:1. Nelson and Mata[45] found it to be about 10:1. In the study by Croucher and colleagues[63] the ratio was about $10^4$:1 in two traffic accident sudden death victims but only 1.2:1 and 10:1 in the two acute heart failure victims

### Table 18
### BACTERIAL FLORA ASSOCIATED WITH THE MUCOSA OF THE LARGE BOWEL

| | |
|---|---|
| *Clostridia* | *Streptococci* |
| *Eubacteria* | *Bifidobacteria* |
| *Peptococci* | Gemmeger |
| *Peptostreptococci* | *Enterococci* |
| *Lactobacilli* | *Enterobacteria* |
| *Bacteroides* | *Staphylococci* |
| *Fusobacteria* | *Micrococci* |
| *Megasphaera* | *Bacilli* |
| *Acidaminococci* | Spiral organisms |
| Veillonella | Yeasts |

### Table 19
### *CLOSTRIDIA* ASSOCIATED WITH THE MUCOSA AT VARIOUS SITES IN THE LARGE BOWEL[58]

| Subject | Site | Clostridial species ($\log_{10}$ No./gm of tissue) | Disease state |
|---|---|---|---|
| 1 | Proximal colon | *C. tertium* (1.6) | Ulcerative colitis |
| | Transverse colon | *C. perfringens* serotype 17, 50 (2.9) | |
| | | *C. paraputrificum* (3.9) | |
| | Distal colon | *C. paraputrificum* (4.0) | |
| | | *C. tertium* (3.4) | |
| | | *C. perfringens serotype* 17, 50 (3.2) | |
| 2 | Proximal colon | *Clostridium sp.* (2.3) | Ulcerative colitis |
| | Transverse colon | Non detected | |
| | Distal colon | *C. perfringens*, nontypable (1.7) | |
| | | *C. perfringens* serotype 30 (1.4) | |
| 3 | Proximal colon | *C. butyricum* (2.1) | Ulcerative colitis |
| | | *C. cochlearium* (2.1) | |
| | Distal colon | *C. carnis* (2.1) | |

studied. There is more agreement about the total numbers of bacteria per gram of large bowel tissue, which achieve a maximum of about $10^8$/g, and about the most frequently found bacteria which are *Bacteroides*. Although most people have found that the mucosal flora at different regions within the large bowel is qualitatively similar, differences do occur. In the study by Edmiston and colleagues[66] *Lactobacilli* were found only in the ascending and sigmoid colon. *Megasphaera* only in the sigmoid colon, *Acidaminococci* and *Veillonella* although present at other sites were absent from the sigmoid and descending colon respectively. Mata and Nelson[45] found *Staphylococci* in the colon area but not in the appendix. Differences in the species of clostridia found at different sites in the same subject were found by Borriello[58] (Table 19). However, due to the small size of samples analyzed in all of these studies the qualitative differences could be a reflection of sampling error.

One of the most important observations from these studies is that there is general agreement that there is no significant difference in the mucosal microbial flora between diseased and adjacent uninvolved tissue.[64,66]

**Table 20**
**PROTOZOA OF THE GASTROINTESTINAL TRACT**[69]

| Nonpathogenic | Rarely pathogenic (symptom) |
|---|---|
| *Chilomastix mesnili* | *Balantidium coli* (diarrhea) |
| *Endolimax nana* | *Dientamoeba fragilis* (? diarrhea) |
| *Entamoeba coli* | *Isopora species* (diarrhea) |
| *E. hartmanni* | |
| *E. polecki* | |
| *Enteromonas hominis* | |
| *Iodamoeba butschlii* | |
| *Trichomonas hominis* | |

### D. Protozoa in the Large Bowel

In general protozoa are not considered to be part of the normal gut flora in health and by most bacteriologists are only expected to be present during gastrointestinal disease. It is only recently that investigators have agreed that the potential pathogen *Entamoeba histolytica*, is usually a commensal of the gut.[68] This is mainly due to the fact that although parasitic infections of the gut are widespread they are mainly restricted to the developing countries. It is important to note that many people with parasitic infections are asymptomatic. The infecting protozoa can live and replicate in the bowel and in this respect behave like the bacterial flora of the gut. A number of the commensal protozoa of the gut are listed in Table 20. Although most of these are nonpathogenic some can cause mild diarrhea.

## IX. CONCLUSIONS

It is evident that the microbial flora of the gastrointestinal tract is extremely complex. Despite its complexity and the problems associated with investigating the flora at various sites within the gastrointestinal tract, a substantial amount of information has been generated. It is obvious from the literature that this learning process is continuing. It is hoped that this overview will serve as a reminder of the difficulties that must be faced in trying to extrapolate from biochemical reactions performed in vitro by single species of bacteria from the gut to the situation as it would occur in vivo. Conversely, there are obvious problems in trying to identify the components of the gut flora involved in a metabolic process noted from in vivo studies of volunteers.

## REFERENCES

1. **Moore, W. E. C. and Holdeman, L. V.,** Human fecal flora: the normal flora of 20 Japanese-Hawaiians, *Appl. Microbiol.*, 27, 961, 1974.
2. **Moore, W. E. C. and Holdeman, L. V.,** Discussion of current bacteriological investigations of the relationships between intestinal flora, diet and colon cancer, *Cancer Res.*, 35, 3418, 1975.
3. **Moore, W. E. C. and Holdeman, L. V.,** Some newer concepts of the human intestinal flora, *Am. J. Med. Technol.*, 41, 427, 1975.
4. **Rall, G. D., Wood, A. J., Wescott, R. B., and Dommert, A. R.,** Distribution of bacteria in feces of swine, *Appl. Microbiol.*, 20, 789, 1970.
5. **Borriello, S. P., Hudson, M., and Hill, M.,** Investigation of the gastrointestinal bacteria flora, in *Clinics in Gastroenterology*, Vol. 7, Russell, R. I., Ed., W. B. Saunders, Philadelphia, 1978, 329.
6. **Crowther, J. S.,** Transport and storage of faeces for bacteriological examination, *J. Appl. Bacteriol.*, 34, 477, 1971.

7. **Calcott, P. H., Lee, S. K., and MacLeod, R. A.,** The effect of cooling and warming rates on the survival of a variety of bacteria, *Can. J. Microbiol.*, 22, 106, 1976.
8. **van Winkelhoff, A. J. and De Graff, J.,** Vancomycin as a selective agent for isolation of *Bacteroides* species, *J. Clin. Microbiol.*, 18, 1282, 1983.
9. **Malnick, H., Thomas, M. E. M., Lotay, H., and Robbins, M.,** Anaerobiospirillum species isolated from humans with diarrhoea, *J. Clin. Pathol.*, 36, 1097, 1983.
10. **Tanner, A. C. R., Badget, S., Lai, C. H., Listgarten, M. A., Visconti, R. A., and Socransky, S. S.,** *Wolinella* gen. nov., *Wolinella succinogenes (Vibrio succinogenes* Wolin et al.) comb. nov., and description of *Bacteroides gracilis* sp. nov., *Wolinella recta* sp. nov., *Campylobacter concisus sp.*, nov., and *Eikenella corrodens* from humans with periodontal disease, *Int. J. Syst. Bacteriol.*, 31, 432, 1981.
11. **Newman, M. G., Sutter, V. L., Pickett, M. J., Blackman, U., Greenwood, J. R., Grinenko, V., and Citron, D.,** Detection, identification and comparison of *Capnocytophaga, Bacteroides ochraceus,* and D. F. I., *J. Clin. Microbiol.*, 10, 557, 1979.
12. **Drasar, B. S., Shiner, M., and McLeod, G. M.,** Studies on the intestinal flora. I. The bacterial flora of the gastrointestinal tract in healthy and achlorhydric persons, *Gastroenterology,* 56, 71, 1969.
13. **Gatehouse, D., Dimock, F., Burdon, D. W., Alexander-Williams, J., and Keighley, M. R. B.,** Prediction of wound sepsis following gastric operations, *Br. J. Surg.*, 65, 551, 1978.
14. **Nichols, R. L., Miller, B., and Smith, J. W.,** Septic complications following gastric surgery: relationship to the endogenous gastric microflora, *Surg. Clin. North. Am.*, 55, 1367, 1975.
15. **Dolby, J. M., Webster, A. D. G., Borriello, S. P., Barclay, F. E., Bartholomew, B. A., and Hill, M. J.,** Bacterial colonization and nitrite concentration in the achlorhydric stomachs of patients with primary hypogammaglobulinaemia or classical pernicious anaemia, *Scand. J. Gastroenterol.*, 19, 105, 1984.
16. **Correa, P., Cuello, C., and Duque, E.,** Carcinoma and intestinal metaplasia of the stomach in Colombian migrants, *J. Natl. Cancer Inst.*, 44, 297, 1970.
17. **Borriello, S. P., Reed, P. J., Dolby, J. M., Barclay, F. E.,Webster, A. D.-B.,** Microbial and metabolic profile of achlorhydric stomach: comparison of pernicious anaemia and hypogammaglobulinaemia, *J. Clin. Pathol.*, 38, 946, 1985.
18. **Skirrow, M. B.,** Session report on taxonomy and biotyping, in *Campylobacter,* Vol. 2, Pearson, A. D., Skirrow, M. B., Rowe, R., Davies, J. R., and Jones, D. M., Eds., Public Health Laboratory Service, London, 1983, 36.
19. **Warren, J. R. and Marshall, B.,** Unidentified curved bacilli on gastric epithelium in active chronic gastritis, *Lancet,* 1, 1273, 1983.
20. **Gordon-Taylor, G. and Whitby, L. E. M.,** Incidence of anaerobic infections in gall bladder, *Br. J. Surg.*, 19, 619, 1932.
21. **Shimada, K., Inamatsu, T., and Yamashiro, M.,** Anaerobic bacteria in biliary disease in elderly patients, *J. Infect. Dis.*, 135, 850, 1977.
22. **Sakaguchi, Y., Murata, K., and Kimura, M.,** *Clostridium perfringens* and other anaerobes isolated from bile, *J. Clin. Pathol.*, 36, 345, 1983.
23. **Imunai, T., Sasaki, T., and Konno, T.,** Antibacterial activity detected in duodenal juice, *Tohoku J. Exp. Med.*, 122, 299, 1977.
24. **Floch, M. H., Binder, H. J., Filburn, B., and Gershengoren, W.,** The effect of bile acids on intestinal microflora, *Am. J. Clin. Nutr.*, 25, 1418, 1972.
25. **Floch, M. H., Gershengoren, W., Elliot, S., and Spiro, H. M.,** Bile acid inhibition of the intestinal microflora — a function for simple bile acids?, *Gastroenterology,* 61, 228, 1971.
26. **Lykkegaard, N. M., Justesen, T., Lenz, K., Nielsen, O. V., and Jensen, S. L.,** Bacterial flora of the small intestine and bile acid metabolism in patients with hepatico-jejunostomy roux-en-Y, *Scand. J. Gastroenterol.*, 12, 977, 1977.
27. **Gorbach, S. L. and Tabaqchali, S.,** Bacteria, bile, and the small bowel, *Gut,* 10, 963, 1969.
28. **King, C. E. and Toskes, P. P.,** Small intestine bacterial overgrowth, *Gastroenterology,* 76, 1035, 1979.
29. **Hersh, T., Floch, M. H., Binder, H. J., Conn, H. O., Prizont, R., and Spiro, H. M.,** Disturbance of the jejunal and colonic bacterial flora in immunoglobulin deficiencies, *Am. J. Clin. Nutr.*, 23, 1595, 1970.
30. **McCarthy, C. F., Austad, W. I., Read, A. E. A.,** Hypogammaglobulinemia and steatorrhea, *Am. J. Dig. Dis.*, 10, 945, 1965.
31. **Hughes, W. S., Cerda, J. J., Holtzapple, P., and Brooks, F. P.,** Primary hypogammaglobulinemia and malabsorption, *Ann. Intern. Med.*, 74, 903, 1971.
32. **Brown, W. R., Savage, D. C., Dubois, R. S., Alp, M. S., Mallory, A., and Kern, F.,** Intestinal microflora of immunoglobulin-deficient and normal human subjects, *Gastroenterology,* 62, 1143, 1972.
33. **Drasar, B. S. and Shiner, M.,** Studies on the intestinal flora. II. Bacterial flora of the small intestine in patients with gastrointestinal disorders, *Gut,* 10, 812, 1969.
34. **Gray, J. D. A. and Shiner, M.,** Influence of gastric pH on gastric and jejunal flora, *Gut,* 8, 574, 1967.

35. **Gorbach, S. L., Plaut, A. G., Nahas, L., Weinstein, L., Spanknebel, G., and Levitan, R.,** Studies of intestinal microflora. II. Micro-organisms of the small intestine and their relations to oral and fecal flora, *Gastroenterology*, 53, 856, 1967.
36. **Gorbach, S. L., Banwell, J. G., Jacobs, B., Chatterjee, B. D., Mitra, R., Sen, N. N., and Muzumder, D. N.,** Tropical sprue and malnutrition in West Bengal. I. Intestinal microflora and absorption, *Am. J. Clin. Nutr.*, 23, 1545, 1970.
37. **Bhat, P., Shantakumari, S., Rajan, D., Mathan, V. I., Kapadia, C. R., Swarnabai, C., and Baker, S. J.,** Bacterial flora of the gastrointestinal tract in Southern Indian control subjects and patients with tropical sprue, *Gastroenterology*, 62, 11, 1972.
38. **Bhat, P., Albert, M. J., Rajan, D., Ponniah, J., Mathan, V. I., and Baker, S. J.,** Bacterial flora of the jejunum: a comparison of luminal aspirate and mucosal biopsy, *J. Med. Microbiol.*, 13, 247, 1980.
39. **Albert, M. J., Bhat, P., Rajan, D., Maiya, P. P., Pereira, S. M., Mathan, M., and Baker, S. J.,** Jejunal microbial flora of Southern Indian infants in health and with acute gastroenteritis, *J. Med. Microbiol.*, 11, 433, 1978.
40. **Jarumilinta, R., Miranda, M., and Villarejos, V. M.,** A bacteriological study of the intestinal mucosa and luminal fluids of adults with acute diarrhoea, *Ann. Trop. Med. Parasitol.*, 70, 165, 1976.
41. **Simon, G. L. and Gorbach, S. L.,** Intestinal microflora, *Med. Clin. North Am.*, 66, 557, 1982.
42. **Tomkins, A. M., Drasar, B. S., and James, W. P. T.,** Bacterial colonisation of jejunal mucosa in acute tropical sprue, *Lancet*, 1, 59, 1975.
43. **Gorbach, S. L., Banwell, J. G., Mitra, R., Chatterjee, B. D., Jacob, B., and Mazumder, D. N.,** Bacterial contamination of the upper small bowel in tropical sprue, *Lancet*, 1, 74, 1969.
44. **Plaut, A. G., Gorbach, S. L., Nahas, L., Weinstein, L., Spanknebel, G., and Levitan, R.,** Studies of intestinal microflora. III. The microbial flora of human small intestinal mucosa and fluids, *Gastroenterology*, 53, 868, 1967.
45. **Nelson, D. P. and Mata, L. J.,** Bacterial flora associated with the human gastrointestinal mucosa, *Gastroenterology*, 58, 56, 1970.
46. **Brown, W. R., Savage, D. C., Dubois, R. S., Alp, M. H., Mallory, A., and Kern, F.,** Intestinal microflora of immunoglobulin-deficient and normal human subjects, *Gastroenterology*, 62, 1143, 1972.
47. **Borriello, S. P. and Larson, H. E.,** Antibiotic and pseudomembranous colitis, *J. Antimicrob. Chemother.*, 7 (Suppl. A), 53, 1981.
48. **Borriello, S. P., Larson, H. E., Welch, A. R., Barclay, F., Stringer, M. F., and Bartholomew, B. A.,** Enterotoxigenic *Clostridium perfringens:* a possible cause of antibiotic-associated diarrhoea, *Lancet*, 1, 305, 1984.
49. **Levison, M. E. and Kaye, D.,** Fecal flora in man: effect of cathartic, *J. Infect. Dis.*, 119, 591, 1969.
50. **Gorbach, S. L., Nahas, L., Plaut, A. G., Weinstein, L., Patterson, J. F., and Levitan, R.,** Studies of intestinal microflora. V. Fecal microbial ecology in ulcerative colitis and regional enteritis: relationship to severity of disease and chemotherapy, *Gastroenterology*, 54, 575, 1968.
51. **Haenel, H. W., Fedheim, W., Muller-Beuthow, W., and Rutloff, H.,** Versuche zur unstimmung der faecalen flora des gesunden erwachsenen, *Zbl. Bakt.*, 188, 70, 1958.
52. **Gorbach, S. L., Banwell, J. G., Jacobs, B., Chatterjee, B. D., Mitra, R., Brigham, K. L., and Neogy, K. N.,** Intestinal microflora in Asiatic cholera. I. "Rice-water" stool, *J. Infect. Dis.*, 121, 32, 1970.
53. **Moore, W. E. C., Cato, E. P., and Holdeman, L. V.,** Anaerobic bacteria of the gastrointestinal flora and their occurrence in clinical infections, *J. Infect. Dis.*, 119, 641, 1969.
54. **Hawksworth, G. M., Drasar, B. S., and Hill, M. J.,** Intestinal bacteria and the hydrolysis of glycosidic bonds, *J. Med. Microbiol.*, 4, 451, 1971.
55. **Gorbach, S. L., Nahas, L., Lerner, P. I., and Weinstein, L.,** Studies of intestinal microflora. I. Effects of diet, age and periodic sampling on numbers of fecal micro-organisms in man, *Gastroenterology*, 53, 845, 1967.
56. **Haenel, H.,** Some roles in the ecology of the intestinal microflora of man, *J. Appl. Bacteriol.*, 24, 242, 1961.
57. **Zubrzycki, L. and Spaulding, E. H.,** Studies on the stability of the normal human fecal flora, *J. Bacteriol.*, 83, 968, 1962.
58. **Borriello, S. P.,** Clostridial Flora of the Gastrointestinal Tract in Health and Disease, Ph.D. thesis, University of London, England, 1981.
59. **Holdeman, L. V. and Moore, W. E. C.,** A new genus, *Coprococcus*, twelve new species, and amended descriptions of four previously described species of bacteria from human feces, *Int. J. Syst. Bacteriol.*, 24, 260, 1974.
60. **Miller, T. L. and Wolin, M. J.,** Stability of *Methanobrevibacter smithii* populations in the microbial flora excreted from the human large bowel, *Appl. Environ. Microbiol.*, 45, 317, 1983.
61. **Tompkins, D. S., Waugh, M. A., and Cooke, E. M.,** Isolation of intestinal spirochaetes from homosexuals, *J. Clin. Pathol.*, 34, 1385, 1981.

62. **Hovind-Hougen, K., Birch-Andersen, A., Henrik-Nielson, R., Orholm, M., Pederson, J. O., Tegibjaerg, P. S., and Thaysen, E. H.,** Intestinal spirochetosis: morphological characterization and cultivation of the spirochete *Brachyspira aalborgi* gen. nov., sp. nov., *J. Clin. Microbiol.*, 16, 1127, 1982.
63. **Croucher, S. C., Houston, A. P., Bayliss, C. E., and Turner, R. J.,** Bacterial populations associated with different regions of the human colon wall, *Appl. Environ. Microbiol.*, 45, 1025, 1983.
64. **Peach, S., Lock, M. R., Katz, D., Todd, I. P., and Tabaqchali, S.,** Mucosal-associated bacterial flora of the intestine in patients with Crohn's disease and in a control group, *Gut*, 19, 1034, 1978.
65. **Hartley, C. L., Neumann, C. S., and Richmond, M. H.,** Adhesion of commensal bacteria to the large intestine wall in humans, *Infect. Immun.*, 23, 128, 1979.
66. **Edmiston, C. E., Jr., Avant, G. R., and Wilson, F. A.,** Anaerobic bacterial populations on normal and diseased human biopsy tissue, *Appl. Environ. Microbiol.*, 43, 1173, 1982.
67. **Seeliger, H. and Werner, H.,** Recherches qualitatives et quantitatives sur la flore intestinale de l'homme, *Ann. Inst. Pasteur*, 105, 911, 1963.
68. **Albach, R. A. and Booden, T.,** Amoeba, in *Parasitic Protozoa*, Vol. 2, Kreier, J. P., Ed., Academic Press, New York, 1978, 455.
69. **Grove, D.,** Amoebic dysentery, intestinal protozoa and helminths, in *Microbes and Infections of the Gut*, Goodwin, C. S., Ed., Blackwell Scientific, Oxford, 1984, 209.

Chapter 2

# FACTORS AFFECTING BACTERIAL METABOLISM

**M. J. Hill**

## TABLE OF CONTENTS

# I. INTRODUCTION

The potential for metabolism of a substrate by the gut bacterial flora is determined by the composition of the flora at the relevant site in the gut, the induction or production of the relevant enzymes, and the activity of those enzymes (Table 1). Each of these are controlled by a complex set of interacting factors and consequently our ability to predict variations in the extent of metabolism is limited and has resulted in many erroneous assumptions.

In studying the factors affecting bacterial metabolism most workers have used animals (usually small rodents) and so there are difficulties in interpreting the results to increase our understanding of the human. However, in recent years there has been a increasing study of humans using various patient groups and this has increased confidence in the validity of the animal studies.

# II. FACTORS CONTROLLING THE COMPOSITION OF THE FLORA

The composition of the flora in an ecosystem is dependent on the supply of specific nutrients, the physicochemical conditions, (e.g., pH, viscosity, $E_h$, etc.), interactions between organisms, (e.g., phage, bacteriocines, antibiotic substances, etc.), interactions with the nonmicrobial environment, (e.g., host factors such as immune systems, surface effects, etc.), the nature of the inoculum, and the time available for the flora to develop.

## A. Nutrient Supply

Although the range of types of intermediate metabolism is limited to a few pathways (principally the Embden-Meyerhof pathway, see Chapter 3) the range of utilizable carbohydrates varies widely between organisms. The ability to utilize a sugar depends on two major factors; the first is the ability to transport the sugar into the cell and the second is the ability to convert the sugar enzymically into a moiety which can be utilized in the pathways of intermediate metabolism. Thus, lactose fermentation is governed by the presence of a lactose permease (which transports lactose into the bacterial cell), and β-galactosidase (which splits lactose into glucose and galactose). A further enzyme system which phosphorylates galactose and converts it to fructose-6-phosphate is necessary for the complete utilization of both monosaccharide components of lactose but almost all bacteria which utilize glucose as an energy source can also utilize galactose. Within the enterobacteria, *Escherichia coli* and *Klebsiella aerogenes* produce β-galactosidase and lactose permease and so ferment lactose; *Salmonella* spp. do not produce either enzyme and so they cannot ferment lactose; *Shigella* spp. produce β-galactosidase but do not produce lactose permease and so cannot utilize lactose. Larger carbohydrates are hydrolyzed by extracellular enzymes (such as amylases) to products which can then be transported into the cell. In a mixed population all components of the flora are able to benefit from this metabolic activity. Thus, an increase in the amount of dietary fiber caused a general increase in the numbers of all genera in the flora of ileostomy fluid,[1] presumably because the major sugars released from the fiber would be utilizable by all of the flora. In contrast, Mallett et al.[2] showed that an increase in the amount of dietary pectin caused an increase in some sentinel enzymes and not in others, indicating a change in the balance of the flora, presumably due to the selective effect of the released mannose which is not utilizable by all organisms. An increase in dietary protein would be expected to favor the more putrefactive organisms at the expense of nonputrefactive species; Fernandez et al.[3] demonstrated that such change has a profound effect on the composition of the flora of ileostomy fluid (Table 2).

## B. Physicochemical Conditions

In in vitro systems the composition of a mixed population depends on a number of physicochemical factors such as pH, $E_h$, oxygen tension, etc., but there is only limited

**Table 1**
**FACTORS AFFECTING THE METABOLIC ACTIVITY OF THE GUT FLORA**

| Effect | Factor |
|---|---|
| Supply of substrate | Nutrient supply |
| Composition of the flora | Nutrient supply |
| | Physicochemical conditions (pH, $E_2$, etc.) |
| | Bacterial interaction (phage, etc.) |
| | Host factors (antibody, lysozyme, etc.) |
| Enzyme activity | Nutrient supply (enzyme induction) |
| | Composition of the flora pH |
| | Supply of cofactors (nutrient supply) |

**Table 2**
**EFFECT OF NUTRIENT SUPPLY ON THE CECAL OR ILEAL FLORA**

| Nutrient | Effect | Ref. |
|---|---|---|
| Cereal fiber | Nonspecific increase in all organisms | 1 |
| Pectin | Change in activity of specific sentinel enzymes | 2 |
| Fat | Increase in anaerobic genera | 3 |
| Protein | Increase in aerobic genera | 3 |

**Table 3**
**FLORA OF GASTRIC JUICE AT VARIOUS pH VALUES**[7]

| | pH of the gastric juice | | |
|---|---|---|---|
| | <4 | 4—5 | 5—8 |
| Total number of samples | 34 | 4 | 19 |
| Those with $>10^5$ bacteria/mℓ | 3% | 25% | 63% |
| Those with fecal *Streptococci* | 9% | 50% | 84% |

evidence available concerning their effects in vivo. This is because it is often difficult to dissociate the effects of different factors.

The effects of extremes of pH are well documented; acid pH values (less than five) are relatively well tolerated by the lactic acid bacteria (*Lactobacilli* and *Streptococci*) Veillonellae and *Micrococci* compared with the *Enterobacteria, Bacteroides* and *Clostridia* while alkaline pH values greater than eight are poorly tolerated by most organisms except *Proteus* spp. and some *Klebsiella* spp. The effects of acid pH values are best documented in the human stomach. The normal stomach in young adults in North America or North West Europe has a pH below 2 except during meals when the food buffers the gastric contents to a pH of 4 to 5[4-6] the normal flora of the stomach is very sparse below pH 4, consists of *Streptococci* and *Micrococci* between pH 4 and 5 while at more alkaline pH values *Enterococci* and *Bacteroides* are often present (Table 3).[7]

The vast majority of species found in the lower gut are strictly anaerobic and are killed by oxygen (Table 4); they require conditions which are sufficiently reducing and with a

**Table 4**
**THE RELATIVE PROPORTIONS OF STRICTLY ANAEROBIC AND MICROAEROPHILIC ORGANISMS AT VARIOUS SITES IN THE GUT OF VARIOUS POPULATIONS**

| | Ratio anaerobes:microaerophils | Ref. |
|---|---|---|
| Gastric juice of patients with pernicious anemia (pH = 7) | 0.1 to 0.001 | |
| Ileostomy fluid | 0.05 | 1 |
| Feces of persons living in | | |
| England | 100 | 9 |
| Scotland | 300 | 9 |
| U.S.A. | 500 | 9 |
| Uganda | 10 | 9 |

sufficiently low oxygen tension. In mixed populations the low oxygen tensions and the low $E_h$ are often created by the aerotolerant organisms; the presence of the aerotolerant organisms thus creates the conditions which allow the strictly anaerobic organisms to grow. In the lumen of the upper gut there are usually sufficient amounts of dissolved oxygen (probably from the air swallowed with food) to prevent the growth of highly oxygen sensitive organisms and only the more robust anaerobes are found. In contrast, the crypts of mucosal surfaces, in dental plaque or any site in which there is stasis and time for anaerobic conditions to be established a more exotic flora with a great range of strictly anaerobic organisms is found, whereas in the lower gut the oxygen tension is much too low to permit the growth of strictly aerobic organisms such as *Pseudomonas* spp.

### C. Interactions between Organisms

Control of the gut flora by microbial interaction is most clearly manifest in colonization resistance. This has been shown to be an active process by the repeated demonstration that, for example, the normal gut flora is highly resistant to colonization by *Staphylococcus aureus* but that Staphylococcal colitis is not a rare consequence of attempts at bowel sterilization using antibiotics.[11] Similarly, high doses of *Salmonella* spp. are needed if the organism is to establish itself in the rat gut in the presence of the normal flora while less than ten are necessary in antibiotic treated rats.[12] People who take prophylactic antibiotics are known to have a higher risk of diarrhea when on holiday than persons who take no medication. The clear failure of attempts to establish a *Lactobacillus* flora by feeding large numbers of *Lactobacillus acidophilus* to persons[13] is a further manifestation of the same effect. The subject has not been widely investigated except by oral bacteriologists.[14,15] Some of the mechanisms thought to be important are the production of bacteriocines, bacteriophages and antibiotic substances. These would serve different purposes since, for example, phage is antagonistic to closely related strains while antibiotics tend to be more active on unrelated strains. There is very little evidence regarding the importance of any of these factors, although the phenomenon of colonization resistance indicates that important and powerful factors are at work.

Bacteria produce a range of substances which help to protect their ecological niche and these can be divided into those which act mainly against organisms of other genera and those whose action is principally against members of the same species. Bacteriocines are antibiotic substances produced by organisms and which are lethal to other strains of the same species. Thus, the action of colicines (produces by *Escherichia coli*) is limited to other strains of *Escherichia coli* and a smaller proportion of strains of other *Escherichia* spp. and some other *Enterobacteria*; pyocines (produced by *Pseudomonas pyocyanaeas*) are only

Table 5
THE GASTRIC BACTERIAL FLORA OF PERSONS WITH PERNICIOUS ANEMIA AND HYPOGAMMAGLOBULINEMIA (PA-HG) COMPARED WITH THAT IN PERSONS WITH PERNICIOUS ANEMIA ALONE (PAS)[8]

| | | | PA-HG | |
|---|---|---|---|---|
| | PA patients Total bacteria | Total nitrate-reducing bacteria | Total bacteria | Total nitrate-reducing bacteria |
| Gastric juice | 7.0[a] | 6.3 | 6.2 | 5.4 |
| Mucosal flora | | | | |
| -antrum | 6.6 | 5.7 | 6.6 | 4.9 |
| -fundus | 6.7 | 6.0 | 6.7 | 6.2 |

[a] Counts expressed as $\log_{10}$ (number of organisms).

active against other members of that species etc. The strains of *Escherichia coli* isolated from the feces of a person usually belong to a small group of colicine types characteristic of that person.

Phages are virus like particles which, like bacteriocines, are only active against a particular species; these also, would act to keep out competing strains of the same bacterial species.

Microbes produce a wide range of antibacterial substances; although very few of those produced by bacteria are of therapeutic value they may nevertheless be of great importance in controlling the composition of the flora in various ecological niches. These may be complex molecules or may be relatively simple metabolites. For example, it has been established that the normal gut flora is resistant to colonization by *Salmonella* spp. because of the fatty acid products of fermentation produced by the nonsporing anaerobic flora.

## D. Interactions with the Nonmicrobial Environment

A major factor in determining the composition of the flora ought to be the ability to resist host factors but this has been difficult to demonstrate. The host factors include bile, immune factors, and host lytic systems. A further host factor is the nature of the body surfaces available for bacterial colonization.

The major immune globulin in the gut is IgA; although the antibody-complement-lysozyme complex is a potent lytic system especially against Gram-negative organisms, that involving IgA is much less powerful than those involving IgG and IgM, and its role in controlling the gut flora is questionable. Patients with hypogammaglobulinemia and also pernicious anemia have a gastric bacterial flora similar to that in pernicious anemia alone (Table 5). Similarly, although the intestinal secretions contain secretory IgA and lysozyme the colonic flora is dominated by the Gram-negative *Bacteroides* spp. In contrast, the intestinal mucosa might be expected to be (and is) dominated by Gram-positive cocci[16] suggesting a possible role for lytic systems in suppressing Gram-negative organisms. Similarly the saliva, which is very rich in lysozyme and in secretory IgA, has a Gram-positive flora dominated by *Streptococci*.

Bile contains potent antibacterial agents, principally the bile acids. Most oral bacteria are sensitive to bile while almost all fecal organisms are resistant to high concentrations of either whole bile or of bile acids. Further, the growth of certain organisms can be stimulated by the presence of bile, principally *B. fragilis*[17] and it is notable that such organisms dominated the flora of western persons (who have high fecal bile acid concentrations). Fernandez et al.,[3] showed that high fat diets cause a big increase in the number of anaerobic bacteria, particularly *Bacteroides* spp. in ileostomy fluid. The bactericidal action of bile and of

**Table 6**
**A COMPARISON OF THE COMPOSITION OF THE FLORA OF ILEOSTOMY EFFLUENT COMPARED WITH THAT OF FECES**

| | Ileostomy effluent[1] | Feces[11] |
|---|---|---|
| Anaerobic bacteria-total | 4.2 | 10.2 |
| *Bacteroides* spp. | 2.7 | 9.8 |
| *Bifidobacterium* spp. | <2.0 | 9.8 |
| *Veillonella* spp. | 3.1 | 4.2 |
| *Clostridium* spp. | 3.7 | 5.7 |
| Microaerophilic organisms | | |
| *Lactobacillus* spp. | 3.1 | 6.5 |
| Facultative organisms - total | 5.4 | 8.0 |
| *Streptococcus* spp. | 5.1 | 7.1 |
| *Esch. coli* | 3.8 | 7.9 |

*Note:* Data expressed as $\log^{10}$ (bacterial numbers).

pancreatic juice is illustrated by the difficulties (still to be overcome) in storing or transporting samples of jejunal juice for bacteriological studies; it is probably part of the explanation for the sparseness of the bacterial flora of the small intestine.

### E. Other Factors

Other factors controlling the composition of a mixed population are the nature and size of the inoculum and the time available for the flora to develop.

The nature of the inoculum is of fundamental importance since clearly no organisms can flourish in an environment to which it was never introduced. The source of the inoculum at various sites in the gut has still to be established. The initial flora in newborn infants is rich in skin organisms (*Coryneforms, Staphlococci*) and respiratory organisms (Gram-positive cocci, Veillonellae) presumably from the maternal skin and breath. Anaerobic organisms, particularly the Gram-positive nonsporing anaerobes in breastfed infants, quickly establish, however their source is unknown.[18,19] In the stomach and the small bowel of adult humans the composition of the flora indicates clearly that the inoculum is from the oral flora. Recent studies of the flora of ileostomy fluid shows clearly that the flora of the large bowel is not derived only from the fluid material entering from above; many organisms present in the fecal flora were absent from the ileostomy fluid (Table 6). The inoculum must be derived from the colonic mucosal flora in addition to ileal fluid; the differences in composition of the flora at the two sites indicates that the mucosa is the more important source of organisms although the upper digestive tract must be the major source of new organisms.

Time is of major importance because it is necessary for the aerobic organisms to scavenge the oxygen to allow growth of anaerobes and removal of simple nutrients also has a major effect on the composition of the flora. The residence time in the stomach is relatively short (2 to 4 hr) and food buffers the gastric contents to a nonbactericidal pH for an even shorter period of time. Thus, the time available for bacterial growth is short and the composition of the inoculum (salivary organisms) has no time to develop. Similarly, transit through the small bowel is rapid and the organisms isolated are mainly salivary. In contrast, the colonic transit time of 30 to 60 hr allows the flora to develop and mature. Plenty of time is available for removal of oxygen and consequently the flora is dominated by anaerobic organisms; readily utilizable nutrients are rapidly removed and so, although they are often of importance in the cecum, they can have no importance in determining the composition of the flora of the sigmoid colon.

## III. INDUCTION AND PRODUCTION OF ENZYMES

Production of enzymes does not simply depend on whether they are constitutive or inducible and on the presence of enzyme inducers in the case of the latter. Utilization of amino acids via the production of deaminases or decarboxylases depends not only on the presence of substrate but also on the pH; the oxygen tension and other factors are equally important with other enzymes. Bacterial amino acid decarboxylases are not produced until the pH falls below the certain value (which is close to pH 5 with *Escherichia coli*) while deaminases are only produced at alkaline pH values. In contrast to those enzymes, whose production is favored by extremes of pH, the production of the cholanoyl 7-dehydroxylase (which yields deoxycholic acid from cholic acid and lithocholic acid from chenodeoxycholic acid) is not produced when the pH falls below 6 in vitro.[20]

*Clostridium perfringens* strains which produce a wide range of bile acid degrading enzymes when grown under strictly anaerobic conditions failed to produce even the constitutive enzymes of that class when grown under less exacting conditions, even though growth of the organism was unimpaired,[23] indicating the importance of strict anaerobiosis in the production of certain enzymes.

Many enzymes are only produced at certain stages in the growth cycle. For example, in vitro enzymes involved in the degradation of the bacterial capsular polysaccharide are usually only produced late in the stationary phase of growth when the pH tends to be extreme, when the nutrients tend to be depleted and when there is a high concentration of the end products of metabolism such as fatty acids, alcohols, and aldehydes. In contrast phosphatases tend to be highly active in the early phase of the growth cycle. A peptidase which utilizes polyglutamic acid as substrate is produced by *C. sporogenes* only during the first few hours of growth and is totally undetectable after overnight incubation of the culture.[21] At a more sophisticated level of study, the rate of enzyme production can be shown to be related to the rate of turnover of cells in chemostat or fermenter studies; this is utilized commercially to optimize enzyme production but might also be of importance in the gut where the turnover time of cells can be altered by dietary manipulation.[22]

The glucose concentration is of importance in determining whether the production of inducible enzymes is repressed. This is of relevance to the situation in saliva, (where the production of virtually all inducible enzymes is repressed) but is unimportant in the lower gut (where the glucose concentration is too low to have any effect on enzyme induction).

## IV. ENZYME ACTIVITY

The major factors known to affect the activity of intestinal enzymes are the pH, substrate concentration, and the presence of co-factors or enzyme inhibitors.

Bacterial bile acid 7$\alpha$-dehydroxylase has a pH-optimum close to neutral and the activity decreases sharply with increasing acidity or alkalinity. Thus, an acid pH may prevent the production of this enzyme (as already described) or may decrease the activity of any enzyme already produced. In attempting to predict the activity of bacterial enzymes in vivo it is important to recognize that the pH-optimum of enzymes in an isolated and purified state may differ grossly from that of the cell-bound enzyme. Thus, the 7-hydroxysteroid oxidoreductase produced by *Escherichia coli* strongly favors the reduction at neutral pH, since the oxidative step has a pH-optimum greater than 9 while that of the reduction step is close to neutral. Nevertheless, when whole cells of *Escherichia coli* are incubated with bile acids at neutral pH it is the oxidative step that is dominant.

Substrate concentration is of importance in certain circumstances. The bile acid 7$\alpha$-dehydroxylase enzyme produced by many genera of gut bacteria is subject to substrate inhibition at high substrate concentrations in vitro[24] and also in vivo.[25]

The effect of enzyme co-factors (NAD, vitamins etc.) on enzyme action is well documented in vitro but is poorly documented in vivo. Bacterial *N*-nitrosation in the urinary bladder is very much greater in mixed infections than in infections due to a single organism (see Chapter 9) and this is probably due to the production of co-factors.

## V. CONCLUSIONS

Because of the complexity of the factors determining the rate of metabolism of substrates by bacteria, it is still necessary to be pragmatic and to avoid the temptation of attempting to predict too closely the fate of a compound. Nevertheless, it is important also not to underestimate the success of workers studying, in particular, the rumen in constructing models of their ecosystems. The success of a model is, in itself, an indication of the level of understanding of a system as well as a means of increasing that understanding. To date those studying the human digestive tract have shown much less interest in the study of bacterial interactions and of consortia.

## REFERENCES

1. **Hori, S., Berghouse, L., Hill, M. J., Hudson, M. J., Rodgers, E., and Lennard-Jones, J. E.,** The effect of dietary fibre on the bacterial flora of ileostomy fluid, *J. Med. Microbiol.*, 16, viii, 1983.
2. **Mallett, A. K., Rowland, I. R., and Wise, A.,** Influence of dietary lactose and age on the metabolic activity of the rat caecal flora, in *Models of Anaerobic Infection*, Hill, M. J., Ed., Martinus Nijhof, Amsterdam, 1984, 250.
3. **Fernandez, F., Kennedy, H., Hill, M., and Truelove, S.,** The effect of diet on the composition of the bacterial flora of the terminal ileum, in *Models of Anaerobic Infection*, Hill, M. J., Ed., Martinus Nijhof, Amsterdam, 1984, 252.
4. **Milton-Thompson, G. J., Lightfoot, N. G., Ahmet, L., et al.,** Intragastric acidity, bacteria, nitrite and N-nitroso compounds before, during and after cimetidine treatment, *Lancet*, i, 1091, 1982.
5. **Kieghley, M. R. B., Youngs, D., Poxon, V., et al.,** Intragastric N-nitrosation is unlikely to be responsible for gastric carcinoma developing after operations for duodenal ulcer, *Gut*, 25, 238, 1984.
6. **Cook, A. R., Meyrick-Thomas, J., Hill, M. J., Misiewicz, J. J., and Woodlings, D. F.,** The effect of ranitidine on the gastric bacterial flora in patients with benign peptic ulcer, *Microecol. Ther.*, 14, 255, 1984.
7. **Bartholomew, B. A., Hill, M. J., Hudson, M. J., Ruddell, W. S. J., and Walters, C. L.,** Gastric bacteria, nitrate, nitrite and nitrosamines in patients with pernicious anemia and in patients treated with cimetidine, in *N-Nitroso Compounds: Analysis, Formation, and Occurrence*, Walker, E. A., Griciute, L., Castagnero, M., and Borzsonyi, M., Eds., International Agency for Research on Cancer, Lyon, 1980, 595.
8. **Dolby, J. M., Webster, A. D. B., Borriello, S. P., Barclay, F. E., Bartholomew, B. A., and Hill, M. J.,** Bacterial colonisation and nitrite concentration in the achlorhydric stomach of patients with primary hypogammaglobinemia or classical pernicious anemia, *Scand. J. Gastroenterol.*, 19, 105, 1984.
9. **Hill, M. J., Drasar, B. S., Aries, V. C., Crowther, J. S., Hawksworth, G. M., and Williams, R. E. O.,** Bacteria and etiology of cancer of large bowel, *Lancet*, i, 95, 1971.
10. IARC Working Party, Dietary fibre, transit time, faecal bacteria, steroids and colon cancer in two Scandinavian populations, *Lancet*, 2, 207, 1977.
11. **Conn, H. O., Leevy, C. M., Vlahcevic, Z. R., Rodgers, J. B., Maddrey, W. C., Seeff, L., and Levy, L. L.,** Comparison of lactulose and neomyn in the treatment of chronic portal-systemic encephalopathy, *Gastroenterology*, 72, 573, 1977.
12. **Meynell, G. G.,** Antibacterial mechanisms in the mouse gut. II. The role of $E_h$ and VFA in the normal gut, *Br. J. Exp. Pathol.*, 44, 209, 1963.
13. **Gilliland, S. E., Speck, M. L., Nauyok, J. R., and Giesbrecht, F. J.,** Influence of consuming nonfermented milk containing *Lactobacillus acidophilus* on faecal flora of healthy males, *J. Dairy Sci.*, 61, 1, 1978.
14. **Mikx, F. M. H. and Van Campen, G. J.,** The microbial community on the teeth, with special reference to necrotizing ulcerative gingivitis in dogs, *Eur. J. Chemother. Antibiot.*, 2, 19, 1982.

15. **Orpin, E. G. C.,** Factors controlling the rumen flora in health and disease, *Eur. J. Chemother. Antibiot.*, 2, 27, 1982.
16. **Hill, M. J. and Hudson, M. J.,** Intestinal microflora in inflammatory bowel disease, in *Etiopathogenesis and Treatment of Inflammatory Bowel Disease*, Japanese MRC, Tokyo, 3, 1984.
17. **Floch, M. J., Binder, H. J., Filburn, B., and Gershengoren, W.,** The effect of bile acids on the intestinal microflora, *Am. J. Clin. Nutr.*, 25, 1418, 1972.
18. **Roberts, A., Harzer, G., Drasar, B. S., and Hill, M. J.,** The effect of diet on the bacterial flora of the infant gut, *J. Med. Microbiol.*, 18, XIV, 1984.
19. **Stark, P. L. and Lee, A.,** The microbial ecology of the large bowel of breast fed and formula fed infants during the first year of life, *J. Med. Microbiol.*, 15, 189, 1982.
20. **Midtvedt, T. and Norman, A.,** Parameters in 7α-dehydroxylation of bile acids by anaerobic lactobacilli, *Acta Path. Microbiol. Scand.*, 72, 313, 1968.
21. **Hudson, M. J., Cohen, B. A., and Hill, M. J.,** The differentiation of *Cl. sporogenes* and *Cl. botulinum* by the enzymatic activities expressed by young cultures, *J. Appl. Bacteriol.*, 55, ix, 1983.
22. **Stephen, A. and Cummings, J. H.,** Mechanism of action of dietary fibre in the human colon, *Nature (London)*, 284, 283, 1980.
23. **Hill, M. J.,** unpublished results.
24. **Aries, V. C. and Hill, M. J.,** Degradation of steriods by intestinal bacteria, *Biochem. Biophys. Acta*, 202, 535, 1970.
25. **Eastwood, M. et al.,** unpublished results, 1974.

Chapter 3

# METABOLISM OF CARBOHYDRATES AND GLYCOSIDES

**M. J. Hill**

## TABLE OF CONTENTS

# I. INTRODUCTION

Although the gut is rich in organisms capable of proteolysis, the vast majority of bacteria in the large intestine derive their energy from saccharolytic reactions. Ultimately these energy-yielding reactions use monosaccharide substrates, but these may enter the gut as free monosaccharides or may be released from glycosides, oligosaccharides, or polysaccharides by bacterial enzymes.

In this chapter the first consideration will be the energy-yielding metabolism of glucose. A wide range of monosaccharides are metabolized by gut bacteria and these will be summarized; many are converted to glucose or fructose by isomerases and are then metabolized as glucose, but some are metabolized by specific pathways. Consideration will be given next to the enzymes responsible for the release of monosaccharides from disaccharides, plant glycosides, and from oligo, polysaccharides, and mucins.

# II. GLUCOSE METABOLISM

The vast majority of gut bacteria derive their energy mainly from glucose metabolism,[1] and most genera that have been examined utilize the Embden-Meyerhoff (E-M) pathway. This is usually detected by assaying fructose 1-6-diphosphate aldolase but may also be demonstrated using labeled substrates and detecting key intermediates. Table 1 lists some of the key organisms in which the E-M pathway has been detected. The only major genus which does not use this pathway is *Bifidobacterium* spp. which utilizes a unique pathway.[2] The E-M pathway yields pyruvate which is converted to a range of short-chain fatty acid end products of fermentation, principally acetic, formic, propionic, butyric, succinic, and lactic acids (Figure 1). Although these end products of fermentation are characteristic of the bacterial genus (Table 2) they do not correspond exactly to those used taxonomically in the identification of anaerobic bacteria (the latter include the products of amino acid deamination). The pathways for the production of the various fatty acids has been reviewed by Miller and Wolin[1] and will be described briefly.

## A. Metabolism of Pyruvate to Fatty Acid End Products

### *1. Lactic Acid*

Lactic acid is formed by the simple reduction of pyruvate using the reverse reaction of lactic acid dehydrogenase.

### *2. Succinic Acid*

The formation of succinic acid has been studied in *B. fragilis* and *B. succinogenes*, both organisms using essentially the same pathway. Phosphoenolpyruvate is carboxylated to oxaloacetic acid which is then reduced to malate followed by dehydration to fumarate and finally reduction to succinate by a pathway involving cytochrome b.

### *3. Propionic Acid*

Propionic acid is produced by a variety of pathways. In *Megasphera elsdenii* it is formed by the reduction of lactic acid while in *Selenomonas* spp and in *Propionibacterium* spp. it is produced by decarboxylation of succinic acid. The latter is the more important pathway in the intestine while the former is more important in ruminants.[1]

### *4. Acetic Acid*

Pyruvate is decarboxylated and dehydrogenated to acetyl Coenzyme A (CoA) by the pyruvate oxidoreductase system. Acetyl CoA is then deacetylated to yield acetic acid in an energy yielding reaction.

## Table 1
## KEY ORGANISMS IN THE GASTROINTESTINAL TRACT IN WHICH THE E-M PATHWAY HAS BEEN DETECTED

| Organism | Approximate counts in human feces | E-M pathway |
|---|---|---|
| *Bacteroides* fragilis | $10^{11}$ | Present |
| *Bacteroides* spp. (other than fragilis) | $10^{11}$ | Present |
| *Propionibacterium* spp. | $10^{11}$ | Present |
| *Eubacterium* spp. | $10^{10}$ | ? |
| *Bifidobacterium* spp. | $10^{11}$ | Not present |
| *Fusobacterium* spp. | $10^{9}$ | Present |
| *Lactobacillus* spp. | $10^{9}$ | Present |
| *Clostridium* spp. | $10^{5}$ | Present |
| *Enterobacteria* | $10^{8}$ | Present |
| *Streptococcus* spp. | $10^{8}$ | Present |

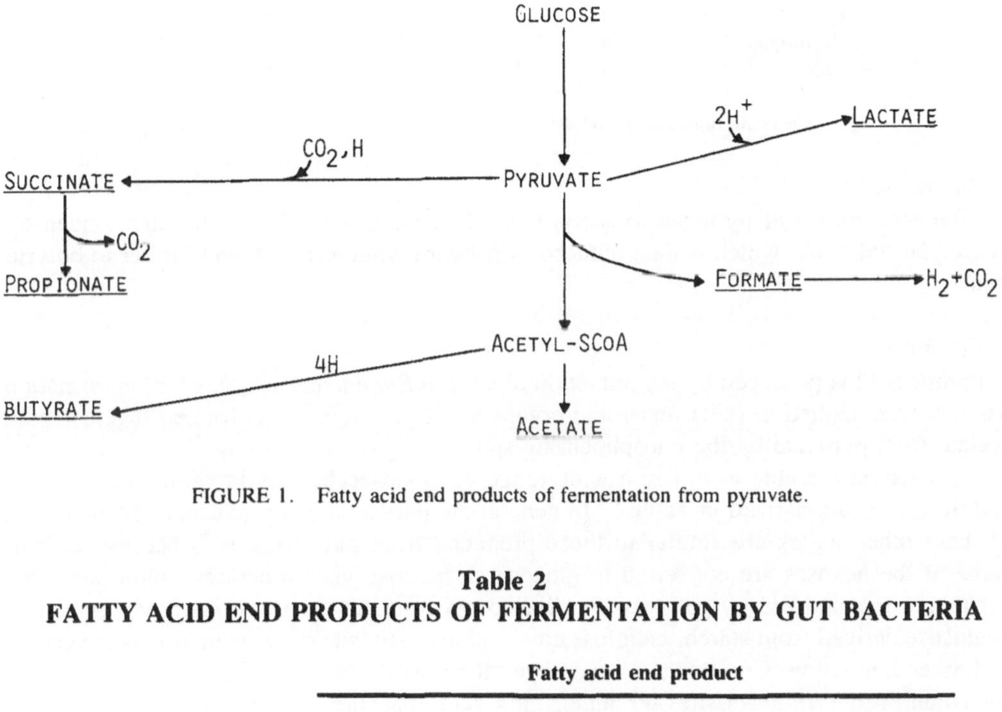


FIGURE 1. Fatty acid end products of fermentation from pyruvate.

## Table 2
## FATTY ACID END PRODUCTS OF FERMENTATION BY GUT BACTERIA

| | Fatty acid end product | | | | | |
|---|---|---|---|---|---|---|
| | Formic | Acetic | Propionic | *n*-butyric | Lactic | Succinic |
| *Bacteroides* spp. | | + | + | | | + |
| *Fusobacterium* spp. | | + | | + | | |
| *Bifidobacterium* spp. | | + | | | + | |
| *Eubacterium* spp. | + | | | | | |
| *Propionibacterium* spp. | | + | + | | | |
| *Lactobacillus* spp. | | + | | | + | |
| *Clostridium* spp. | | + | + | + | + | + |

**Table 3**
**UTILIZATION OF MONOSACCHARIDES BY GUT BACTERIA**

| | Arabinose | Fructose | Mannose | Rhamnose | Ribose | Xylose | Galactose |
|---|---|---|---|---|---|---|---|
| *Peptococcus* spp. | − | − | − | − | − | − | − |
| *Peptostreptococcus* spp. | − | + | − | − | − | − | + |
| *Bacteroidis fragilis* | − | + | + | − | − | + | + |
| *B. ovatus* | + | + | + | + | + | + | + |
| *Eubact. rectale* | + | + | + | − | + | + | + |
| *Eub. Lentum* | − | − | − | − | − | − | − |
| *Propionibact acnes* | − | + | + | − | − | − | + |
| *Lactobacillus plantarum* | + | + | + | − | + | + | + |
| *Bifidobacterium bifidum* | − | + | − | − | − | − | + |
| *Bifidobacterium adolescentis* | + | + | + | − | + | + | + |

**Table 4**
**VOLATILE FATTY ACIDS AT VARIOUS SUBSITES IN THE COLON**

| | Acetic acid[a] | Propionic acid | Butyric acid |
|---|---|---|---|
| Cecum | 81 | 18 | 36 |
| Colon | 73 | 22 | 26 |
| Rectosigmoid | 68 | 19 | 21 |
| Feces | 46 | 17 | 14 |

[a] Concentrations are in mmol/ℓ.

*5. Butyric Acid*

After metabolism of pyruvate to acetyl CoA, two molecules of the latter are condensed to acetylacetyl CoA, which is then reduced to β-hydroxybutyryl CoA and further to butyric acid.

*6. Formic Acid*

Formic acid is produced by the reduction of $CO_2$ in *Ruminicoccus albus* but its formation has not been studied in other anaerobic organisms. In coliforms it is formed together with acetate from pyruvate by the phosphoclastic split.

Gut bacteria are able to utilize a wide range of monosaccharides in addition to glucose and these are summarized in Table 3. In general the fatty acid end products of fermentation of these other sugars are similar to those produced from glucose; this is because at least some of the hexoses are converted to glucose or fructose via isomerases, while the other sugars are metabolized by pathways leading to pyruvate. In vivo, the main energy source is glucose derived from starch, cellulose etc., and the main site of fermentation is the cecum and ascending colon. Cumming[3] have assayed the volatile fatty acids at various subsites in the colon (Table 4); at all sites the major fatty acid is acetic, and the molar ratio of acetic to propionic to butyric acid remains fairly constant at 60:25:15. Other fatty acids are present only in small amounts and in particular, lactic acid is rarely found in the feces of adults although it is normally present in newborn infants. This suggests that lactate is utilized as a fermentable substrate by some component of the gut flora and it is known that this is the main energy source for *Veillonella* spp. Similarly ethanol is rarely present in stools although it is a major end product of fermentation in vitro of many species. The production of short-chain fatty acids in the gut has been reviewed by Cummings.[4]

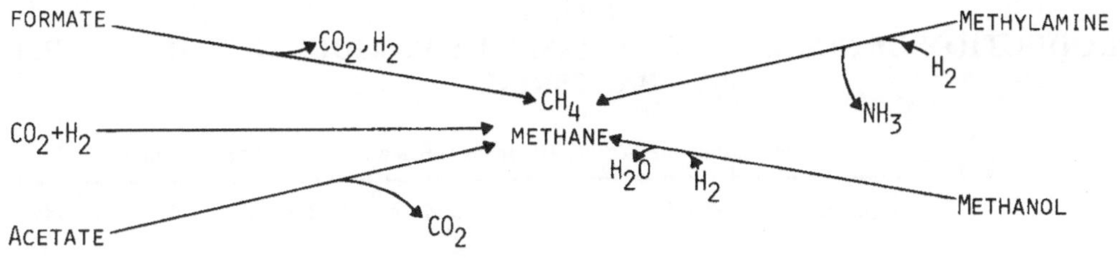


FIGURE 2. Pathways for methanogenesis by intestinal bacteria.

**Table 5**
**GLYCOSIDASES PRODUCED BY GUT BACTERIA**

| Enzyme | Substrates |
|---|---|
| β-galactosidase | Lactose, lactulose |
| β-glucuronidase | Glucuronides secreted in bile (see Chapter 10) |
| β-glucosidase | Iso-maltose, plant glucosides, small oligosaccharides |
| α-glucosidase | Sucrose |
| Cellobiase | Cellobiose, small oligosaccharides |

In addition to the short-chain fatty acids, the other major waste product of glucose fermentation is gas, principally hydrogen, carbon dioxide, and methane. The first two are produced as byproducts of a wide range of reactions but methane production was thought to be produced by only a limited range of pathways (Figure 2). In the rumen the major pathway utilizes carbon dioxide and hydrogen and requires a highly reducing environment. This pathway is carried out by *Methanobacterium* spp. which has only been isolated rarely from human feces and then only in small numbers. Methane is produced in the colon in most persons[5] and is produced in amounts sufficient for it to be present in the breath of 30 to 40% of human adults (see Chapter 13); thus, although the failure to detect *Methanobacterium* spp. more often may be due to inadequate techniques it is unlikely that this is the major pathway of methanogenesis. Further, methanogenesis can be demonstrated in feces under conditions of anaerobiosis that would certainly be insufficient to sustain methanobacteria.[5] To date the mechanism of methane production has still to be determined. There is clear substrate specificity in methanogenesis and this has been investigated by Tadesse and Eastwood[6] who demonstrated increased breath methane as a result of the metabolism of pectin but not following the metabolism of lactulose. This illustrates the substrate preferences of the different genera of anaerobes, which has been studied by Russell[7] and will be discussed later.

## III. THE RELEASE OF MONOSACCHARIDES BY BACTERIAL GLYCOSIDASES

Since monosaccharides are so well absorbed from the upper gastrointestinal tract the main source of fermetable saccharide is that released from conjugated compounds, oligo, and polysaccharides. To release these, bacteria produce a range of glycosidases (which remove terminal sugars) and also enzymes able to hydrolyze glycosidic bonds in the middle of a polymeric chain. These are listed in Table 5 and will only be described briefly.

### A. β-Galactosidase

This enzyme will hydrolyze β-galactosidic bonds involving any substrate, in contrast to intestinal β-galactosidase which is specific for lactose (glucose-β-D-galactopyranoside). Thus,

**Table 6**
**PRODUCTION OF SOME GLYCOSIDASES BY VARIOUS GENERA OF GUT BACTERIA[7]**

| | Mean enzyme activity (μ moles substrate degraded/hr/$10^8$ cells) | | | | |
|---|---|---|---|---|---|
| | α-galactosidase | β-galactosidase | α-glucosidase | β-glucosidase | β-glucuronidase |
| *E. coli* | 12.7 ± 2.1 | 42.4 ± 3.2 | 5.9 ± 0.5 | 5.8 ± 2.5 | 24.7 ± 2.1 |
| *S. faecalis* | 20.8 ± 4.3 | 53.8 ± 6.0 | 14.0 ± 1.1 | 192.7 ± 19.5 | 2.9 ± 0.6 |
| *Lactobacillus* spp. | 97.7 ± 12.1 | 90.6 ± 10.7 | 26.6 ± 3.5 | 26.0 ± 7.4 | 1.6 ± 0.2 |
| *Bifidobacterium* spp. | 28.2 ± 3.8 | 39.1 ± 4.7 | 20.7 ± 3.0 | 29.3 ± 6.0 | 1.9 ± 0.8 |
| *Bacteroides* spp. | 24.0 ± 4.8 | 50.7 ± 4.9 | 9.8 ± 2.0 | 35.1 ± 4.8 | 6.0 ± 3.5 |
| *Clostridium* spp. | 53.1 ± 8.1 | 13.7 ± 2.7 | 30.1 ± 6.4 | 22.1 ± 5.0 | 11.3 ± 2.3 |

**Table 7**
**THE CALCULATED SOURCE OF β-GALACTOSIDASE IN THE TERMINAL ILEUM AND FECES[7]**

| | | Terminal ileum | | | Colon | | |
|---|---|---|---|---|---|---|---|
| Organism | Enzyme activity | No. of organisms | Activity × $10^5$ | % of total activity | No. of organisms | Activity | % of total activity |
| *E. coli* | 42.4 | $5 \times 10^3$ | 212 | 6.9 | $10^8$ | 42.4 | — |
| *S. faecalis* | 53.8 | $5 \times 10^4$ | 2690 | 87.4 | $10^5$ | 0.05 | — |
| *Lactobacillus* spp. | 90.6 | $1 \times 10^3$ | 91 | 3.0 | $10^7$ | 9.1 | — |
| *Bacteroides* spp. | 50.7 | $5 \times 10^2$ | 25 | 0.8 | $10^{11}$ | 50700 | 56.4 |
| *Bifidobacterium* spp. | 39.1 | $1 \times 10^2$ | 4 | 0.1 | $10^{11}$ | 39100 | 43.5 |
| *Clostridium* spp. | 13.7 | $4 \times 10^3$ | 55 | 1.8 | $10^6$ | 0.1 | — |
| Total activity | | | 3077 | | | 89,852 | |

the bacterial enzyme will hydrolyze lactose, lactulose (fructose-β-D-galactopyranoside) as well as a range of synthetic galactosides the hydrolysis of which releases chromophores, (e.g., phenolphthalein, α-naphthol, *p*-nitrophenol) useful in enzyme assay. Strains of most species tested produce β-galactosidase;[8] although the enterobacteria produce enzymes of the highest activity (Table 6) because of their numerical dominance the major source of β-galactosidase in the large bowel is the nonsporing anaerobic bacteria (Table 7) and *Streptococci* in the terminal ileum. The enzyme is inducible in *E. coli* since very little lactose is fermented in the oral cavity it is likely that in most organisms this is so; however it is constitutive in *S. faecalis*.[8] The major substrate for β-galactosidase in vivo is lactose.

### B. β-Glucuronidase

A wide range of foreign compounds (usually lipophilic) undergoes detoxification in the liver by conjugation as the β-glucuronides and this is usually followed by biliary secretion. These may be hydrolyzed in the colon by bacterial β-glucuronidase. The enzyme is produced by strains of all genera, those with the highest mean activity being *E. coli* and related organisms (Table 6). Although *E. coli* are also a major source of the enzyme in the terminal ileum, the numerical dominance of the nonsporing anaerobic bacteria and of the lacto bacilli result in coliforms being only a minor source of colonic β-glucuronidase. The role of the enzyme in the enterohepatic circulation of xenobiotics has been described elsewhere in this chapter, it is important in the antimicrobial activity of chloramphenicol, the laxative action

**Table 8**
**α AND β-GLUCOSIDASE SUBSTRATES USED IN BACTERIAL IDENTIFICATION**

| Trivial name | Composition |
|---|---|
| Amygdalin | Benzaldehyde cyanhydrin (β-D-glucoside)$_2$ |
| Salicin | O-hydroxymethylphenyl-β-D-glucoside |
| Aesculin | 6,7-dihydroxycoumarin-β-D-glucoside |
| Iso-maltose | 6-0-αglucosyl-β-D-glucoside |
| Trehalose | αglucosyl-α-D-glucoside |
| Cellobiose | 4-O-glucosyl-β-D-glucoside |
| Melibiose | 6-O-α-Dgalactosyl-α-D-glucose |
| Gentiobiose | 6-O-β-glucosyl-β-D-glucoside |
| Arbuten | Hydroquinoe-β-D-glucoside |

**Table 9**
**LOSS OF LACTASE DURING CHILDHOOD IN TWO POPULATIONS % LACTOSE — INTOLERANT IN VARIOUS AGE BANDS (YEARS)**

| | 1—3 | 3—4 | 4—5 | 5—7 | 7—11 | 11—15 | 21—65 |
|---|---|---|---|---|---|---|---|
| Finland | 0% | | | 6% | | 12% | 15% |
| Pima Indians (U.S.A.) | 0% | 40% | 71% | 92% | | | 95% |

of phenolphthalcin, and in the variability of the half-life of, for example, morphine and stilboestrol within the body.

### C. β-Glucosidase

This enzyme will hydrolyze the β-glucoside linkage to a wide range of aglycones; the very wide range in substrate specificity contrasts with the very narrow range of substrates hydrolyzed by the β-glucosidase (iso-maltase) present in the intestinal mucosal layer. The diet is a rich source of β-glucosides of plant origin; many plants cannot be used unprocessed as food sources because of the presence of β-glucosides which yield toxic aglycones. β-glucosidase is a constitutive enzyme although the activity is increased in the presence of the substrate. Some of the β-glucosides used in bacterial identification are listed in Table 8. Although strains of all bacterial genera tested produce a β-glucosidase the mean activity is highest in *Streptococci* and *Lactobacilli* (Table 6). *Streptococci* are the major source of the enzyme in the lower ileum but in the colon, as with the other glycosidases, the major source of the enzyme is the nonsporing anaerobic bacteria because of their numerical dominance (Table 7).

### D. Cellobiase

Cellobiose consists of 3 glucose molecules linked by β (1 - 4)-bonds and so can be degraded by β-glucosidase. However, this is a relatively slow hydrolysis and the enzyme cellobiase hydrolyses, chains of 3 to 8 β (1 - 4) linked glucose molecules much more rapidly. This enzyme is produced by many species of *Lactobacillus, Bifidobacterium, Clostridium* and *Bacteroides* but not by most species of *Eubacterium, Propionibacterium, Fusobacterium* or the anaerobic cocci (Table 9).

### E. Other Glycosidases

A number of other glycosidases have been assayed in bacterial strains. Hawksworth et

**Table 10**
**OTHER GLYCOSIDASES PRODUCED BY GUT BACTERIA**

| Enzyme | Organisms producing the enzyme at highest activity | Ref. |
|---|---|---|
| α-Glucosidase | *Lactobacilli, Clostridia* | 8 |
| α-Galactosidase | *Lactobacilli, Clostridia, Bifidobacteria* | 8 |
| α-Mannosidase | Very low activity in feces | 9 |
| α-Fucosidase | High activity in feces; no details on organisms | 9 |
| β-Xylosidase | High activity in feces; no details on organisms | 9 |

al.[8] assayed α-glucosidase (sucrase) and α-galactosidase in addition to the enzymes described above (Table 10). Salyers[9] reported assays of α-mannosidase, β-xylosidase, and α-fucosidase in fecal samples from four volunteers; the former was not present but the other two enzymes were present at activities similar to that of β-glucosidase.

## IV. OTHER SACCHAROLYTIC ENZYMES

### A. Starch Hydrolysis

Starch contains two major components: amylose and amylopectin. Amylose, which accounts for approximately 25% of most starch sources consists of linear chains of glucose molecules linked by α(1 - 4)-bonds. Amylopectin, which accounts for the remaining 75% of starch, consists of amylase with branches of α(1 - 6)-linked glucose chains.

Both classes of molecule are degraded by two types of amylase (both of which hydrolyze α(1 - 4)-linkages. α-Amylase attacks starch molecules randomly in the interior parts of the chain to yield maltose and maltotriose from amylose, while from amylopectins the same products together with oligosaccharides containing 4 to 8 glucose units and the α(1 - 6) linkages. The maltose produced by α-amylase has an α-configuration (β-amylase yields products with a β-configuration). In addition, there is an iso-amylase which hydrolyses the α(1 - 6)-linkages.

Most bacteria capable of starch hydrolysis produce extracellular α-amylases. The hydrolysis of starch by anaerobic bacteria is summarized in Table 9.

### B. Metabolism of Nonstarch Polysaccharide

The metabolism of dietary fiber components has been reviewed by Salyers[9] and will be considered in more detail in the next chapter. Cellulose and other β-glucans are hydrolyzed by the enzyme complex cellulase to soluble oligosaccharides of 3 to 8 glucose units; these are then further hydrolyzed to glucose by β-glucosidase and cellobiase.

## V. METABOLISM OF GLYCOSIDES AND DISACCHARIDES IN THE COLON

The diet of human adults contains two major disaccharides, namely sucrose and lactose, both of which are consumed in much greater amounts in Western countries than in Africa, Asia, and South America. Both of these are efficiently digested in the small intestine of Western persons and only trace amounts reach the bacterially colonized regions of the large bowel; exceptions are persons with disaccharidase deficiency or with disease states accompanied by very severe malabsorption. In such persons disaccharide metabolism by the gut flora causes disease.[10]

Plants produce a wide range of β-glucosidases and many of these yield pharmacologically active glycosides. The hydrolysis of such glycosides is not a major source of fermentable carbohydrate to the gut flora but with disaccharides, can produce clinical symptoms which can be severe.

## A. Disaccharide Metabolism

### *1. Lactose*

Lactose is a major component of the diets of all human infants but in most parts of the world, ceases to be of major importance post-weaning because of the lack of dairy products in the diet. In North and Western Europe and in populations derived from those regions (North Americans, South Africans, Australians) dairy products are an important part of the diet and are well tolerated except in a small minority of persons with lactose intolerance. The digestion of lactose has been reviewed by Dahlqvist,[11] who describes four classes of lactose intolerance. These are impaired digestion of lactose, abnormal permeability of the gastric wall to lactose, impaired transport of galactose, and impaired metabolism of galactose. Only impaired lactose digestion involves bacterial metabolism; under such conditions the dietary lactose reaches the bacterially colonized region of the large bowel where it is rapidly metabolized to short-chain fatty acid and gas, thereby causing flatulence, gut pain and distension, and fermentative or osmotic diarrhea. There are three causes of impaired lactose digestion, these are congenital lactase deficiency, adult lactase deficiency, and secondary lactase deficiency. In congenital lactase deficiency, a very rare disease, the brush border lactase is absent from birth[12] and the infant must be fed lactose-free milk, (i.e., where the lactose is prehydrolyzed or using an artificial formulation containing another sugar) from the time of diagnosis. Adult lactase deficiency, in contrast, is very common (Table 8) and indeed, it is clear that the vast majority of persons around the world lose their brush border lactase in late childhood (Table 9). The widespread nature of adult lactase deficiency was not realized until 1970,[13-16] and provided an explanation for some previously inexplicable observations regarding the uses made of dried milk powder used in the WHO famine relief programs, (e.g., as whitewash in Guatamala, as a mild purgative in the Phillipines).

Although the lactose tolerance test uses the equivalent of four glasses of milk, studies of American Indians[17] who were lactose intolerant showed that they could tolerate one to one and one-half glasses of milk, when taken with a meal, with no obvious ill effects. This would be expected since the symptoms of lactose intolerance are caused by excess acid and gas; the smaller amount of acid produced from the smaller dose is evidently buffered by the indigestible components of the meal.

Secondary lactose intolerance is due to an altered mucosal surface with leaf-like or stunted villi which have a very much smaller surface area than the usual finger-like villous surface. Since lactase is in the brush border at the surface, the reduced surface area results in a decreased lactase activity and lactose intolerance. Lactose intolerance of this type can be secondary to gastroenteritis, tropical sprue, gluten-sensitive enteropathy, and possibly also to inflammatory bowel disease. When the primary disease is short-lived (as in, for example, enteritis) the secondary lactase deficiency is also transient.

### *2. Sucrose-Isomaltose*

Congential sucrase-isomaltase deficiency is a rare disease which is usually manifest at weaning but may not become symptomatic until adulthood.[18] The symptoms are similar to those of lactase deficiency. Unlike lactase, both sucrase and maltase can be induced; the enzyme activity rapidly increases in normal persons in response to either sucrose or fructose but not to glucose, galactose, or lactose.[19]

### B. Plant Glucosides

#### *1. Cathartic Glucosides*

A number of plant glucosides, including senna and cascara sagrada, have cathartic activity. The active agents in senna are the β-glucosides of a family of compounds related to emodin; the glucosides have no cathartic activity but the aglycones are highly active. However, administration of the aglycone is of no value since it is inactivated during transit through the stomach and the small bowel. The activity of senna is therefore dependent on the protective action of the β-glucoside against inactivation, followed by the activating effect of bacterial β-glucosidase in releasing the active aglycone. A similar situation applies with cascara sagrada.

#### *2. Cyanogenic Glucosides*

There is a family of cyanogenic glucosides including linamarin (formyl cyanhydrin β-D-glucoside), and amygdalin (acetylcyanhydrin β-D-glucoside). These are hydrolyzed by bacterial β-glucosidase (they are resistant to the brush border enzyme) to release the cyanhydrin which spontaneously hydrolyzes to release hydrogen cyanide. Amygdalin is present in almonds (and its breakdown is responsible for the characteristic smell) and in cassava. The latter is a major starch source in many tropical areas but can only be used for human consumption after exhaustive aqueous extraction of the amygdalin (which is otherwise present in lethal amounts).

#### *3. Other Glucosides*

The carcinogenic glucoside cycasin (methylazoxymethanol β-D-glucoside) is produced by the Cycadaceae; these latter are used as a major starch source in many parts of South-East Asia and Polynesia, the cycasin having first been removed by aqueous extraction. If cycasin is ingested it is hydrolyzed by the gut bacteria β-glucosidase to release methylazoxymethanol which is hepatotoxic and, in rodents, is also carcinogenic. The carcinogenicity of cycasin has been reviewed by Laqueur and Spatz.[20]

A number of plant glucosides are cardioactive, for examples the Lanatosides. Their cardioactivity is not affected by hydrolysis but the aglycone may be responsible for some of the abdominal side-effects of their use.

## VI. CONCLUSIONS

Bacterial hydrolysis of glycosides is of great value in the identification of bacterial species and many of the enzymes are important in determining human lifestyle. Dairy products are consumed in large amounts by those populations with intestinal β-galactosidase, but populations which have lost the enzyme do not consume such foods because the metabolism of the lactose by the gut bacteria causes gut pains, flatus, and diarrhea. A number of plants cannot be used as foods because they contain glucosides which are hydrolyzed by the gut flora to toxic products.

## REFERENCES

1. **Miller, T. L. and Wolin, M. J.,** Fermentation by saccharolytic intestinal bacteria, *Am. J. Clin. Nutr.*, 32, 164, 1979.
2. **De Vries, W. and Stouthamer, A. H.,** Pathway of glucose fermentation in relation to the taxonomy of Bifidobacteria, *J. Bacteriol.*, 93, 574, 1967.
3. **Cummings, J. H.,** Fiber and short chain fatty acids, in *2nd George Washington Univ. Int. Symp. Diet. Fibre*, Vohouny, G. and Dritchevsky, D., Eds., Plenum Press, New York, 1984, in press.

4. **Cummings, J. H.,** Fermentation in the human large intestine: evidence and implications for health, *Lancet,* i, 1206, 1983.
5. **Bond, J. H., Engel, R. R., and Levitt, M. D.,** Factors influencing pulmonary methane excretion in man, an indirect method of studying *in situ* metabolism of the methane producing bacteria, *J. Exp. Med.,* 133, 572, 1971.
6. **Tadesse, K. and Eastwood, M. A.,** Metabolism of dietary fibre components in man assessed by breath hydrogen and methane, *Br. J. Nutr.,* 40, 393, 1978.
7. **Russell, E. G.,** Types and distribution of anaerobic bacteria in the large intestine of pigs, *Appl. Env. Microbiol.,* 37, 187, 1979.
8. **Hawksworth, G. M., Drasar, B. S., and Hill, M. J.,** Gut bacteria and the hydrolysis of glycosidic bonds, *J. Med. Microbiol.,* 4, 451, 1971.
9. **Salyers, A. A.,** Energy sources of major intestinal fermentative anaerobes, *Am. J. Clin. Nutr.,* 32, 158, 1979.
10. **Neale, G.,** Disaccharidase deficiencies, *J. Clin. Pathol.,* 5, 22, 1971.
11. **Dahlqvist, A.,** Digestion of lactose, in *Milk Intolerances and Rejection,* Delmont, J., Ed., S. Karger, Basel, 1983, 11.
12. **Levin, B., Abraham, J. M., Burgess, E. A., and Wallis, P. G.,** Congenital lactose malabsorption, *Arch. Dis. Child.,* 45, 173, 1970.
13. **Bolin, T. D., Crane, G. G., and Davis, A. L. E.,** Lactose intolerance and its various ethnic groups in South East Asia, *Aust. Ann. Med.,* 17, 300, 1968.
14. **Chung, M. H. and McGill, D. B.,** Lactase deficiency in orientals, *Gastroenterology,* 54, 225, 1968.
15. **Gundman-Hoyer, E. and Jarnum, S.,** Lactose malabsorption in Greenland Eskimos, *Acta Med. Scand.,* 186, 235, 1969.
16. **Reddy, V. and Pershad, J.,** Lactase-deficiency in Indians, *Am. J. Clin. Nutr.,* 25, 114, 1972.
17. **Newcomer, A. D., McGill, D. B., Thomas, P. J., and Hofmann, A. F.,** Tolerance to lactose among lactose-deficient American Indians, *Gastroenterology,* 74, 44, 1978.
18. **Pink, I. J.,** Diarrhoea due to sucrase and isomaltose deficiency, *Gut,* 8, 373, 1967.
19. **Rosensweig, N. S. and Herman, R. H.,** Control of jejunal sucrase and maltase activity by dietary sucrose or fructose in man. A model for the study of enzyme regulation in man, *Am. J. Clin. Nutr.,* 22, 99, 1969.
20. **Laqueur, G. L. and Spatz, M.,** Toxicology of cycasin, *Cancer Res.,* 28, 2262, 1968.

Chapter 4

# BACTERIAL METABOLISM OF DIETARY FIBER

**M. J. Hill**

## TABLE OF CONTENTS

## I. INTRODUCTION

Fermentation of dietary fiber in the colon is of major importance both to the gut bacterial flora and to the host. Most of the organisms in the gut are saccharolytic, obtaining their energy mainly from the fermentation of carbohydrate; since the small intestine is so efficient in the absorption of mono- and disaccharide, dietary fiber and intestinal mucin are overwhelmingly the most important sources of fermentable carbohydrate in the large intestine. This is despite the fact that it is now clear that some starch reaches the cecum.[1] The main fermentation products are short-chain fatty acids (SCFA's) gases (hydrogen, carbon dioxide, and methane) and energy for bacterial growth. Only a small fraction of the SCFA released is lost in feces, much being absorbed from the colon and a proportion being converted into bacterial cell mass; the SCFA absorbed from the colon contributes about 1 to 3% of the total energy yield from the diet.[2] In addition, a byproduct of intestinal fermentation of fiber is an increase in fecal bulk[3] and this has important consequences for intestinal physiology since intestinal transit time is related to fecal bulk.[4] The physiological effects of fiber fermentation are thought to protect against constipation, hemorrhoids, and diverticular disease as well as a host of other diseases.[5-7]

In a description of the bacterial metabolism of dietary fiber it is first necessary to define dietary fiber and to consider its composition and analysis. The evidence that dietary fiber is metabolized during colonic transit will then be considered; this evidence is from in vivo and in vitro studies. The enzymes involved in fiber metabolism will be described and finally the consequences to the host of fiber metabolism will be discussed.

## II. COMPOSITION AND ANALYSIS OF DIETARY FIBER

Interest in dietary fiber is cyclical; it is currently very high but the subject was ignored during the 1950s and 1960s, as it was during the 1920s. There have been many definitions of dietary fiber, (the variations depending on the particular interests of the investigator) and this has led to much confusion. It has been defined as that part of the plant cell wall which is not digested by small intestinal enzymes,[7] but this definition assumes a degree of certainty about the efficiency of digestion which is not at present warranted. For example, potato starch is more readily digested when cooked than when raw, while other plant components are not readily digested after heating; this leads to the considerable uncertainty associated with a definition of dietary fiber based on physiological factors. In contrast, definitions based on analytical characteristics are much more certain, but are more difficult to interpret.

The earliest analytical definition was of crude fiber, determined by the Weende method in Germany in the early years of the 19th century. Crude fiber is the fraction remaining insoluble after treatment with hot petroleum ether, hot acid, and hot alkali; it consists of lignin and cellulose but does not contain any noncellulosic polysaccharides (NCP). In 1967 Van Soest[8] described two fiber fractions, neutral detergent fiber (NDF) and acid detergent fiber (ADF). ADF is the fraction remaining insoluble after refluxing with hot detergent in sulfuric acid; it corresponds to crude fiber. NDF is the insoluble fraction of the plant cell wall after refluxing with detergent at neutral pH and corresponds to crude fiber plus the hemicelluloses contaminated with some starch but not containing any soluble NCP. The dietary fiber according to the AOAC method is analyzed gravimetrically after removal of the starch and after correcting for any protein not removed; it corresponds to the dietary fiber as defined physiologically, but does not subdivide the dietary fiber into its components. This method is based on that described by Asp.[9] The most useful definition is based on the analytical procedure described by Englyst et al.[10] In this procedure the starch etc., is first removed from the plant cell wall, the insoluble residue is then subdivided into cellulose, soluble NCP, and insoluble NCP, each being assayed by gas chromatography after prior

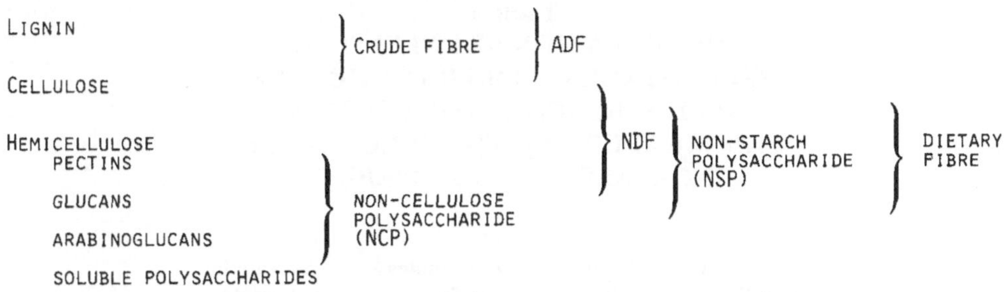


FIGURE 1. The various definitions of dietary fiber and the components included in each.

hydrolysis. The total dietary fiber on this procedure is termed nonstarch polysaccharide (NSP) and does not include lignin. A concensus appears to be emerging that dietary fiber should be defined as lignin plus NSP. These various definitions are described in Figure 1. When hydrolyzed the various NCP fractions yield glucose, galactose, mannose, arabinose, and xylose together with a range of uronic acids. A major cause of error in dietary fiber analysis is the incomplete removal of starch (especially in very starchy foods such as rice) and incorrect correction of gravimetric assays for residual protein (especially with high protein foods such as soya). At present the method of Englyst et al.[10] appears to overcome these problems most effectively.

## III. IN VIVO METABOLISM OF DIETARY FIBER

It is now widely recognized that dietary fiber is extensively degraded during transit through the gut and that this is the cause of many of the physiological effects of dietary fiber. Early studies by Williams and Olmstedt[11] sought an explanation for the effect of dietary fiber on stool bulk. They showed that in three young healthy male adults lignin was virtually undegraded while cellulose and hemicellulose (NCP) were extensively fermented. They showed that the fecal bulk was proportional to the extent of digestion of the fiber and suggested that the stool bulking effect was due to the production of short-chain fatty acids (SCFA's) which had an osmotic effect. Hummel et al.[12] studied 18 children aged 4 to 12 years for a period of 30 to 225 days on a diet containing 4 to 6 g fiber (cellulose + hemicellulose); an average of 71% (range 3 to 98%) of the cellulose and 67% (17 to 83%) of the hemicellulose was digested. They suggested that the range in digestion might be due to variations in the particle size of the fiber and physiological factors such as peristalsis. The proportion digested increased with the age of the child. Many similar studies were carried out in the period 1930 to 1945 while dietary fiber was fashionable as a laxative and in the general maintenance of good health. During the 1950s and early 1960s there was a loss of interest in dietary fiber but this was revived by the evangelical teachings of Burkitt[5] and Trowell[7] that many diseases of Western civilization were caused by the consumption of excessively refined food and the consequent deficiency in consumption of dietary fiber. This thesis was broadly in agreement with that of Cleave[13] who suggested that the cause of the diseases was excessive consumption of refined carbohydrate. The diseases concerned were mainly gastrointestinal (hemorrhoids, constipation, diverticular disease, appendicitis, colorectal cancer, hiatus hernia, peptic ulcer) but included other diseases such as coronary heart disease, diabetes, varicose veins, and obesity. A wide range of different properties of dietary fiber were incriminated in the causation of these diseases including stool bulking, stool softening, more rapid transit of stools, cholesterol binding, affects on glucose absorption and gastric emptying, changes in the gut bacterial flora. This led to an outburst of activity by groups investigating the effect of dietary fiber on the colon and its fate during colonic transit.

**Table 1**
**THE RELATION BETWEEN THE PERCENTAGE OF FIBER IN THE DIET WHICH REACHES THE RECTUM UNDEGRADED AND THE PERCENTAGE INCREASE IN STOOL WEIGHT**

| Source of dietary fiber | Amount in the diet (g) | % remaining in feces | % increase in fecal weight |
|---|---|---|---|
| Wheat bran | 20 | 50 | 127 |
| Cabbage | 20 | 10 | 69 |
| Carrot | 20 | 0 | 59 |
| Guar gum | 20 | 0 | 20 |

**Table 2**
**THE BREAKDOWN DURING COLONIC TRANSIT OF VARIOUS FIBER COMPONENTS IN FIBER FROM A NUMBER OF SOURCES**

| | Degradation of dietary fiber components | | | |
|---|---|---|---|---|
| Source | Ligin | Cellulose | Hemi-cellulose | Ref. |
| Wheat bran | 0% | 29% | 35% | 11 |
| | 0% | 23% | 50% | 4 |
| | 0% | 63% | 42% | 4 |
| Cabbage | 0% | 55% | 80% | 11 |
| | 0% | 81% | 53% | 14 |
| Carrot | 0% | 67% | 84% | 11 |

Some of the most relevant studies were carried out by Stephen and Cummings,[3] who studied the mechanisms by which dietary fiber increased stool bulk. They noted that when 18 g wheat fiber was added to the diet of young healthy volunteers, 11.6 g reached the rectum undegraded and the remaining 6.4 g (36%) must have been degraded during colonic transit. When a similar amount of cabbage fiber was added to the diet only 1.6 g was excreted in feces so that 16.4 g (91%) must have been degraded by the bacteria. Similar results were obtained with fiber from apple, carrot, pectin, and guar (Table 1). These results, using modern and reliable methods of fiber analysis, were in agreement with the early results already described. The differences between bran and the other fiber sources in the extent of metabolism were due to the differences in the composition of the fibers, cellulose being less degraded than the NCP (Table 2). When purified components of dietary fiber were added to the diet, cellulose (as Solka floc) was only 20% degraded while pectin was almost completely degraded,[3] as was guar gum.[3] In contrast there is general agreement that lignin is almost totally undegraded during colonic transit.[11]

The major products of fiber metabolism are the SCFA's (acetic, propionic, and butyric acids), gases ($CO_2$, $H_2$, and $CH_4$) and energy. The production of SCFA's in the human colon has been reviewed by Cummings.[4] They are the predominant anions in the colon of all mammalian species and the major proportion of the SCFA's synthesized in the colon is absorbed across the colonic mucosa with a small proportion lost in feces. The molar ratio of the principal SCFA's in feces is acetic to propionic to butyric 60:24:16 and it has been

**Table 3**
**THE METABOLISM OF PLANT POLYSACCHARIDES BY HUMAN GUT BACTERIA[16,18]**

| Polysaccharide | Composition | Bacterial species able to degrade the polysaccharide |
|---|---|---|
| Cellulose | Linear β (1,3) glucan | *Bacteroides* spp. (various) |
| Noncellulosic | Linear β (1,3) glucans or mixed β (1,3) and β (1,4) glucans | *Bacteroides* spp. (various) |
| Xylan | β (1,4) xylan with arabinose side-chains | *Bacteroides* spp., *Bifidobacterium* spp. |
| Arabinogalactans | β (1,4) galactan with arabinose side-chains | *Bacteroides* spp., *Bifidobacterium* spp. |
| Galactomannans | β (1,4) mannan with galactose side-chains | *Bacteroides* spp., *Ruminococcus* spp. |
| Pectins | Linear β (1,4) polymers containing galacturonic acid | *Bacteroides* spp., *Eubacterium* spp. |

calculated that more than 200 mmol (possibly considerably more) are absorbed from the colon daily[4] in addition to the 7 to 20 mmol/day excreted in feces.

## IV. IN VITRO METABOLISM OF DIETARY FIBER

Salyers and colleagues have carried out an extensive study of the metabolism of plant polysaccharides by gut bacteria. They tested the ability of 25 strains of colonic bacteria to ferment various nonstarch polysaccharides; the results were reviewed by Salyers[16] and are summarized in Table 3. Most of the species able to degrade NSP belonged to the genera *Bacteroides* and *Bifidobacterium*; some species of *Bacteroides* were able to degrade a wide range of different NSP containing a range of sugars and linkages. Ford et al.[17] assayed the enzymes responsible for NSP breakdown in fecal pellets (using nitrophenyl glucosides as substrates) and found high activity of β-galactosidase, β-glucosidase, β-xylosidase, and β-fucosidase.

There has been some work on the location of NSP-degrading enzymes; many of those produced by *Bacteroides* spp. appear to be cell-associated.[16,18] The products of such cell-associated enzymes appear to be small molecular weight (mono- or disaccharide) and so all of the product would be available to the organism; if such enzymes were extracellular the major proportion of the products would be utilized by competing organisms. The enzymes involved in the degradation of guar gum by *Bacteroides* spp. include extracellular enzymes whose products are high molecular weight but noncolloidal; the released products can then be digested by cell associated enzymes.[16]

The in vitro degradation of wheat bran and pea fiber in continuous cultures of mixed fecal organisms has been studied by Fadden et al.[19] who demonstrated the production of large amounts of SCFA under such conditions.

## V. THE CLINICAL SIGNIFICANCE OF DIETARY FIBER METABOLISM

The major consequences of bacterial fermentation of dietary fiber are fecal bulking, more rapid transit of the colonic contents, increased nitrogen utilization in the gut, and the formation of SCFA's usable by the gut mucosa.

It has been suggested that fecal bulking may be important in the prevention or treatment of constipation, diverticular disease, or hemorrhoids. Constipation is one of the commonest diseases in Western civilizations, especially in the elderly, and has been reviewed by Godding[20] who divided the disease into primary and secondary types. Primary constipation includes selfinduced and environmental types while secondary constipation includes idiopathic, psychiatric, genetic, geriatric, obstetric and pathogenic, or iatrogenic types; clearly constipation has a multifactorial etiology and the search for a single cause is likely to be fruitless. Bran fiber is widely used in the treatment of constipation and undoubtedly has had considerable success (presumably due to its stool bulking action) although many prove to be extremely resistant to its effects.[21] The use of wheat bran in the treatment of diverticular disease was strongly and persuasively promoted by Burkitt[6] and by Painter et al.[22,23] and its value in uncontrolled trials has been extensively documented. Double blind trials have been few; one by Brodribb[24] showed that bran was superior to placebo in the alleviation of symptoms of the disease but Ornstein et al.[25] using the same dose of bran and the same formulation for the placebo found that the beneficial effects of the two were equal. The beneficial effects of placebo suggest a psychosomatic component in the etiology of the disease that benefits from placebo treatment but more so from bran given by a convinced clinician (hence the results of the uncontrolled trials).

Although it is clear that a more rapid transit of colonic contents is achieved by bran therapy there is little evidence that this, in itself, has any clinical significance.

Stephen and Cummings[3] demonstrated that one of the major effects of fiber degradation was the synthesis of new bacterial mass and the consequent utilization of nitrogen in the colon. There is a urea cycle, in which blood urea diffuses into the colon to be hydrolyzed by bacterial urease; the released ammonia is absorbed from the colon, returns to the liver via the portal blood where it is metabolized to urea and secreted into the blood for recycling. Part of the ammonia released in the colon may be utilized in the synthesis of bacterial protein and the amount utilized in this way is increased in the presence of readily fermentable carbohydrate. This may be important in patients with renal failure (since it can help to remove excess urea from the blood) and may be very important in patients with hepatoportal encephalopathy. Indeed, Stephen and Cummings[3] suggested that this may be the reason for the success in the treatment of such patients with lactulose.

The SCFA's produced as a result of the degradation of fiber are absorbed from the colon and it has been demonstrated by Roediger[26] that butyrate is utilized by the intestinal mucosal cells as an energy source; indeed it is used in preference to glucose or other potential energy sources reaching the mucosa from the vascular system. Thus, there is a symbiosis between the mucosa and the flora in which the mucosa provides attachment sites for the flora, the host provides nutrients and fermentable carbohydrate to the gut lumen and the flora provides butyrate which is apparently important to the health and well being of the mucosa. Roediger[26] has gone on to hypothesize that inflammatory bowel disease is a disease of the gut mucosa caused by a deficiency in available butyrate. This is an intriguing hypothesis that has had support particularly from Japanese workers and is discussed later in Chapter 15.

## VI. CONCLUSIONS

Dietary fiber is equivalent to that fraction of the diet that used to be referred to as unavailable carbohydrate, and both terms have major shortcomings. Dietary fiber, to many, is a pictorial term which means the structural polysaccharides of plant cell walls — the parts that are fibrous — and yet include components that are readily water soluble. This can be justified by extending the analogy — pectins and gums can be regarded as the equivalent of the plaster on the walls. Unavailable carbohydrate suggests that the carbohydrate is truly unavailable and yet Stephen and Cummings[3] showed that many sources of dietary fiber were almost totally degraded during colonic transit and that they were extremely available to the gut bacteria.

The evidence that dietary fiber is degraded by the gut bacterial flora is very strong and copious, although there have been relatively few investigations of how the degradation takes place. The location of the enzymes and the nature of the products are known for a few strains and with a few substrates but much more needs to be known about how fiber metabolism affects the ability of the flora to utilize and metabolize other substrates.

## REFERENCES


1. **Stephen, A. M., Haddad, A. C., and Phillips, S. F.,** Passage of carbohydrate into the colon. Direct measurements in humans, *Gastroenterology,* 85, 589, 1983.
2. **McNeil, N. I.,** The contribution of the large intestine to energy supplies in man, *Am. J. Clin. Nutr.,* 39, 338, 1984.
3. **Stephen, A. M. and Cummings, J. H.,** Mechanism of action of dietary fibre in the human colon, *Nature (London),* 284, 283, 1980.
4. **Cummings, J. H.,** Short chain fatty acids in the human colon, *Gut,* 22, 763, 1981.
5. **Burkitt, D. P.,** Related disease-related cause?, *Lancet,* 2, 1229, 1969.
6. **Burkitt, D. P.,** Some diseases characteristic of western civilisation, *Br. Med. J.,* 1, 274, 1973.
7. **Trowell, H.,** Definition of dietary fiber and hypotheses that it is a protective factor in certain diseases, *Am. J. Clin. Nutr.,* 29, 417, 1976.
8. **Van Soest, P. J.,** Development of a comprehensive system of food analysis and its application to forages, *J. Anim. Sci.,* 26, 119, 1967.
9. **Asp. N.,** Rapid enzymatic assay of insoluble and soluble dietary fibre, *J. Agric. Food Chem.,* 31, 746, 1983.
10. **Englyst, H., Wiggins, H. S., and Cummings, J. H.,** Determination of the non-starch polysaccharides in plant foods by gas-liquid chromatography of constituent sugars as alditol acetates, *Analyst,* 107, 307, 1982.
11. **Williams, R. D. and Olmstedt, W. H.,** The effect of cellulose, hemi-cellulose and lignin on the weight of a stool: a contribution to the study of laxation in man, *J. Nutr.,* 11, 433, 1936.
12. **Hummel, F. C., Sheperd, M. L., and Macy, I. G.,** Disappearance of cellulose and hemicellulose from the digestive tract of children, *J. Nutr.,* 25, 59, 1943.
13. **Cleave, T. L.,** *The Saccharine Disease,* R. Wright, Bristol, 1974.
14. **Van Soest, P. J., Robertson, J. D., Roe, D. A., Rivers, J., Lewis, B. A., and Hackler, I. R.,** The role of dietary fibre in human nutrition, in Proc. 1978 Cornell Nutr. Conf. Feed Manuf., 1978, 5.
15. **Cummings, J. H., Southgate, D. A. T., Branch, W. J., Wiggins, H. S., Houston, H., Jenkins, D. J. A., Jivraj, T., and Hill, M. J.,** The digestion of pectin in the human gut and its effects on calcium absorption and large bowel function, *Br. J. Nutr.,* 41, 477, 1979.
16. **Salyers, A. A.,** Energy sources of major fermentative anaerobes, *Am. J. Clin. Nutr.,* 32, 58, 1979.
17. **Ford, J. R., Nunley, J. A., Li, Y. T., Chambers, R. P., and Cohen, W.,** A continuously monitored spectrophotometric assay of glycosidases with nitrophenyl glycosides, *Anal. Biochem.,* 54, 120, 1973.
18. **Salyers, A. A.,** Digestion of larch arabinogalactan by a strain of human colonic *Bacteroides* growing in continuous culture, *J. Agric. Food Chem.,* 29, 475, 1980.
19. **Fadden, K., Owen, R. W., Mason, A., and Hill, M. J.,** The effect of wheat bran fibre on the anaerobic metabolism of cholic acid by mixed faecal bacteria, *Trans. Biochem. Soc.,* 12, 860, 1984.
20. **Godding, E. W.,** Physiological yardsticks for bowel function and the rehabilitation of the constipated bowel, *Pharmacology,* 20 (Suppl. 1), 88, 1980.
21. **Brodribb, A. J. M.,** Dietary fibre as a tool of the clinician, in *Dietary Fibre,* Birch, G. G. and Parker, K. J., Eds., Applied Science, London, 1983, 195.
22. **Painter, N. S.,** Diverticular disease of the colon; a disease of this century, *Lancet,* ii, 586, 1969.
23. **Painter, N. S., Almedia, A. Z., and Colebourne, K. W.,** Unprocessed bran in the treatment of diverticular disease, *Br. Med. J.,* 2, 137, 1972.
24. **Brodribb, A. J. M.,** Treatment of symptomatic diverticular disease with a high fibre diet, *Lancet,* i, 664, 1977.
25. **Ornstein, M. H., Littlewood, E. R., Baird, I. M., Fowler, J., North, W. J., and Cox, A. G.,** Are fibre supplements really necessary in diverticular disease of the colon? A controlled clinical trial, *Br. Med. J.,* 282, 1353, 1981.
26. **Roediger, W. E. W.,** The colonic epithelium in ulcerative colitis: an energy-deficiency disease, *Lancet,* ii, 712, 1980.

27. **Heller, S. N.,** The Effect of Particle Size of Dietarty Wheat Bran on Colomic Function in Young Adult Men, Ph.D. thesis, Cornell University, Ithiaca, N.Y., 1977.
28. **Farrell, D. J., Girle, L., and Arthur, J.,** Effects of dietary fiber on the apparent digestibility of major food components and on blood lipids in men, *Austr. J. Exp. Biol. Med. Sci.*, 56, 469, 1978.

Chapter 5

# THE METABOLISM OF BILE ACIDS

**Robert W. Owen**

## TABLE OF CONTENTS

# I. INTRODUCTION

The intestinal tract of human adults contains approximately 1 kg of bacteria equivalent to $10^{14}$ organisms/g representing at least 400 distinct species.[1,2] Over 99% of the organisms are obligate anaerobes. The stomach of humans is essentially sterile but a transient flora can be detected after a meal due to washout from the mouth. Subjects with deficient acid secretion (achlorhydria) however can maintain a relatively rich flora in their stomachs. The jejunum is essentially sterile and the ileum can contain small numbers of bacteria. Thus, from the stomach arborally there is an increase in the gradient of bacteria present until bacteria represent the bulk of the solid material of the colon. Diet has a major influence upon the mass of bacteria present in the colon and it has been shown that acid-resistant dietary fiber escaping from the stomach constitutes a major proportion of the carbohydrate required to maintain the colonic flora at the levels described earlier.[3] Thus, a wide spectrum of bacterial species exists within the colon encompassing obligate anaerobes and facultative anaerobes (aerobes) which have the capacity to metabolize both exogenous and endogenous substances entering the intestine.

Cholesterol enters the intestine from both exogenous (dietary cholesterol) and endogenous sources (hepatic cholesterol) while bile acids are produced endogenously in the liver from cholesterol. The metabolism of cholesterol and plant sterols is discussed elsewhere in this book (see Chapter 6) and will not be referred to in this chapter. Here the metabolism of bile acids only by bacteria, will be addressed.

Prior to describing the numerous reactions undergone by bile acids in relation to bacteria a brief summary of the role of the entero-hepatic circulation (EHC) is outlined below.

The EHC of bile acids in man is a system by which bile acids are recycled and conserved. Bile acids are synthesized and conjugated exclusively in the liver and excreted through the canaliculi into the biliary system. In addition to conjugated bile acids, bile contains cholesterol, phospholipids, bile pigments (mainly bilirubin), and inorganic salts. Bile is concentrated 5- to 10-fold in the gall bladder where it is stored until required. After a meal, on the appearance of food in the duodenum, specialized cells of the duodenum secrete a hormone cholecystokinin (CCK) into the blood; this simultaneously causes contraction of the gall bladder and relaxation of the sphincter of Oddi. Thus, bile flows from the gall bladder into the duodenum and jejunum where the bile salts play a major role along with cholesterol and lecithin in the digestion and absorption of lipids. More than 95% of the bile salts are actively reabsorbed in the terminal ileum and returned to the liver via the hepatic portal vein. The remaining bile salts pass into the large intestine where they undergo bacterial degradation. The degradation products and some unchanged bile acids are mainly excreted in feces but a proportion is passively absorbed from the colon and returned to the liver for reconjugation and resecretion into bile. I shall now describe the various degradative sequences that bile acids may be subjected to, in vivo and also in vitro experiments which have increased our knowledge of bile acid metabolism by bacteria.

# II. METABOLISM OF BILE ACIDS

Bile acid metabolism by bacteria can be studied in a variety of ways including

1. In vivo administration of radio-labeled steroids.
2. Animal model systems; in which germfree animals either mono-, di-, or polyassociated with specific organisms can be compared in metabolic capability to conventional animals.
3. Indirectly by the analysis of bile, urine, serum, and feces.
4. Metabolism of bile acids by mixed fecal bacteria and pure cultures in vitro.

In this chapter I should like to concentrate on the metabolism of bile acids by bacteria in both in vivo and in vitro situations. However, appropriate studies from the other facets described above will be drawn upon to emphasize particular points.

The metabolism of bile acids by bacteria can be conveniently segregated under six main headings;

1. Hydrolysis of conjugated bile acids.
2. Oxidation of 3α-, 7α-, and 12α-hydroxyl groups.
3. Oxido-reduction of 3α-, 7α-, and 12α-hydroxyl groups.
4. 7-Dehydroxylation.
5. Formation of unsaturated bile acids.
6. Hydrolysis of bile acids sulfates.

**A. Hydrolysis of Conjugated Bile Acids**

The major biliary bile acids in man are cholic acid (CA), chenodeoxycholic acid (CDCA), and deoxycholic acids (DCA). The conjugated bile acids are released from the gall bladder by hormonal response to CCK. The conjugates remain intact in the distal ileum but the proportion which escapes absorption, passes into the colon where they are deconjugated by bacteria. Deconjugation in the gut is very thorough and was once thought to go to completion[4,5] however, a proportion of conjugates can be detected in feces of normal healthy people.[6] Under abnormal conditions resulting in bacterial overgrowth of the small intestine, such as the blind-loop syndrome, conjugates are split in the small bowel and fat absorption is impaired because of a reduction of conjugated bile salts below the level necessary for normal fat absorption.[7,8]

The discovery that bile salts are important in maintenance of a healthy colon led to an upsurge in studies related to bile acid physiology and bacterial metabolism. Deconjugation in particular was studied in detail in the late 1960s.[9-11] It is now known that bile salt hydrolysis may be caused by many intestinal organisms including *Bacteroides, Clostridium, Eubacterium, Lactobacillus, Streptococcus,* and *Bifidobacterium.*[12-14]

The enzyme bile acid hydrolase (cholylglycine hydrolase) from *C. perfringens*[15] and *B. fragilis*[16] has been purified and characterized and is now used routinely for the analysis of bile acid conjugates in a variety of biological samples. Among deconjugating bacteria there is considerable enzyme specificity for taurine vs. glycine conjugated bile acids. For example, a bile acid hydrolase from *Peptostreptococcus intermedius* displays a specificity for taurine conjugates while the enzyme from *S. faecalis* and *L. brevis* will hydrolyze chiefly glycine conjugates.[17] In contrast the *C. perfringens* enzyme will hydrolyze both conjugate types and is thus marketed as a commercial enzyme.[15]

Thus, the deconjugation of bile acids by bacteria is well documented and while this reaction has physiological importance in the small bowel it does not appear to exert influence in the large bowel.

**B. Oxidation of 3α-, 7α-, and 12α-Hydroxyl Groups**

The oxidation of bile acid hydroxyl functions to ketone groups is inherently linked to oxido-reduction because hydroxysteroid dehydrogenase (HSDH) enzymes are invariably reversible. The reverse reaction for α-oriented hydroxyl groups is generally a reversion back from the ketonic intermediate (Figure 1); however reduction can also give rise to β-oriented hydroxyl groups. In this section I shall emphasize studies which have concentrated upon the formation of ketone groups by oxidation of bile acid hydroxy groups. Reversible oxidation and epimerization will be dealt with separately later on in this chapter.

Ketonic bile acids have been identified in many biological samples including bile,[18] feces,[19] serum,[20] urine,[21] and gastric contents.[22] Ketonic bile acids while they may represent a

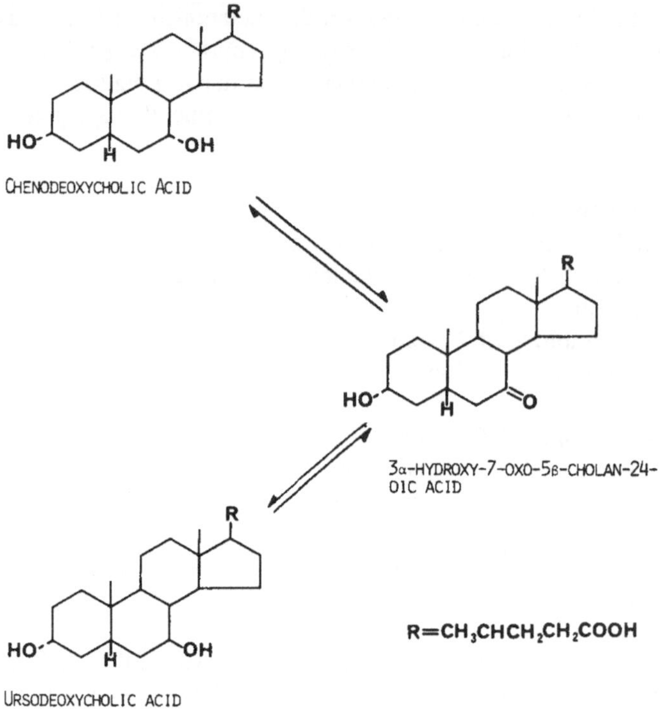


FIGURE 1. 7α-Hydroxysteroid dehydrogenation of chenodeoxycholic acid.

considerable proportion of the fecal bile acid pool in some instances, are rarely detectable in bile. This is because ketonic bile acids, if they are absorbed from the intestine and returned to the liver are reduced primarily to the α-hydroxy conformation by cytochrome P450 mediated hepatic enzymes.[18]

Cell free preparations of 3α-, 7α-, and 12α-HSDH were first reported by Aries and Hill[23] who studied these enzymes from strains of *Clostridium, Bacteroides, Bifidobacterium,* and *Enterobacterium,* isolated from human fecal specimens. 7-HSDH was isolated from two strains of *Clostridium* and *Bacteroides* and one strain of *E. coli*. The enzymes from the *Clostridium* spp. were $NADP^+$-dependent while those from *E. coli* and *Bacteroides* strains utilized $NAD^+$. Compared with the $NADP^+$-dependent enzyme of the *Clostridia*, the $NAD^+$-dependent enzyme of the *Bacteroides* strains was more stable to heat and storage; was subject to inhibition by excess bile acid substrate and had a greater affinity for CDCA than CA. The pH optimum for the oxidation reaction of the *Bacteroides* enzyme was 8.8 to 9.3 while reaction for the *Clostridial* enzymes was 9.8 to 10.2. Subsequently HSDH enzymes have been studied by many groups[24-36] and the following points are pertinent. HSDH are $NAD^+$ or $NADP^+$ dependent and are elaborated by bacteria only. A given organism may elaborate a single enzyme acting solely on a single hydroxy function while others may elaborate several HSDH's acting upon hydroxy functions at 3, 7, and 12. The HSDH enzymes usually have different pH optima for oxidation and reduction; alkaline conditions favor oxidation while pH optima below 7 favors reduction. It is apparent that human fecal bacteria favor the elaboration of 7 HSDH enzymes because bile acids carrying 7-ketonic groups are the major products of co-incubating bile acids with mixed fecal bacteria.

### C. Oxido-Reduction of 3α-, 7α-, and 12α-Hydroxyl Groups

Oxido-reduction of 3α-hydroxy and 7α-hydroxy groups is a common occurrence; 12α-hydroxy oxido-reduction somewhat less so. This section is to be limited to oxido-reductase reactions which involve epimerization only, i.e., specifically the oxidation of a α-orientated hydroxy group to a ketone group with subsequent reduction of the ketone to a β-orientated hydroxyl group.

Evidence for the epimerization of 3α- and 7α-hydroxyl groups of the primary bile acids in man, CDCA and CA has been substantiated by both in vivo and in vitro observations. The transformation of the 3α- and 7α-hydroxyl groups of CDCA and CA to 3β- and 7β-epimers by intestinal microorganisms has been indicated by the presence of such bile acids in the feces of laboratory animals[37] and ursodeoxycholic acid (UDCA) in the bile of man[38] and bears.[39] Ursocholic acid (UC) the 7β-epimer of CA has also been demonstrated in the bile of bears.

The mechanism of 3α-hydroxyl epimerization has been described by Hayakawa[40] who proposed that 3α-OH epimerization occurred by oxidation of the 3α-hydroxyl group to a 3-ketone, followed by subsequent reduction of the ketone to a 3β-hydroxyl group. An analogous mechanism probable gives rise to the 7β-epimer of 7α-hydroxyl groups. These mechanisms have been corroborrated by in vitro studies with individual and mixed fecal cultures of bacteria.

3α-Hydroxyl epimerization has recently been demonstrated by Hirano et al.[41] who showed that 35 strains of *C. perfringens* oxidized the 3α-hydroxy function of CDCA, CA, and DCA to a ketone group, and furthermore 32 of these strains reduced the 3-ketone to the 3β-configuration. This study supported the observations of Macdonald et al.[42] who showed that *C. perfringens* could also carry out the reverse reaction 3β-OH → 3-ketone → 3α-OH. This indicates that *C. perfringens* species contain both 3α-HSDH and 3β-HSDH and shows unequivocally that interconversion involves a 3-keto intermediate.

7α-Hydroxy epimerization is well documented and such activity has been demonstrated in among others *C. absonum*,[43] lecithinase-lipase negative *Clostridia*,[44] and *Eubacterium aureofaciens*.[45] Epimerization of CDCA to UDCA by human intestinal lecithinase-lipase negative *Clostridia* was demonstrated by Edenharder and Knaflic.[44] There were six clostridial strains isolated from human feces which oxidized CDCA to 3α-hydroxy-7-keto-5β-cholanoic acid and epimerized the intermediate to UDCA. Additionally 3α hydroxy-7-keto-5β-cholanoic acid (7-keto-lithocholic acid) was reduced by growing cultures of all six strains to both CDCA and UDCA with the latter reaction being favored.

Macdonald et al.,[43] showed that eight strains of *C. absonum* were capable of forming UC from CA and UDCA from CDCA. Additionally the 7-keto intermediates were also formed. Macdonald et al.[43] have thus proposed that the mechanistic reactions of 7α-epimerization involves the following sequence CA ⇌ 7-KDCA ⇌ UC and CDCA ⇌ 7-KLCA ⇌ UDCA. Further studies by Sutherland and Macdonald[46] using *C. absonum* have shown that at relatively low redox potential (Eh) values high yields of urso-bile acids accrue (60 to 75%). If, however the Eh of the culture was allowed to rise above approximately — 100 mV the 7-keto bile acid would tend to dominate (75%). It is proposed that the formation of excess amounts of 7-keto bile acid is a manifestation of oxygen toxicity and that it could be mediated by an increasing intracellular NAPD:NADPH ratio.

The role of 7-keto-lithocholic acid (7-KLCA) as an intermediate in the epimerization of CDCA to UDCA has been substantiated by Fromm et al.[47] The formation of UDCA from CDCA and the role of 7-KLCA as an intermediate in this biotransformation was studied in vitro in fecal incubations as well as in vivo in the human colon. [24$^{14}$C]-labeled 7-KLCA was instilled into the human colon and rapid conversion of 7-KLCA to UDCA and to a lesser extent CDCA was observed. When [24$^{14}$C]-labeled CDCA was instilled into the colon various amounts of 7-KLCA and UDCA were detected. These results were confirmed using mixed fecal cultures in vitro. Furthermore isotope dilution studies in which trace amounts of labeled 7-KLCA were incubated with unlabeled CDCA indicated 7-KLCA to be a major intermediate in the intestinal bacterial conversion of CDCA to UDCA.

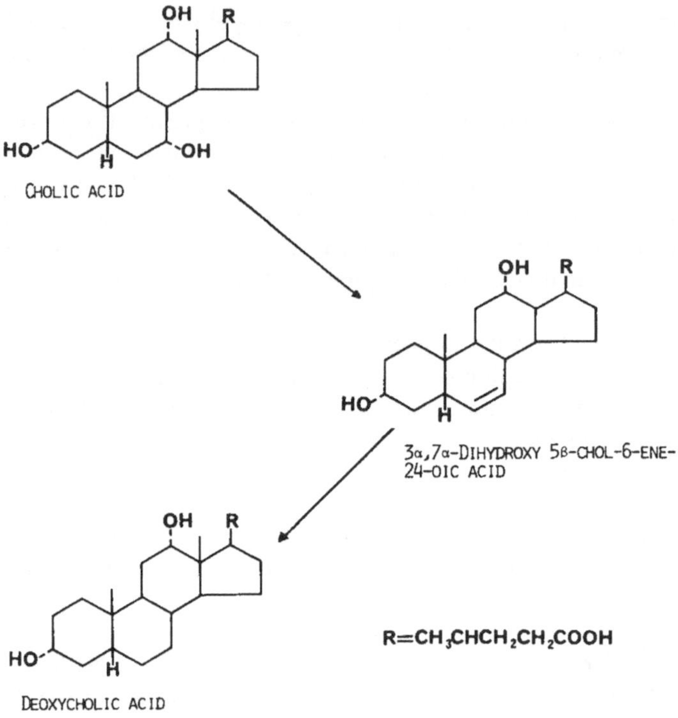


FIGURE 2. 7α-Dehydroxylation of cholic acid.

Epimerization of the 7α-hydroxy group of bile acids has also been observed using consortia of fecal bacteria with 7α- and 7β-HSDH activity respectively.[48] An unidentified Gram-positive anaerobic organism capable of dehydrogenating the 7α-hydroxy group of UDCA was isolated from human feces. By combined action of this organism with known 7α-dehydrogenating organisms such as *E. coli* and *Bacteroides* the consortium was able to epimerize CDCA to UDCA and vice versa.

Thus, the evidence indicates that in the epimerization of 7-hydroxy functions a keto intermediate must be involved in the same way as described for 3-hydroxy functions by Hayakawa.[40]

### D. 7-Dehydroxylation

7α-Dehydroxylation of the primary bile acids CA and CDCA to DCA and LCA respectively is a major bile acid transformation occurring in the human intestine. Samulelsson[49] has proposed that the first step of the 7α-dehydroxylation reaction is the diaxial transelimination of water from the C6-C7 position of CA yielding a C6-C7 unsaturated intermediate which is subsequently reduced to DCA (Figure 2). In man DCA is efficiently conserved and thus, undergoes EHC and can account for about 20% of biliary bile acids. In contrast to DCA, LCA is not efficiently conserved in the EHC and accounts for up to 2% of biliary bile acids.[50] LCA is known to be hepatoxic in several animal species,[51] but in man this is overcome by sulfation,[52] rendering LCA nontoxic. Sulfated LCA ($LASO_4$) can be excreted via the urine but a proportion is known to be released into the small intestine via the gall bladder.

The isolation of fecal bacteria capable of transforming bile acids by 7α-dehydroxylation in vitro was first reported by Gustafsson et al.[53] There were eight strains with this capacity isolated from rat feces and one from human feces and a later investigation by Midvedt[54] revealed that these organisms belonged to the genus *Lactobacillus*.

Later studies by Bokkenheuser et al.[55] and Aries and Hill[23] described the isolation of a

7α-hydroxylating strain of *Bacteroides*. Attempts were made by Aries and Hill to purify the *Bacteroides* enzyme but this was only partially successful.

7α-Dehydroxylase activity has more recently been demonstrated in cultures of *C. bifermentans*,[56] *C. leptum*[57,58] a *Eubacterium* species,[59] and a variety of other anaerobes including *Clostridium* and nonclostridial isolates from human feces and sewers.[60]

Recently important advances have been made in purifying 7α-dehydroxylase enzymes especially from species of the genus *Eubacterium*. Hylemon et al.[61] studied the 7α-dehydroxylation of cholic acid by cell extracts of *Eubacterium* species V.P.I. 12708. 7α-Dehydroxylase was induced 9-fold by CA in growing cultures of this organism. Assays of 7α-dehydroxylase activity in the presence of various co-factors revealed that optimal activity was obtained in the presence of $NAD^+$ and $FADH_2$. The pH optimum of 7α-dehydroxylase activity in whole cells and cell extracts being 7.0. These results suggested a relationship between 7α-dehydroxylase and NADH:flavin oxido-reductase.

These studies were furthered by White et al.[62] who studied the co-factor requirements for 7α-dehydroxylation of CA and CDCA by the same organism. The characteristics of the 7α-dehydroxylase were determined using dialyzed cell extracts of *Eubacterium* sp. V.P.I. 12708. Induction of 7α-dehydroxylase in this organism resulted in the differential synthesis of at least five new polypeptides with molecular weights of 77,000, two at 56,000, 27,000 and 23,500 as determined by both one- and two-dimensional sodium dodecyl sulfate polyacrylamide gel electrophoresis. The relative molecular weight of 7α-dehydroxylase activity was estimated by Biogel A 1.5 *M* gel filtration chromatography to be 114,000. $NAD^+$ was the only co-factor to consistently stimulate 7α-dehydroxylase activity in cell free extracts.

Later studies[63] by the same group showed that NADH inhibited 7α-dehydroxylase activity. Saturation kinetics and double reciprocal plots of NADH inhibition were consistent with negative co-operativity. 7α-Dehydroxylase activity was modulated by the molar ratio of $NAD^+$-NADH with maximal activity at a $NAD^+$ mole fraction of 0.75 to 0.85. Thus, NADH was stimulatory at low concentrations ($<0.15$ m*M*) and inhibitory at high concentrations ($>0.5$ m*M*). Moreover, kinetics for the $NAD^+$ saturation curve did not obey the Michaelis-Menton equation in the presence of NADH (but did so in its absence) suggesting allosteric binding of NADH to the enzyme. The purified enzyme 7-dehydroxylated CA, CDCA, and UDCA but the reaction with UDCA was somewhat slower. This observation is in agreement with the findings of Federowski et al.[64] who showed that dehydroxylation of CDCA was fivefold faster than UDCA in mixed fecal cultures. However in vivo studies show that intestinal transit time is sufficient to allow a similar degree of dehydroxylation of both CDCA and UDCA in normal subjects undergoing dihydroxy bile acid treatment.[65] The dehydroxylase enzymes studied so far appear to be specific in their need for a free $C_{24}$-carboxyl group. The organisms studied by White et al.[62,63] did not produce enzymes that reacted with conjugated, methylated, or unsaturated bile acids which confirm the findings of Aries and Hill.[23]

### E. Formation of Unsaturated Bile Acids

The detection of unsaturated bile acids in vivo has met with very little success. However, it is generally regarded that unsaturated bile acids are produced in the human intestine but their existence is (at most) transient.

To gain an insight into the capability of intestinal bacteria to desaturate bile acids and bile acid type molecules much of the work has involved in vitro studies. Aries et al.[66] and Goddard et al.[67] have shown that lecithinase-negative *Clostridium* spp. contain potent nuclear dehydrogenase (NDH) enzymes. This work concentrated upon the metabolism of 5β-androstan-3,17-dione which has an identical nucleus to an oxo bile acid. NDH *Clostridia* are capable of introducing a double band at $C_4$-$C_5$ into the nucleus of this substrate in the presence of a suitable electron acceptor such as menadione. Later work by Owen and Hill[68]

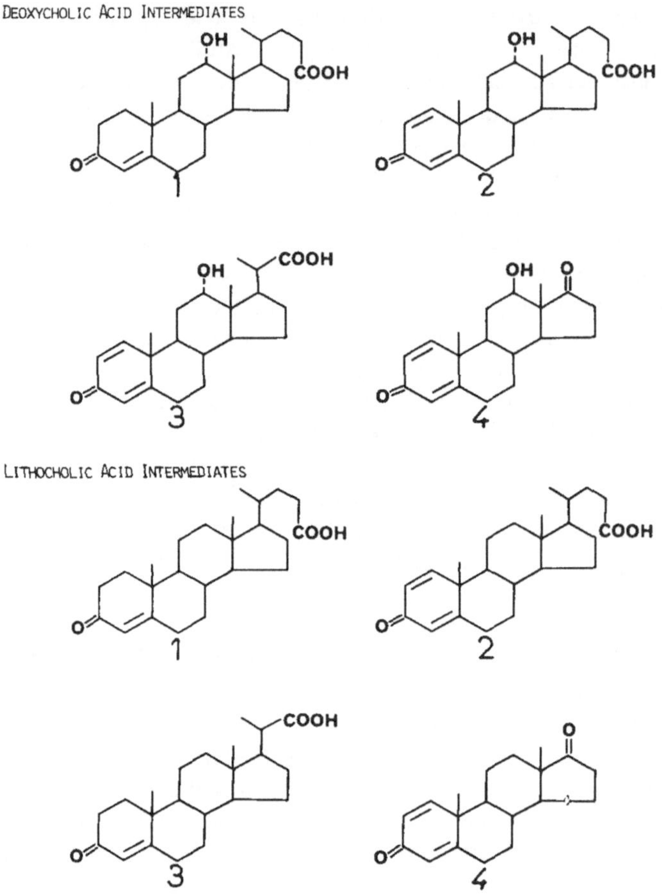


FIGURE 3. Major unsaturated intermediates of deoxycholic acid and lithocholic acid degradation by *Pseudomonas* sp. NCIB 10590 under anaerobic conditions. Deoxycholic acid intermediates: (1) 12α-hydroxy-3-oxochol-4-one-24-oic acid, (2) 12α-hydroxy-3-oxochola-1,4-diene-24-oic acid, (3) 12α-hydroxy-3-oxopregna-1,4-diene-20-carboxylic acid, and (4) 12β-hydroxyandrosta-1,4-diene-3,17-dione. Lithocholic acid intermediates: (1) 3-oxochol-4-ene-24-oic acid, (2) 3-oxochola-1,4-diene-24-oic acid, (3) 3-oxopregna-4-ene-20-carboxylic acid, and (4) androsta-1,4-diene-3,17-dione.

extended these studies to the bile acids and it was shown that *Clostridia* are highly efficient at introducing a double bond at $C_4$-$C_5$ of the major 3-oxo bile acids under anaerobic conditions. The production of unsaturated bile acids has also been demonstrated in *E. coli* (Tenneson et al.[69]), *Bacteroides* species (Owen et al.[70]), and *Bifidobacterium* species (Owen[71]). However apart from *E. coli* and the *Clostridia* the yield of unsaturated bile acids is extremely low and studies involving the more common colonic bacteria has yielded a small range of unsaturated intermediates.

Another approach has been adopted by Owen et al.[72-76] using *Pseudomonas* species NCIB 10590 as a model system. *Pseudomonas* sp. NCIB 10590 although an obligate aerobe which can utilize bile acids as a sole carbon source and can also be induced to metabolize bile acids under anaerobic conditions. Simulation of intestinal Eh in this way has shown that the Pseudomonad in the presence of potassium nitrate as terminal electron acceptor can produce a range of unsaturated intermediates. The metabolism of DCA[72] and LCA[73] gives rise to predominantly 4-ene-3-one and 1,4-dienone intermediates of the cholanoyl, pregnane, and androstane series (Figure 3). The metabolism of CDCA[74] and CA[75] on the other hand yields

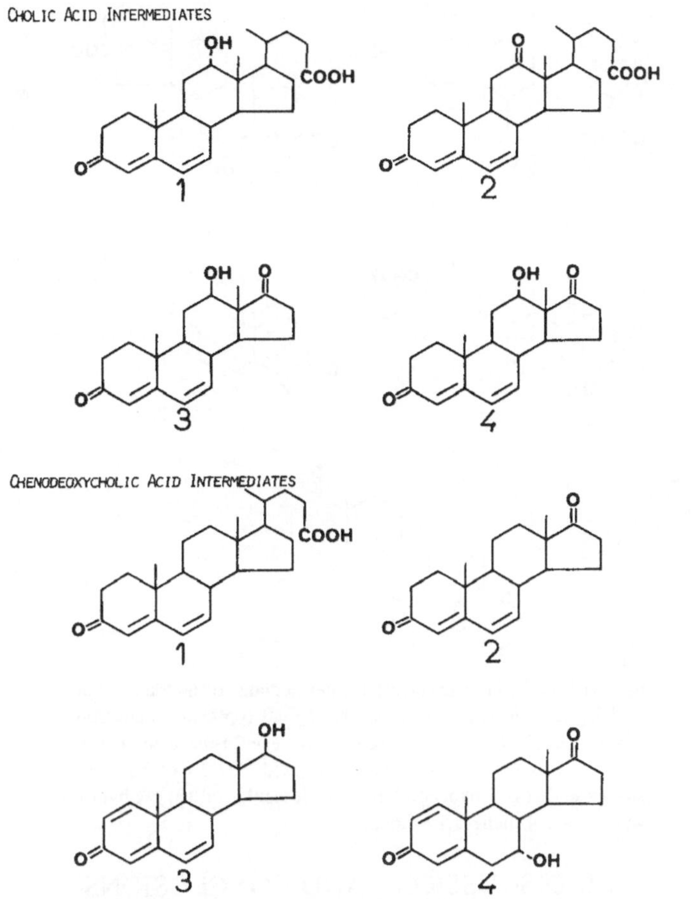


FIGURE 4. Major unsaturated intermediates of cholic acid and chenodeoxycholic acid degradation by *Pseudomonas* sp. NCIB 10590 under anaerobic conditions. Cholic acid Intermediates: (1) 12α-hydroxy-3-oxochola-4,6-diene-24-oic acid, (2) 3,12-dioxochola-4,6-diene-24-oic acid, (3) 12β-hydroxyandrosta-4,6-diene-3,17-dione, and (4) 12α-hydroxyandrosta-4,6-diene-3,17-dione. Chenodeoxycholic acid intermediates: (1) 3-oxochola-4,6-diene-24-oic acid, (2) androsta-4,6-diene-3,17-dione, (3) 17β-hydroxyandrosta-1,4,6-triene-3-one and (4) 7α-hydroxyandrosta-1,4-diene-3,17-dione.

predominantly 4,6-dienone intermediates due to the elaboration of a 7α-dehydroxylase enzyme (Figure 4). Furthermore the metabolism of hyodeoxycholic acid (HDCA)[76] induces a 6α-dehydroxylase enzyme such that two distinct types of unsaturated intermediates accrue (Figure 5).

This work has enabled a comprehensive library of mass spectra of unsaturated bile acids to be lodged at the Bacterial Metabolism Research Laboratory for reference in ongoing studies into the detection of unsaturated bile acids in feces as part of a major investigation into the etiology of colo-rectal cancer.

### F. Hydrolysis of Bile Acid Sulfates

It has been shown that the sulfate ester group of $LASO_4$ can be hydrolyzed by mixed fecal cultures. Kelsey et al.[77] and Borriello and Owen[78] revealed that such a reaction gave rise to LCA, isolithocholic acid, 3-oxo-5β-cholanoic acid, 5β-cholanic acid and $\Delta^3$-cholenic acid. The physiological implications of such reactions have yet to be realized.

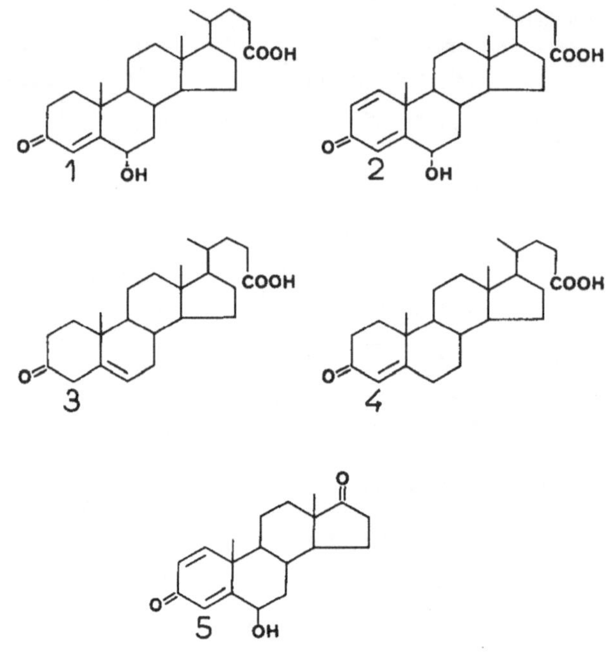


FIGURE 5. Major unsaturated intermediates of hyodeoxycholic acid degradation by *Pseudomonas* sp. NCIB 10590 under anaerobic conditions: (1) 6α-hydroxy-3-oxochol-4-ene-24-oic acid, (2) 6α-hydroxy-3-oxo-chola-1,4-diene-24-oic acid, (3) 3-oxochol-5-ene-24-oic acid, (4) 3-oxochol-4-ene-24-oic acid, and (5) 6α-hydroxyandrosta-1,4-diene-3,17-dione.

## III. DISCUSSIONS AND CONCLUSIONS

Bacteria indigenous to the human large intestine are capable of many and varied bile acid transformations. In this chapter I have categorized broadly the major enzymes elaborated by intestinal and fecal bacteria, and the resulting bile acid transformations which accrue. We now have extensive knowledge about these bile acid enzymes but know little about their physiological significance. For instance there is very little evidence to suggest that bile acid transformations are either beneficial or detrimental to the organisms or to the host. Nevertheless a stable and rich bacterial flora is essential for efficient excretion of intestinal steroids,[79] and there is some evidence to suggest that elaboration of 7α-dehydroxylase enzymes are beneficial to certain bacteria.

To date the bacteria/bile acid interface of the human large intestine has led several research groups[80,81] to suggest that the metabolism of bile acids may be detrimental to the host.

Epidemiological studies[82-86] have shown that continents with a higher per capita intake of dietary fat such as North America and Western Europe have a higher incidence of colorectal cancer (CRC) than continents with low intakes (North Africa, Asia, and South America). Higher intakes of dietary fat and lack of dietary fiber combine to increase colonic bile acid concentrations which in turn positively correlate with the incidence of CRC.[87]

Several studies indicate that bacteria such as *Clostridia*[66,88] may exacerbate the onset of CRC because certain types can produce unsaturated bile acids in vitro.[68] However, although unsaturated bile acids may resemble proximate carcinogens such as 3-methyl cholanthrene and benzo[a]pyrene, to date unsaturated bile acids have not proven to be either mutagenic or carcinogenic.[89]

The epidemiological and circumstantial evidence to my mind favors a role for the bile acids as promotors of CRC. LCA and DCA as the major colonic bile acids have been shown to be co-mutagenic and/or co-carcinogenic in several animal studies.[90,91] Therefore it is possible that these bile acids may augment as yet unidentified colonic carcinogens or mutagens in the etiology of CRC. However because LCA and DCA are the predominant native bile acids in the colonic contents of all populations some other factors must be involved. Recent studies by Owen et al.[92,93] have shown that susceptibility to CRC maybe dependent upon the ratio of LCA to DCA in colonic contents. In a British case-control study[92] cancer patients exhibited a fecal LCA:DCA ratio of 1.43 as opposed to 0.72 in healthy subjects; 75% of the cancer patients excreting more LCA than DCA with 86% of the healthy subjects excreting more DCA and LCA. These observations have been supported by a case-control study in Denmark (Owen et al.[93]) with cancer patients again having a much higher LCA:DCA ratio than controls.

A possible reason for the difference in the fecal LCA:DCA ratio between CRC subjects and controls is a disproportionate synthesis of the primary bile acids in the liver. CDCA is a precursor of LCA, while CA is a precursor of DCA indicating that CRC subjects synthesis more CDCA than CA, the reverse being true for controls. A possible explanation of such differences occurring within a population has been demonstrated by Uchida et al.[94] In animal studies it has been shown that diets rich in cholesterol increases the synthesis of the hepatic CDCA pool. Therefore subjects whose intake of cholesterol is high may change the ratio of primary bile acids in bile, which in turn would lead to a higher colonic and fecal LCA:DCA ratio. The consequences of such changes may lead to an increased risk of contracting CRC, a conclusion which is supported by the recent studies of Wilpart et al.[95] who have shown, especially, that co-incubating high ratios of LCA:DCA in the presence of 1,2-dimethylhydrazine causes the latter to become mutagenic in the Ames Salmonella mutagenicity assay.

In conclusion bacterial metabolism of bile acids in the human large intestine is extensive and our knowledge of these reactions has increased steadily by in vitro experiments. It is essential that areas in which the metabolism of bile acids are implicated in gastrointestinal disease, are studied more fully.

## ACKNOWLEDGMENTS

The author is grateful to the Cancer Research Campaign for financial support.

## REFERENCES

1. **Holdeman, L. V., Good, I. J., and Moore, W. E. C.,** Human faecal flora: a variation in bacterial composition within individuals and a possible effect of emotional stress, *Appl. Environ. Microbiol.*, 31, 359, 1976.
2. **Moore, W. E. C., Cato, C. P., and Holdeman, L. V.,** Some current concepts in intestinal bacteriology, *Am. J. Clin. Nutr.*, 31, 33, 1978.
3. **Stephen, A. M., Haddad, A. C., and Phillips, S. F.,** Passage of carbohydrate into the colon: direct measurements in humans, *Gastroenterology*, 85, 589, 1983.
4. **Hofmann, A. F.,** The enterohepatic circulation of bile acids in man, in *Bile Acids Clinics in Gastroenterology*, Vol. 6, Paumgartner, G., Ed., W. B. Saunders, Philadelphia, 1979, 3.
5. **Carey, M. C.,** The enterohepatic circulation, in *The Liver, Biology and Pathobiology*, Arias, I., Popper, P., Schachter, D., and Shafritz, D. A., Eds., Raven Press, New York, 1982, 429.
6. **Owen, R. W., Thompson, M. H., and Hill, M. J.,** Analysis of metabolic profiles of steroids in faeces of healthy subjects undergoing chenodeoxycholic acid treatment by liquid-gel chromatography and gas-liquid chromatography — mass spectrometry, *J. Steroid Biochem.*, 21, 593, 1984.
7. **Kim, Y. S., Spritz, N., Blum, M., Terz, J., and Sherlock, P.,** The role of altered bile acid metabolism in the steatorrhoea of experimental blind loop, *J. Clin. Invest.*, 45, 956, 1966.
8. **Tabaqchali, S., Hatzionnou, J., and Booth, C. C.,** Bile-salt deconjugation and steatorrhoea in patients with stagnant-loop syndrome, *Lancet*, 2, 12, 1968.

9. **Drasar, B. S., Hill, M. J., and Shiner, M.,** The deconjugation of bile salts by human intestinal bacteria, *Lancet,* 1, 1237, 1966.
10. **Hill, M. J. and Drasar, B. S.,** Degradation of bile salts by human intestinal bacteria, *Gut,* 9, 22, 1968.
11. **Midvedt, T. and Norman, A.,** Anaerobic bile acid transforming microorganisms in rat intestinal contents, *Acta Pathol. Microbiol. Scand.,* 72, 337, 1968.
12. **Dickinson, A. A., Gustafsson, B. E., and Norman, A.,** Determination of bile acid conversion potencies of intestinal bacteria by screening in vitro and subsequent establishment in germfree rats, *Acta Pathol. Microbiol. Scand.,* 79, 691, 1971.
13. **Mahony, D. E., Meier, C. E., Macdonald, I. A., and Holdeman, L. V.,** Bile salt degradation by nonfermentative Clostridia, *Appl. Environ. Microbiol.,* 34, 419, 1977.
14. **Gilliland, S. E. and Speck, M. L.,** Deconjugation of bile acids by intestinal Lactobacilli, *Appl. Environ. Microbiol.,* 33, 15, 1977.
15. **Nair, P. P., Gordon, M., and Reback, J.,** The enzymatic cleavage of the carbon-nitrogen bond in 3α,7α,12α tri-hydroxy-5β-cholan-24-oyl glycine, *J. Biol. Chem.,* 243, 7, 1967.
16. **Stellwag, E. J. and Hylemon, P. B.,** Purification and characterisation of bile salt hydrolase from *Bacteroides fragilis* sub sp *fragilis, Biochim. Biophys. Acta,* 452, 165, 1976.
17. **Kobashi, K., Nishizawa, I., Yamada, T., and Hase, J.,** A new hydrolase specific for taurine conjugates of bile acids, *J. Biochem. (Toyko),* 84, 495, 1978.
18. **Hepner, G. W., Hofmann, A. F., Malagelada, J. R., Szczepanik, P. A., and Klein, P. D.,** Increased bacterial degradation of bile acids in Cholecystectomised patients, *Gastroenterology,* 66, 556, 1974.
19. **Eneroth, P., Gordon, B., Ryhage, R., and Sjövall, J.,** Identification of mono- and dihydroxy bile acids in human faeces by gas-liquid chromatography and mass spectrometry, *J. Lipid Res.,* 7, 511, 1966.
20. **Bjorkhem, I., Angelin, B., Einarsson, K., and Ewerth, S.,** Fasting levels of monoketonic bile acids in human peripheral and portal circulation, *J. Lipid Res.,* 23, 1020, 1982.
21. **Alme, B., Bremmelgaard, A., Sjövall, J., and Thomassen, P.,** Analysis of metabolic profiles of bile acids in urine using a lipophilic anion exchanger and computerized gas-liquid chromtography-mass spectromatry, *J. Lipid Res.,* 18, 339, 1977.
22. **Clayton, P. T., Muller, D. P. R., and Lawson, A. M.,** The bile acid composition of gastric contents from neonates with high intestinal obstruction, *Biochem. J.,* 206, 489, 1982.
23. **Aries, V. C. and Hill, M. J.,** Degradation of steroids by intestinal bacteria. II. Enzymes catalysing the oxido-reduction of the 3α-, 7α-, and 12α-hydroxyl groups in cholic acid, and the dehydroxylation of the 7-hydroxyl group, *Biochim. Biophys. Acta,* 202, 535, 1970.
24. **Macdonald, I. A., Webb, G. R., and Mahony, D. E.,** Faecal hydroxysteroid dehydrogenase activities in vegetarian Seventh Day Adventists, control subjects, and bowel cancer patients, *Am. J. Clin. Nutr.,* 31, 233, 1978.
25. **Macdonald, I. A., Singh, G., Mahony, D. E., and Meier, C. E.,** Effect of pH on bile salt degradation on mixed faecal cultures, *Steroids,* 32, 245, 1978.
26. **Macdonald, I. A., Meier, C. E., Mahony, D. E., and Costain, G. A.,** 3α-, 7α-, and 12α-hydroxysteroid dehydrogenase activities from *Clostridium perfringens, Biochim. Biophys. Acta,* 450, 142, 1976.
27. **Macdonald, I. A., Forrest, T. P., Costain, G. A., and Rao, B. G.,** Identification of 7α-, 12α-dehydroxy-3-oxo-cholanoic acid as the major degradation product from cholic acid by *C. perfringens, J. Steroid Biochem.,* 9, 353, 1978.
28. **Macdonald, I. A., Mahony, D. E., Jellet, J. F., and Meier, C. E.,** NAD-dependent 3α- and 12α-hydroxysteroid dehydrogenase activities, *Biochim. Biophys. Acta,* 489, 466, 1977.
29. **Macdonald, I. A., Jellet, J. F., Mahony, D. E., and Holdeman, L. V.,** Bile salt 3α- and 12α-hydroxysteroid dehydrogenases from *Eubacterium lentum* and related organism, *Appl. Environ. Microbiol.,* 37, 992, 1979.
30. **Haslewood, E. S. and Haslewood, G. A. D.,** The specificity of a 7α-hydroxysteroid dehydrogenase from *Escherichia coli, Biochem. J.,* 157, 207, 1976.
31. **Hylemon, P. B. and Sherrod, J. A.,** Multiple forms of 7α-hydroxysteroid dehydrogenase in selected strains of *Bacteroides fragilis, J. Bacteriol.,* 122, 418, 1975.
32. **Macdonald, I. A., Williams, C. N., Mahony, D. E., and Christie, W. M.,** NAD- and NADP-dependent 7α-hydroxysteroid dehydrogenases from *Bacteroides fragilis, Biochim. Biophys. Acta,* 384, 12, 1975.
33. **Hirano, S. and Masuda, N.,** Transformation of bile acids by *Eubacterium lentum, Appl. Environ. Microbiol.,* 42, 912, 1981.
34. **Harris, J. N. and Hylemon, P. B.,** Partial purification and characterisation of NADP-dependent 12α-hydroxysteroid dehydrogenase from *Clostridium leptum, Biochem. Biophys. Acta,* 528, 148, 1978.
35. **Hylemon, P. B. and Stellwag, E. J.,** Bile acid biotransformation rates of selected Gram-positive and Gram-negative intestinal anaerobic bacteria, *Biochem. Biophys. Res. Comm.,* 69, 1088, 1976.
36. **Hirano, S. and Masuda, N.,** Characterisation of NADP-dependent 7β-hydroxysteroid dehydrogenases from *Peptosteptococcus productus* and *Eubacterium aureofociens, Appl. Environ. Microbiol.,* 43, 1057, 1982.

37. **Elliot, W. H.,** Allo bile acids, in *The Bile Acids,* Vol. 1, Nair, P. P. and Kritchevsky, D., Eds., Plenum Press, New York, 1971, 47.
38. **Igimi, H.,** Ursodeoxycholic acid — a common bile acid in gallbaldder bile of Japanese subjects, *Life Sci.,* 18, 993, 1976.
39. **Kurozumi, K., Harano, T., Yamasaki, K., and Ayak, Y.,** Studies on bile acids in bear bile, *J. Biochem.,* 74, 485, 1973.
40. **Hayakawa, S.,** Microbial transformation of bile acids, *Adv. Lipid Res.,* 11, 143, 1973.
41. **Hirano, S., Masuda, N., Oda, H., and Mukai, H.,** Transformation of bile acids by *Clostridium perfringens, Appl. Environ. Microbiol.,* 42, 394, 1981.
42. **Macdonald, I. A., Hutchison, D. M., Forrest, T. P., Bokkenheuser, V. D., Winter, J., and Holdeman, L. V.,** Metabolism of primary bile acids by *Clostridium perfringens, J. Steroid Biochem.,* 18, 94, 1983.
43. **Macdonald, I. A., Hutchison, D. M., and Forrest, T. P.,** Formation of urso- and ursodeoxycholic acids from primary bile acids by *Clostridium absonum, J. Lipid Res.,* 22, 458, 1981.
44. **Edenharder, R. and Knaflic, T.,** Epimerization of chenodeoxycholic acid to ursodeoxycholic acid by human intestinal lecithinase-lipase-negative Clostridia, *J. Lipid Res.,* 22, 652, 1981.
45. **Macdonald, I. A., Rochon, Y. P., Hutchison, D. M., and Holdeman, L. V.,** Formation of ursodeoxycholic acid from chenodeoxycholic acid by a 7β-hydroxysteroid dehydrogenase-elaborating *Eubacterium aerofociens* strain when co-cultured with 7β-hydroxysteroid dehydrogenase-elaborating organisms, *Appl. Environ. Microbiol.,* 44, 1187, 1982.
46. **Sutherland, J. D. and Macdonald, I. A.,** The metabolism of primary, 7-oxo, and 7-hydroxy bile acids by *Clostridium absonum, J. Lipid Res.,* 23, 841, 1983.
47. **Fromm, H., Sarva, R. P., and Bazzoli, F.,** Formation of ursodeoxycholic acid from chenodeoxycholic acid in the human colon: studies of the role of 7-ketolithocholic acid as an intermediate, *J. Lipid Res.,* 24, 841, 1983.
48. **Hirano, S. and Masuda, N.,** Epimerization of the 7-hydrosy group of bile acids by the combination of two kinds of microorganisms with 7α- and 7β-hydroxysteroid dehydrogenase activity, respectively, *J. Lipid Res.,* 22, 1060, 1981.
49. **Sammuelsson, B.,** Bile acids and steroids. 96: on the mechanism of the biological formation of deoxycholic acid from cholic acid, *J. Biol. Chem.,* 235, 361, 1960.
50. **Cowen, A. E., Kotman, M. G., Hofmann, A. F., Cass, O. W., and Coffin, S. B.,** Metabolism of lithocholic acid in healthy man. II. Enterohepatic circulation, *Gastroenterology,* 69, 67, 1975.
51. **Palmer, R. H.,** Bile acids, liver injury and liver disease, *Arch. Intern. Med.,* 130, 606, 1975.
52. **Palmer, R. H.,** The formation of bile acid sulphates. A new pathway of bile acid metabolism in humans, *Proc. Natl. Acad. Sci. U.S.A.,* 58, 1047, 1967.
53. **Gustafsson, B. F., Midvedt, T., and Norman, A.,** Isolated faecal microorganisms capable of 7α-dehydroxylating bile acids, *J. Exp. Med.,* 123, 413, 1966.
54. **Midvedt, T.,** Properties of anaerobic Gram-positive rods capable of 7α-dehydroxylating bile acids, *Acta Pathol. Microbiol. Scand.,* 71, 147, 1967.
55. **Bokkenheuser, V., Hoshita, T., and Mosbach, E. H.,** Bacterial 7-dehydroxylation of cholic acid and allocholic acid, *J. Lipid Res.,* 10, 421, 1969.
56. **Ferrari, A. and Beretta, L.,** Activity on bile acids of a *Clostridium bifermentans* cell-free extract, *FEBS Lett.,* 75, 163, 1977.
57. **Stellwag, E. J. and Hylemon, P. B.,** 7α-dehydroxylation of cholic acid and chenodeoxycholic acid by *Clostridium leptum, J. Lipid Res.,* 20, 325, 1979.
58. **Stellwag, E. J. and Hylemon, P. B.,** Characterisation of 7α-dehydroxylase in *Clostridium leptum, Am. J. Clin. Nutr.,* 31, 243, 1978.
59. **White, B. A., Lipsky, R. L., Fricke, R. J., and Hylemon, P. B.,** Bile acid induction specificity of 7α-dehydroxylase activity in an intestinal *Eubacterium* species, *Steroids,* 35, 103, 1980.
60. **Hirano, S., Nakama, R., Tomaki, M., Masuda, N., and Oda, H.,** Isolation and characterisation of thirteen intestinal microorganisms capable of 7α-dehydroxylating bile acids, *Appl. Environ. Microbiol.,* 41, 737, 1981.
61. **Hylemon, P. B., Cacciapuoti, A. F., White, B. A., Whitehead, T. R., and Fricke, R. J.,** 7α-Dehydroxylation of cholic acid by all extracts of *Eubacterium* species V.P.I. 12708, *Am. J. Clin. Nutr.,* 33, 2507, 1980.
62. **White, B. A., Cacciapuoti, A. F., Fricke, R. J., Whitehead, T. R., Mosbach, E. H., and Hylemon, P. B.,** Cofactor requirements for 7α-dehydroxylation of cholic and chenodeoxycholic acid in cell extracts of the intestinal anaerobic bacterium, *Eubacterium* species V.P.I. 12708, *J. Lipid Res.,* 22, 891, 1981.
63. **White, B. A., Paone, D. A. M., Cacciaputoi, A. F., Fricke, R. J., Mosbach, E. H., and Hyleman, P. B.,** Regulation of bile acid 7-dehydroxylase activity by $NAD^+$ and NADH in cell extracts of *Eubacterium* species V.P.I. 12708, *J. Lipid Res.,* 24, 20, 1983.
64. **Federowski, T., Salen, G., Tint, G. S., and Mosbach, E. H.,** Transformtion of chenodeoxycholic acid and ursodeoxycholic acid by human intestinal bacteria, *Gastroenterology,* 77, 1068, 1979.

65. **Dodo, M., Owen, R. W., Thompson, M. H., and Hill, M. J.,** A comparison of the effects of chenodeoxycholic and ursodeoxycholic acid treatment on faecal bile acid profiles in healthy subjects, *Biochem. Soc. Trans.*, 12, 862, 1984.
66. **Aries, V. C., Goddard, P., and Hill, M. J.,** Degradation of steroids by intestinal bacteria. III. 3-oxo-5β-steroid $\Delta^4$-dehydrogenase and 3-oxo-5β-steroid $\Delta^{1,4}$-dehydrogenase, *Biochim. Biophys. Acta*, 248, 482, 1972.
67. **Goddard, P., Fernandez, F., West, B., Hill, M. J., and Barnes, P. J.,** The nuclear dehydrogenation of steroids by intestinal bacteria, *J. Med. Microbiol.*, 8, 429, 1975.
68. **Owen, R. W.,** Biotransformation of bile acids by Clostridia, *J. Med. Microbiol.*, 20, 233, 1985.
69. **Tenneson, M. E., Owen, R. W., and Mason, A. N.,** The anaerobic side-chain cleavage of bile acids by *Escherichia coli* isolated from human faeces, *Biochem. Soc. Trans.*, 5, 1711, 1977.
70. **Owen, R. W., Bilton, R. F., and Tenneson, M. E.,** The degradation of cholic acid and deoxycholic acid by *Bacteroides* species under strict anaerobic conditions, *Biochem. Soc. Trans.*, 5, 1711, 1977.
71. **Owen, R. W.,** The degradation of bile acids by *Bifidobacterium* species, in the *Anaerobic degradation of steroids*, Ph.D. thesis, Council for National Academic Awards 1980, Liverpool Polytechnic, England, 121.
72. **Owen, R. W., Mason, A. N., and Bilton, R. F.,** The anaerobic degradation of deoxycholic acid by *Pseudomonas* sp. NCIB 10590, *Tetrahedron*, 40, 421, 1984.
73. **Owen, R. W. and Bilton, R. F.,** Bioconversion of lithocholic acid under anaerobic conditions by *Pseudomonas* sp NCIB 10590, *Appl. Environ. Microbiol.*, 48, 606, 1984.
74. **Owen, R. W., Hill, M. J., and Bilton, R. F.,** Biotransformation of chenodeoxycholic acid by *Pseudomonas* sp NCIB 10590 under anaerobic conditions, *J. Lipid Res.*, 24, 1109, 1983.
75. **Owen, R. W. and Bilton, R. F.,** The degradation of cholic acid by Pseudomonas sp NCIB 10590 under anaerobic conditions, *Biochem. J.*, 216, 641, 1983.
76. **Owen, R. W. and Bilton, R. F.,** The transformation of hyodeoxycholic acid by Pseudomonas sp NCIB 10590 under anaerobic conditions, *J. Steroid Biochem.*, 19, 1355, 1983.
77. **Kelsey, M. I., Muschik, G. M., and Sexton, G. A.,** The metabolism of lithocholic acid-3α-sulphate by human intestinal microflora, *Lipids*, 13, 152, 1978.
78. **Borriello, S. P. and Owen, R. W.,** The metabolism of lithocholic acid and lithocholic acid-3α-sulphate by human faecal bacteria, *Lipids*, 17, 477, 1982.
79. **Burczak, J., Ghnassia, J., Waggoner, C., Vernon, M., Griscelli, C., and Kellog, T.,** Steroid excretion of children in the germfree and conventional state, *Clinical and Experimental Gnotobiotics*, Fliedner et al., Eds., Gustav Fischer, New York, 1979, 285.
80. **Mastromarino, A., Reddy, B. S., and Wynder, E. L.,** Metabolic epidemiology and colon cancer: enzymic activity of faecal flora, *Am. J. Clin. Nutr.*, 29, 1455, 1976.
81. **Aries, V. C., Crowther, J. S., Drasar, B. S., Hill, M. J., and Williams, R. E. O.,** Bacteria and the aetiology of cancer of the large bowel, *Gut*, 10, 334, 1969.
82. **Armstrong, B. and Doll, R.,** Environmental factors and cancer incidence and mortality in different countries, with special reference to dietary practices, *Int. J. Cancer*, 15, 617, 1975.
83. **Burkitt, D. P.,** Large bowel carcinogenesis: an epidemiologic jigsaw puzzle, *J. Natl. Cancer Inst.*, 54, 3, 1975.
84. **Haenszel, W., Berg, J. W., Segi, M., Kurihara, M., and Locke, F. B.,** Large bowel cancer in Hawaiian Japanese, *J. Natl. Cancer Inst.*, 51, 1765, 1973.
85. **Wynder, E. L.,** The epidemiology of large bowel cancer, *Cancer Res.*, 35, 3388, 1975.
86. **Wynder, E. L. and Shigematsu, T.,** Environmental factors of cancer of the colon and rectum, *Cancer*, 20, 1520, 1967.
87. **Hill, M. J., Drasar, B. S., Aries, V. C., Crowther, J. S., Hawksworth, G. M., and Williams, R. E. O.,** Bacteria and aetiology of cancer of the large bowel, *Lancet*, 1, 95, 1971.
88. **Hill, M. J., Drasar, B. S., Williams, R. E. O., Meade, T. W., Cox, A. G., Simpson, J. E. P., and Morson, B. C.,** Faecal bile acids and Clostridia in patients with cancer of the large bowel, *Lancet*, 1, 535, 1975.
89. **Mckillop, C. A., Owen, R. W., Bilton, R. F., and Haslam, E. A.,** Mutagenicity testing of steroids obtained from bile acids and cholesterol, *Carcinogenesis*, 4, 1179, 1983.
90. **Narisawa, T., Magadia, N. E., Weisberger, J. H., and Wynder, E. L.,** Promoting effect of bile acids on colon carcinogenesis after intrarectal instillation of MNNG in rats, *J. Natl. Cancer Inst.* 53, 1093, 1974.
91. **Silverman, S. J. and Andrews, A. W.,** Bile acids; co-mutagenic activity in the Salmonella-mammalian-microsome mutagenicity test. Brief communication, *J. Natl. Cancer Inst.*, 59, 1557, 1977.
92. **Owen, R. W., Dodo, M., Thompson, M. H., and Hill, M. J.,** The faecal ratio of lithocholic acid to deoxycholic acid may be an important aetiological factor in colo-rectal cancer, *Biochem. Soc. Trans.*, 12, 861, 1984.
93. **Owen, R. W., Thompson, M. H., and Jensen, O. M.,** unpublished observations.

94. **Uchida, K., Takase, H., Nomura, Y., Takeda, K., Takeuchi, N., and Ishikawa, Y.,** Changes in biliary and faecal bile acids in mice after treatment with diosgenin and β-sitosterol, *J. Lipid Res.*, 25, 236, 1984.
95. **Wilpart, M., Mainguet, P., Maskens, A., and Roberfroid, M.,** Structure-activity relationship amongst biliary acids showing co-mutagenic activity towards 1,2-dimethyl-hydrazine, *Carcinogenesis*, 4, 1239, 1983.

Chapter 6

# METABOLISM OF NEUTRAL STEROIDS

**M. H. Thompson**

## TABLE OF CONTENTS

# I. INTRODUCTION

The complex populations of predominantly anaerobic bacteria that colonize the digestive tract, especially the terminal ileum and large bowel, exhibit a wide range of metabolic activities. In particular these organisms can metabolize a variety of steroids including those encountered in vivo such as cholesterol from dietary and bodily sources, phytosterols from plants, and steroid hormones from endogenous sources. In general it may be assumed that the profile of fecal steroids mirrors that found in the descending colon but is probably qualitatively different from that found in the cecum. All of the steroids passing into the colon come into intimate contact with the colonic flora and may therefore be considerably degraded by these organisms to yield a complex mixture of substrates. Some of the minor components may be derived from metabolites absorbed from the large bowel undergoing further circulation and degradation.[1]

Studies with human volunteers, primates, and rodents have confirmed that the bacteria residing in the gastrointestinal tract are actively involved in the degradation of steroids entering the colon. Both in vitro and in vivo investigations have identified a number of degradative pathways associated with these organisms including deconjugation, nuclear- and hydroxysteroid-dehydrogenation, epimerization of hydroxyl groups, isomerization of double bonds, dehydroxylation, oxido-reduction and cleavage of side chains. In vitro studies have demonstrated that cultures of mixed fecal organisms derived from humans or rodents can degrade many steroids. Similar metabolic activities have frequently proved difficult to induce in pure cultures of fecal organisms, the incubation requiring carefully controlled conditions and specially supplemented media.[2]

This review will therefore examine the role that intestinal bacteria play in the metabolism of cholesterol, phytosterols, and steroid hormones. Evidence for these activities may be derived by reference to indirect data from human and animal studies as well as from in vitro studies with fecal bacterial.

# II. DEGRADATION OF CHOLESTEROL AND PHYTOSTEROLS

## A. Human Studies

In man a proportion of the cholesterol entering the digestive tract bypasses absorption in the small intestine and enters the colon. This cholesterol is derived from biliary and dietary sources and is supplemented by material from desquamated intestinal cells. Depending upon the source of the cholesterol it may be present in the lumen as the free steroid,[3] complex lipid esters,[4] or as cholesterol sulfate.[5] Under normal circumstances healthy subjects can excrete up to 800 mg/day of cholesterol and its major degradation product coprostanol (Figure 1). Feces from adults also contain a number of minor cholesterol related steroids including coprostanone, cholestanol, epicoprostanol, cholest-4-ene-3-one, lanosterol, dihydrolanosterol, cholest-7-ene-3-ol and methostenol (Figure 2).[6,7] Reddy et al.[8] have also reported detectable levels of cholestane-3β, 5α, 6β-triol presumably derived from cholesterol-5α,6α-epoxide.

The role that intestinal bacteria play in forming cholesterol degradation products in the colon has been deduced from a number of studies of the steroid profiles in feces from different human groups. In infants very little fecal coprostanol is formed during the early weeks of life[9] with some degradation being observed at 12 months and an adult-like profile being established by the age of 4.[10] In a further study of young subjects maintained under germfree conditions[11] cholesterol degradation was suppressed when compared to control subjects. It may therefore be inferred from these observations that the degradation of cholesterol in the lumen is associated with the establishment and activation of the luminal flora in the previously sterile infant bowel.

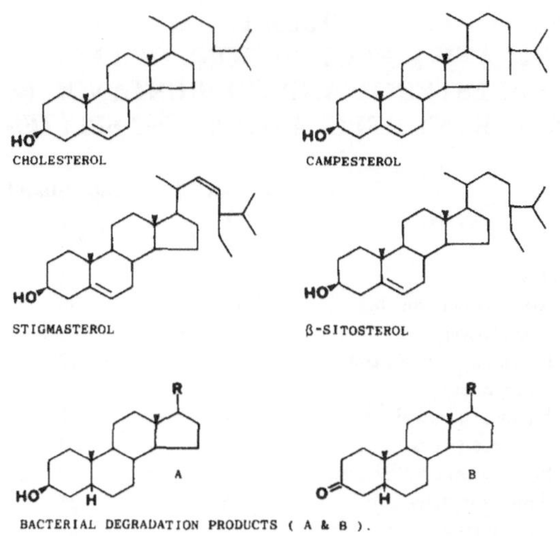


FIGURE 1. Neutral sterols present in feces. (R = CH $(CH_3)$ $CH_2CH_2CH_2CH(CH_3)_2$, A = coprostanol, B = coprostanone; R = $-CH$ $(CH_3)CH_2CH_2CH$ $(CH_3)CH(CH_3)_2$, A = campestanol, B = campestanone; R = $-CH(CH_3)CH{=}CH{\cdot}CH(CH_2CH_3)CH(CH_3)_2$, A = stigmastanol, B = stigmastanone; R = $-CH(CH_3)CH_2CH_2CH(CH_2CH_3)CH(CH_3)_2$, A = sitostanol, B = sitostanone).

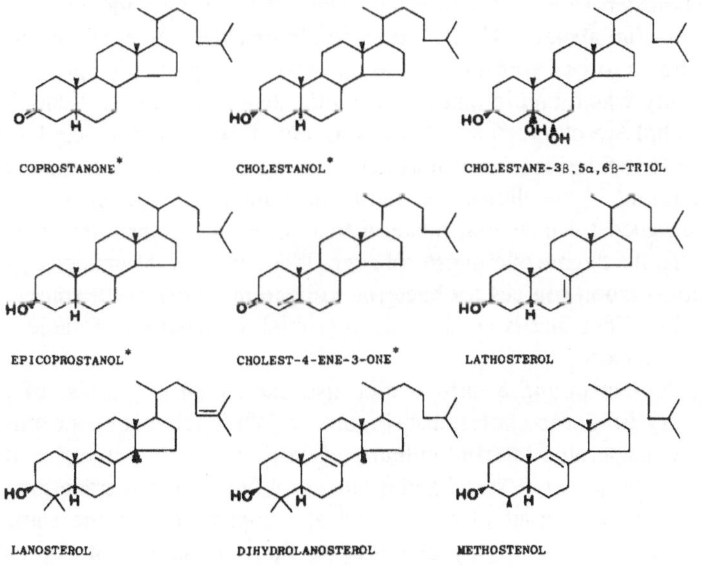


FIGURE 2. Minor cholesterol derivatives present in feces. (*Known microbial degradation products.)

In healthy adults more than 70% of the fecal cholesterol is normally converted to coprostanol although in some population studies[3,12] it has been shown that a small proportion of the subjects do not excrete metabolized cholesterol (Table 1). The degree of cholesterol degradation is also suppressed in subjects taking antibiotics that may alter the activity of the fecal flora. Korpela et al.[13] have recently demonstrated such an effect when the deg-

**Table 1**
**RELATIVE PROPORTIONS OF FECAL CHOLESTEROL AND COPROSTANOL IN DIFFERENT POPULATIONS AND SPECIES**

| Species | Cholesterol (%) | Coprostanol (%) |
|---|---|---|
| Human | | |
| North American[3](high converter) | 12 | 88 |
| North American (low converter) | 88 | 12 |
| Japanese (Hawaii)[25] | 51 | 49 |
| Japanese (Rural) | 36 | 64 |
| Finns (Control)[(13)] | 3 | 97 |
| Finns (Oxytetra-cyc-line treated) | 75 | 25 |
| Rats[(20)] | | |
| Conventional (Male) | 61 | 39 |
| Germfree (Male) | 100 | 0 |
| Pig[22] | 27 | 73 |
| Pigeon[23] | 100 | 0 |
| Alligator[24] | 62 | 38 |

radation of cholesterol in feces of subjects taking the wide spectrum antibiotic oxytetracycline fell from approximately 80 to 10%. In additional studies where adult volunteers ingested $^{14}$C-labeled cholesterol[7] assay of the distribution of radioactivity in the fecal steroids has separated the cholesterol derived substrates (labeled) from the endogenously derived cholesterol precursors (unlabeled). Under these conditions the major cholesterol degradation products were shown to be coprostanol, coprostanone, and probably epicoprostanol (Figure 2). No radioactivity was found in lanosterol, methostenol, or cholest-7-ene-3-ol confirming that they were cholesterol precursors, nor was any label found in cholestanol which is normally assumed to be derived from cholesterol. In some subjects participating in cholesterol balance studies not all of the cholesterol can be accounted for, suggesting that a significant proportion is degraded by intestinal bacteria to nonsteroidal substrates.[14] It has also been hypothesized[15] that the degree of cholesterol degradation may be influenced by the availability of more accessible carbon sources for bacterial utilization. When these sources are depleted, such as when a low fiber diet is consumed, microbial degradation of cholesterol and other steroids may be enhanced.

Healthy subjects consuming a normal diet also excrete plant sterols, of dietary origin, with structures very similar to cholesterol (Figure 1). While the fecal concentrations of these substrates show considerable interindividual and daily variations the excretion of the main plant sterols stigmasterol, campesterol and β-sitosterol, with their degradation products, does not normally excede the output of cholesterol and coprostanol in the same subject.[16] In general fecal phytosterols are highly degraded to the homologues of coprostanol and coprostanone (Figure 1).[17] Unlike cholesterol most evidence indicates that the plant sterols and their degradation products, with the exception of β-sitosterol,[18] are neither well absorbed nor secreted in the human gut.[14] Therefore as these substrates are not absorbed in the host their degradation is more clearly associated with the metabolic activity of the intestinal bacteria. In infants β-sitosterol has been detected in the feces of 4-week-old subjects although the microbial reduction product 24α-ethylcoprostanol does not appear until the subjects are 1-year-old.[10] In some subjects a proportion of the phytosterols cannot be accounted for and may also be degraded to nonsteroidal substrates.[14]

The development of phytosterol metabolism in the human gut is therefore very similar to that for cholesterol and is strongly associated with the development of the gut flora.

### B. Animal Studies

Analysis of the feces from different animals has revealed that cholesterol is degraded in the intestine of these species to coprostanol and coprostanone (Table 1). The profile of cholesterol degradation products in these animals is qualitatively similar to that in man although considerable interspecies and interindividual variation is observed.[19] Where degradation is not observed, as in some pigeon species,[23] this may be attributed to the gut flora being different to that found in animals able to reduce cholesterol. In germfree rats[20] and rats treated with penicillin[21] the degradation of cholesterol is suppressed when compared to controls. It may therefore be inferred from these observations that the intestinal flora in various animals is intimately associated with cholesterol metabolism in the intestine.

As in man, rat feces contain a complex profile of cholesterol associated derivatives (Figures 1 and 2) although the relative proportions of cholesterol precursors is higher.[26] Prelabeling the rats with [$^{14}$C]-cholesterol similarly revealed that the major cholesterol degradation products in the intestine are coprostanol, coprostanone, and cholestanol.[7] More detailed investigations[27] have demonstrated that the rat cecum is the primary site of cholesterol degradation as this activity is lost within 2 days of the cecum being surgically removed. In germfree rats intestinal cholesterol metabolism may be re-established by introducing a Bacterium*(Eubacterium 21408)* known to possess this activity, into the gut.[28] The carriage of the bacterium requires stabilization by the presence of a second intestinal organism *(Clostridium paraputrificum).* As observed with conventional rats this artificially established activity is lost when the cecum is removed and the carriage of the *Eubacterium* is suppressed. Although these surgical procedures alter the overall function of the rat bowel these data are further evidence that intestinal bacteria are closely associated with cholesterol degradation in the gut. It has also been demonstrated that rats consuming varying amounts of the common plant sterols excrete stigmasterol, β-sitosterol and campesterol as well as the homologous degradation products of coprostanol and coprostanone.[7] The degradation of phytosterols is suppressed in germfree rats and lost in cecotamized rats, paralleling similar effects on cholesterol degradation in these animal models.

In general the data from animal studies have confirmed the human observations that the intestinal bacteria can degrade cholesterol and plant sterols to reduced products. The evidence from germ free animals confirms these deductions with the additional observation that the major site of microbial degradation of steroids, in the rat at least, is the cecum.

### C. In Vitro Studies

Additional information on the mechanism of microbial degradation of cholesterol and phytosterols has been derived from the incubation of these substrates with intestinal bacteria. Cholesterol is reduced to coprostanol in cultures of mixed fecal organisms derived from freshly voided feces.[2,29] Cholesterol and coprostanol may also be extensively esterified in mixed fecal culture and, in some cases, degraded to nonsteroidal substrates.[30] β-Sitosterol is similarly metabolized in cultures of mixed fecal organisms where the extent of reduction, esterification, and destruction is similar to that for cholesterol in the same incubation.[3] When fecal samples from different subjects are used to provide the cultures for these incubations considerable variation in the extent of the reactions is observed. This is attributed in part to each fecal sample having a different profile of bacteria present which alters the metabolic activity of the incubation.

In contrast considerable difficulty has been encountered in identifying the organisms that reduce cholesterol in vivo and in carrying out this reaction with pure cultures in vitro. Using a medium supplemented with homogenized pig brain Snog-Kjaer et al.[2] demonstrated that

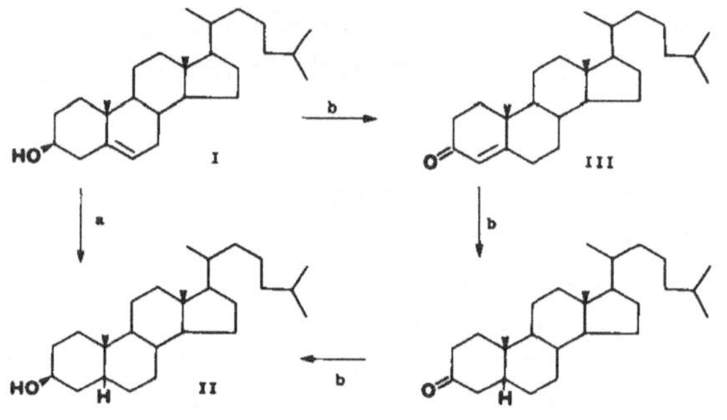

FIGURE 3. Degradation of cholesterol (I) by intestinal organisms: (a) direct reduction to coprostanol (II), (b) indirect reduction via the formation of 4-cholesten-3-one (III).

a proportion of the anaerobic bacteria isolated from human feces can reduce cholesterol although the organisms were not identified. Utilizing the same medium Crowther et al.[31] noted that many intestinal anaerobes, including strains of *Clostridia, Bifidobacteria,* and *Bacteroides,* may reduce cholesterol to coprostanol. The fastidiousness of this reaction is evidenced by the failure of subsequent investigations to confirm this activity in commonly found intestinal organisms[32] although it has been demonstrated in some strains of *Eubacteria.*[33]

The mechanism of the microbial reduction of cholesterol to coprostanol may involve one of two pathways involving either the direct reduction of cholesterol or the formation of an oxidized intermediate, cholest-4-ene-3-one, which is subsequently reduced (Figure 3). Early investigation utilizing [3α-$^{3}$H] labeled cholesterol incubated with human fecal homogenates revealed that the bulk of the label was retained in the 3α position of the product.[34] As the label did not appear to have been lost from the ring position adjacent to the hydroxyl group the reaction was assumed to proceed via the direct reduction pathway. In a subsequent study when [3α-$^{3}$H, 4-$^{14}$C] labeled cholesterol was incubated with a pure strain of *Eubacterium* only 35% of the tritium label was retained in the coprostanol, mainly at C-3.[35] Furthermore when [3α-$^{3}$H, 4-$^{14}$C] labeled coprostanol was incubated in the presence of *Eubacteria* up to 40% of the tritium was lost. Finally when [4β-$^{3}$H, 4-$^{14}$C] cholesterol was metabolized in this system the coprostanol formed retained 80% of the label although most of the tritium had migrated to the C-6 position. These data indicates that microbially mediated cholesterol reduction proceeds via oxidation of the hydroxy group at C-3 and does not support a reaction pathway involving direct reduction.[34] Interpretation is confounded by the fact that tritium may be lost from coprostanol subsequent to its formation. Furthermore there is strong evidence that the reaction proceeds with isomerization of the double bond associated with the intramolecular shift of tritium from C-3 to C-6.[35] Björkhem and Gustaffson observed similar losses and transfer of tritium when labeled cholesterol was incubated with homogenates of rat cecal organisms[36] and that cholest-4-ene-3-one is efficiently reduced to coprostanol under similar conditions.[37] Much of these data strongly support a multi-step reduction process for the conversion of cholesterol to coprostanol but the vagaries of using [$^{3}$H-]labeled substrates cannot eliminate the possibility of a single step direct reduction occurring or that both pathways can occur simultaneously.

Cholesterol may also be degraded by intestinal bacteria with loss of side-chain. Intracecal administration of [4-$^{14}$C] cholesterol in guinea pigs was accompanied by the conversion of a small proportion of this substrate to estradiol.[38] Pretreatment of the animals with a mixture of antibiotics (clindamycin, gentamycin, and nystatin) suppressed the formation of estradiol indicating that the intestinal organisms were associated with the formation of these substrates. Although cholesterol undergoes loss of side-chain when incubated with bacteria such as *Pseudomonas testosteroni*[39] there is little evidence that pure strains of intestinal bacteria can carry out this reaction in vitro. However Owen et al.[40] have demonstrated that a wild strain of *E. coli* can metabolize cholesterol to unsaturated products including cholest-4-ene-3-one, cholesta-1,4-diene-3-one, androst-4-ene-3,17-dione and androst-1,4-diene-3,17-dione. Finally in a unique report it was noted that when incubated with cultures of *E. coli* [$^3$H-] cholesterol is bound to the cellular DNA indicating that the steroid may be metabolized to an active intermediate capable of binding to cellular biopolymers.[41]

Phytosterols and allocholesterol are also reduced by the intestinal organisms such as *Eubacterium* (ACTC 21408).[42] Under anaerobic incubation conditions campesterol is reduced to 5β-campestanol, β-sitosterol to 5β-sitostanol, stigmasterol to 5β-stigmastanol and allocholesterol (4-cholesten-3β-ol) to coprostanol (Figure 1). Epicholesterol is not reduced to epicoprostanol under these conditions nor is the side chain double bond of stigmasterol. All of these reductions yield metabolites homologous to coprostanol and it may be deduced that the mechanism of this reaction follows a similar pattern to that for the microbial degradation of cholesterol, involving either direct reduction or the formation of an oxidized intermediate.

## III. DEGRADATION OF STEROID HORMONES

### A. Human Studies

Compared to the excretion of bile acids, cholesterol, and phytosterols relatively small amounts of steroid hormones escape enterohepatic circulation and enter the colon and are subsequently excreted.[1] At birth meconium contains a large number of sulfate or glucuronide conjugates of steroid hormones including estrogens, androstanes, and pregnanes from both maternal and infant sources.[43] As the child develops and the bowel is colonized by metabolically active bacteria the pattern of fecal steroids is substantially altered. Thus, meconium contains relatively high concentrations of 5-androstene-3β , 16β , 17α-triol whereas infant feces contain the microbially derived epimer 5-androstene-3β , 16α, 17β-triol and traces of 5-androstene-3β, 16α, 17α-triol and 5-androstene-3β, 16β, 17β-triol.[44] Beyond 6 months the major fecal steroids excreted are the sulfate conjugates of 5-pregnene-3β, 20α-diol and 5-pregnene-3β, 12α, 20α-triol.

In adults the fecal excretion pattern of steroid hormones escaping enterohepatic circulation is again complex. Taylor[44] has compiled a list of more than 50 steroid hormones that have been detected in the feces of adult males and females. The complexity of these excretion patterns is compounded by the possibility that the steroids may be excreted as disulfates, monosulfates, or unconjugated substrates. Adults excrete low concentrations of androstane and pregnane derivatives, the fecal steroid profile varying between men, women, and pregnant women. The major differences in the fecal steroid profiles is that men tend to excrete mainly unconjugated substrates with some monosulfates whereas in women more sulfated material is found. In pregnant subjects the fecal loss of steroid hormones particularly is markedly increased the pregnane derived substrates, such as pregnane diols and 17α- or 17β-pregnanolones. Compared to the pattern of biliary steroids it is important to note that the fecal steroid hormones are frequently present in reduced forms, that no glucuronide substrates are found and many of the sulfated hormones found in bile are partially or fully hydrolyzed in feces. The deconjugation processes are attributed in part to the activity of intestinal bacteria.

Supportive evidence for the role of intestinal bacteria in determining the profile and levels of fecal estrogens has been derived from observing the effects of antibiotics.[45] Administration of oral ampicillin suppresses the urinary excretion of estrogens in pregnant women[46] while simultaneously raising the relative proportions of conjugated estrogens in feces. Under these conditions the antibiotic is clearly interfering with the enteroheptic circulation and excretion of estrogens by suppressing the deconjugating activity of the intestinal bacteria. In a further study[47] it was demonstrated that progesterone was formed in vivo from deoxycorticosterone sulfate probably via the sequential deconjugation and dehydroxylation of the precursor by the intestinal flora during enterohepatic circulation.

The development of the fecal steroid hormone profile in babies and adults is partially determined by major changes in the endogenous synthesis of these substrates as well as the development of an active intestinal flora. It may therefore be inferred from this evidence that in humans steroid hormones entering the colon may undergo considerable microbial degradation.

### B. Animal Studies

More direct evidence for the role of intestinal bacteria in the metabolism of steroid hormones has been derived from the comparison of steroid excretion patterns in germfree and conventional rats. The fecal steroid hormone profiles of these groups of animals exhibit quantitative differences attributable to the metabolic activity of the intestinal bacteria,[48] although account has to be taken of the possible differences in the cellular activities of the intestinal wall. In particular no steroids of the pregnane series hydroxylated in the side-chain C-21 are excreted in conventional rats suggesting that dehydroxylation at this site may be carried out by the intestinal flora.[49] Similarly only steroids with a 3β-hydroxy conformation are excreted in germfree rats suggesting that the intestinal bacteria can epimerize a 3α-hydroxy substitutent to the 3β-isomer.[44]

In general conventional male rats excrete unconjugated steroid hormones whereas in female rats they are excreted as monosulfates, although these differences are by no means absolute. In germfree rats treated with [$^{14}$C]-pregnenalone or [$^{14}$C]-corticosterone these substrates are excreted in the feces as sulfates whereas in conventional animals they are excreted in the unconjugated form indicating that the intestinal bacteria are able to desulfate these substrates.[50] Although glucuronide conjugates are excreted in the bile they are not observed in the feces of germfree or conventional animals. Therefore while these compounds probably enter the colon their degradation cannot be particularly associated with microbial metabolism.

Many other minor variations have been detected in the fecal steroid profiles of germfree and conventional rats.[44] Some of these differences may be attributed to variations in enterohepatic circulation of these steroids or their metabolism in the intestinal tissue, as well as to the influence of the intestinal bacteria. Therefore the degradation of hormonal steroids cannot be as unambiguously attributed to the flora as can the metabolism of bile acids and cholesterol.

### C. In Vitro Studies

As with cholesterol and phytosterols it has been demonstrated that steroid hormones may be degraded in vitro by intestinal bacteria. Many different metabolic activities have been identified including side-chain cleavage, deconjugation, reduction of keto groups and double bonds, dehydroxylation, and epimerization of hydroxyl groups. Steroids of the pregnane, androstane, and estrogen series enter the bowel, generally as sulfate or glucuronic acid conugates and are degraded by a combination of these metabolic pathways. Glucuronides are generally deconjugated by cellular enzymes although Graef et al.[51] have isolated a steroid specific glucuronidase from *E. coli*. In contrast bacterial sulfatases are widely distributed in the intestinal flora.[48,52,53]

Bokkenheuser and Winter[54] have extensively investigated the microbial degradation of corticosteroids in pure and mixed cultures of fecal organisms. The substrates are commonly reduced in ring A and undergo side-chain dehydroxylation either independently or sequentially. Thus, cortisol is metabolized to tetrahydrocortisol, 21-deoxycortisol and tetrahydro-21-deoxycortisol (Figure 4)[55] whereas corticosterone is metabolized to tetrahydrocorticosterone, 21-deoxycorticosterone and the 3 α-hydroxy- or 3β-hydroxy-epimers of tetrahydro-21-deoxycorticosterone (Figure 5).[56] Deoxycorticosterone is similarly reduced or dehydrox-

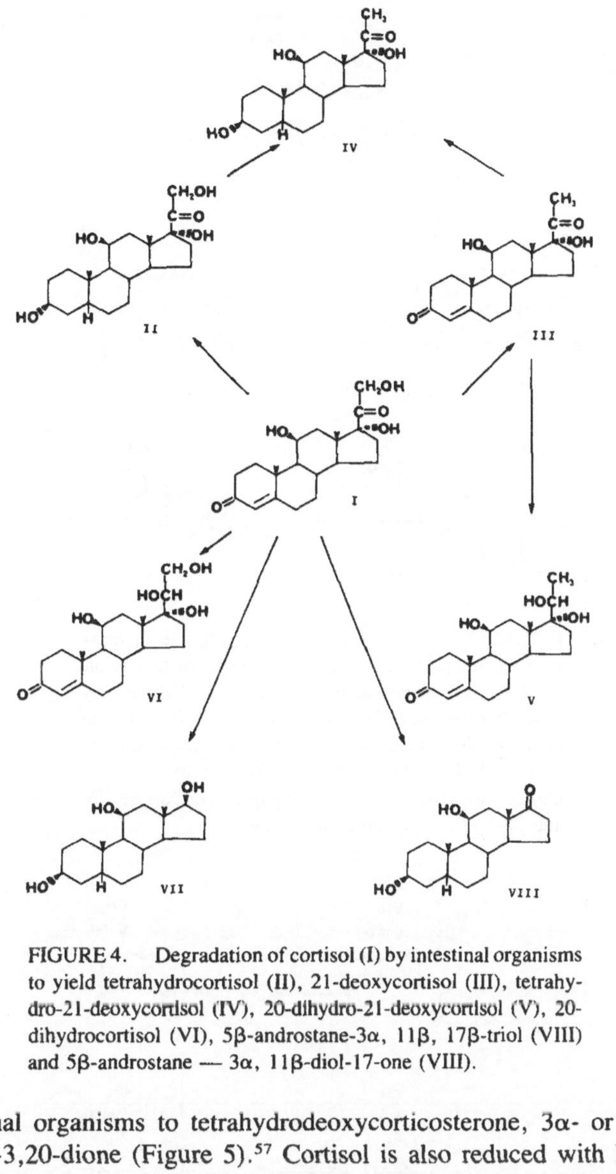

FIGURE 4. Degradation of cortisol (I) by intestinal organisms to yield tetrahydrocortisol (II), 21-deoxycortisol (III), tetrahydro-21-deoxycortisol (IV), 20-dihydro-21-deoxycortisol (V), 20-dihydrocortisol (VI), 5β-androstane-3α, 11β, 17β-triol (VIII) and 5β-androstane — 3α, 11β-diol-17-one (VIII).

ylated by intestinal organisms to tetrahydrodeoxycorticosterone, 3α- or 3β-pregnanolone and 5α-pregnane-3,20-dione (Figure 5).[57] Cortisol is also reduced with loss of side-chain to 5β -androstane-3α, 11β, 17β-triol and 5β-androstane-3α,11β-diol-17-one. Finally the C-20 keto group of cortisol, costicosterone, and deoxycorticosterone may be reduced to yield C-20-hydroxylated substrates (Figures 4 and 5) such as 20-dihydrocortisol and 20-dihydro-21-deoxycortisol.

Progesterones are also degraded by intestinal organisms with the substrates undergoing 16α-dehydroxylation associated with epimerisation of the side-chain as well as ring A reduction (Figure 6).[58] Thus, 16α-hydroxy-progesterone yields 17α-pregnanolone when incubated with mixed fecal organisms under anaerobic conditions. Epimerization of the side-chain indicates that the reaction proceeds via the formation of a Δ16 double bond which is subsequently reduced. In germfree rats 16α-hydroxy progesterone is excreted in the urine whereas conventional animals excrete 17α-pregnanolone, confirming that the intestinal bacteria are involved in the formation of this substrate.[44] It is not clear if 17α-pregnanolone may be further metabolized by these organisms.

Androstanes are also metabolized by fecal organisms when incubated under anaerobic conditions with pure or mixed cultures of fecal organisms. Typically 1,4-androstadiene-

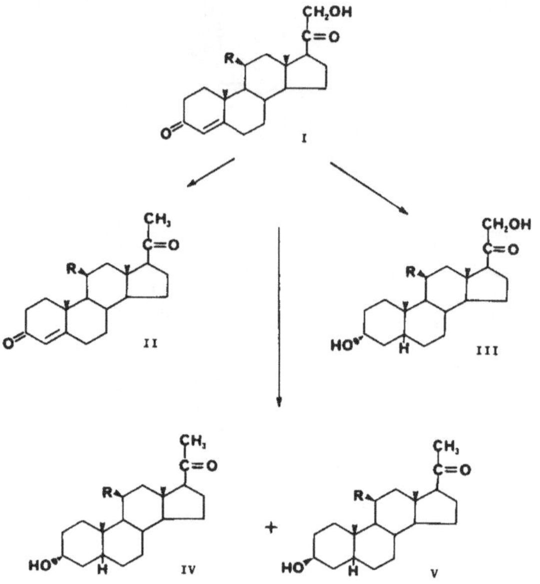


FIGURE 5. Degradation of corticosterone or deoxycorticosterone by intestinal organisms: corticosterone (I, R = OH) yields deoxycorticosterone (II), tetrahydrocorticosterone (III) and 3α- or 3β-epimers of tetrahydro-21-deoxycorticosterone (IV, V). Deoxycorticosterone (I, R = H) yields 3α, 21-dihydroxy-5β-pregnan-20-one (III), 3α-hydroxy-5β-pregnan-20-one (IV) and 3β-hydroxy-5β-pregnan-20-one (V).

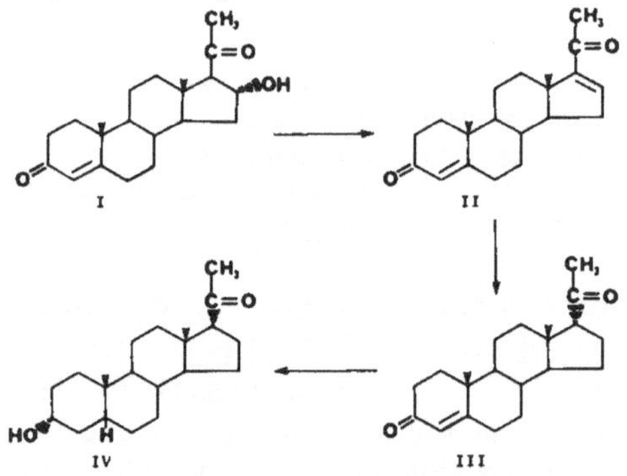


FIGURE 6. Degradation of 16α-hydroxyprogesterone (I) by intestinal organisms to yield 16-dehydroprogesterone (II), 17α-progesterone (III), 17α-pregnenolone (IV), and 3β-hydroxy-5β-pregnan-20-one (V).

3,17-dione is sequentially reduced in mixed fecal culture to 4-androstene-3,17-dione, 5β-androstane-3,17-dione, 3α-hydroxy-5β- androstane-17-one and 5β-androstane-3α,17β-diol (Figure 7).[59] This reduction has also been investigated in pure cultures of *Clostridium paraputrificum* where Glass et al.[60] demonstrated that 4-androsten-3,17-dione is also reduced to 3α-hydroxy-5β-androstan-3,17-dione probably via the intermediate formation of 5β-androstane-3,17-dione. Furthermore polyunsaturated androstanes, such as 1,4,6-androstatriene-

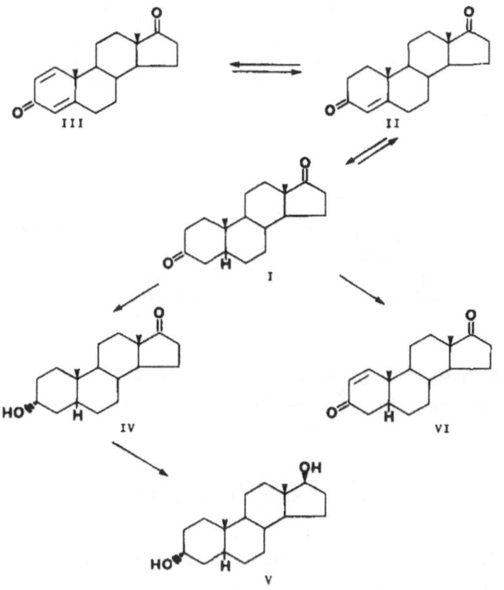


FIGURE 7. Degradation of androstanes by *Clostridia*. 5β-androstane-3,17-dione (I) may be oxidized to 4-androstene-3,17-dione (II), 1,4-androstadiene-3,17-dione (III) or 1-androstene-3,17-dione (VI) and reduced to 3α-hydroxy-5β-androstane-17-one (IV) or 3α,17β-dihydroxy-5β-androstane (V).

3,17-dione and 4,6-androstadiene-3,17-dione, are also reduced by *Clostridium para-putrificum*.[61] Oxidative degradation of androstanes has been observed under less rigorous anaerobic culture conditions whereby 5β -androstane-3,17-dione is metabolized to 4-androsten-3,17-dione in mixed fecal culture[62] or in pure cultures of *Clostridia*.[63] Some strains of *Clostridia* also possess a 3-oxo-5β-steroid-$\Delta^1$-dehydrogenase whereby 5β-androstan-3,17-dione may be metabolized to 1-androsten-3,17-dione. The oxidative reactions do not occur under rigorous anaerobic incubation conditions and in vitro require the presence of cofactors, such as menadione, to yield oxidized products.[60]

Estrogens, like other steroids, circulate in the body as sulfate or glucuronic acid conjugates. Estriol may enter the bowel as estriol-3-sulfate-16-glucuronide which is rapidly hydrolyzed by the gut flora. Microbial estrogen sulfatase activity has been demonstrated in organisms isolated from the cecum of rats.[53] Similarly estrogen sulfatase and glucuronidase activities have been identified in mixed fecal cultures of human organisms.[62] Depending upon the incubation conditions employed estrogens may be metabolically reduced or oxidized. Under anaerobic conditions estrone and 16α-hydroxyestrone are reduced to estradiol and estriol respectively in mixed fecal culture (Figure 8). In the presence of oxidizing agents and low concentrations of fecal organisms this mechanism may be reversed with estriol being oxidized to estrone.[62] More recently Järvenpää et al.[64] have confirmed these observations and demonstrated further that in anaerobic mixed fecal culture 16-oxo-estradiol and 15α-hydroxyestrone may be reduced to 16-epiestriol and 15α-hydroxyestradiol respectively.

Finally synthetic steroid hormones undergo enterohepatic circulation, enter the bowel as conjugates when they may be degraded and excreted.[65,66] Bokkenheuser et al.[67] have demonstrated that a number of progestins are reduced both in cultures of mixed fecal organisms or of specific strains of *Clostridia*. Under these conditions synthetic progestins, as well as naturally occurring progesterone, undergo reduction in ring A whereby the 4-ene-3-one structure is reduced to saturated 3α-hydroxy or 3β-hydroxy epimers.

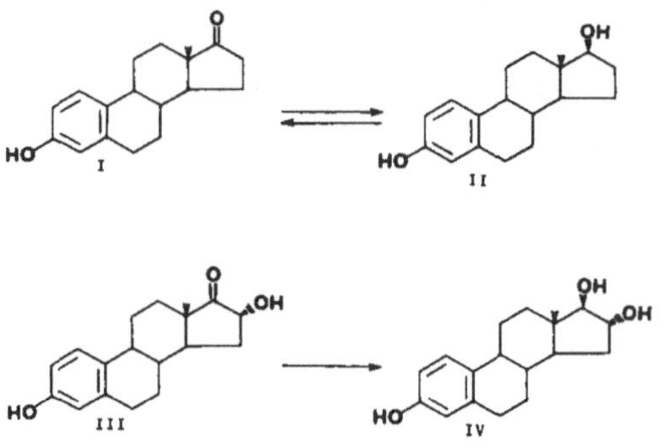

FIGURE 8. Degradation of estrone (I) to estradiol (II) and 16α-hydroxyestrone (III) to estriol (IV) by intestinal organisms.

## IV. CONCLUSION

It may be concluded from these observations that steroids can be extensively metabolized by intestinal bacteria. Variations in the fecal steroid profiles from humans and animal models have shown that cholesterol and the homologous phytosterols are extensively reduced by the luminal flora. In vitro experiments have demonstrated that these steroids can be similarly modified in cultures of mixed fecal organisms although it has proved difficult to establish which particular organisms carry out the reduction. In contrast it has not proved so easy to determine whether the intestinal bacteria modify steroid hormones by examining the fecal profiles of these substrates from humans and conventional or germfree animals. This is partly due to many of the metabolites being absorbed and undergoing urinary as opposed to fecal excretion. In vitro studies, however, have demonstrated that steroid hormones are extensively degraded by intestinal organisms. Corticosteroids, progesterones, androstanes and estrogens are all deconjugated, reduced, and dehydroxylated by gut bacteria in mixed or pure cultures. A degree of interconversion has also been observed particularly in those reactions where corticosteroids lose the side-chain substitutent to form androstanes and pregnanes. Under normal incubation conditions reduction products are formed although the oxidative introduction of double bonds into the steroid nucleus can occur. The metabolic pathways are generally sensitive to pH and the redox potential of the incubation and particular steps may be expressed by defined strains of bacteria. It is unlikely that strict comparisons and correlations can be made between these in vitro observations and the complex situation pertaining in the bowel although some insight may be gained into the in vivo mechanisms of steroid degradation.

It is important to note that both the extent of steroid metabolism in the human gut and the overall concentration of the substrates may be of importance. Many studies of the role of cholesterol in heart disease have required accurate cholesterol balance studies to be peformed. A major route of excretion of cholesterol is in the feces and it is therefore necessary to quantify the fecal loss of this substrate and its metabolites. In some cases it has been concluded that microbial degradation of cholesterol to nonsteroidal products in the bowel may be a major source of error in these determinations.[14] These studies have also attempted to use the poorly absorbed phytosterols, such as β-sitosterol, as a marker for cholesterol degradation although there is some evidence that this substrate may also be degraded to nonsteroidal products. The conversion of cholesterol to nonabsorbable coprostanol may be of importance in reducing bodily levels of cholesterol.

It has also been hypothesized that cholesterol may be associated with the risk of developing colon cancer.[68] Although it is difficult to separate the effects primarily due to cholesterol from other associated factors, in rats with pre-induced colon tumors enhancing the intake of dietary cholesterol is associated with a significant increase in the tumor load of the animals.[69] It is not clear whether the degree of cholesterol degradation in the intestine of these animals can influence the incidence of tumors. In this context Hill and co-workers[70] have hypothesized that subjects at risk of developing large bowel cancer excrete higher concentrations of fecal bile acids. These subjects also carry more bacteria capable of degrading bile acids to ring unsaturated substrates (the nuclear dehydrogenating *Clostridia*). This activity appears to be quite general and is currently assayed by quantifying the carriage of organisms capable of degrading androstane -3,17-dione to androst-4-ene-3,17-dione (Chapter 16). Thus, subjects and populations at high risk of developing colon cancer appear to have a generally enhanced potential to degrade steroids to unsaturated derivatives. This phenomenon may be directly associated with the in vivo generation of precarcinogens from steroids or may be a more generalized expression of the oxidative potential of microbial populations in the intestine of at risk subjects.

With respect to the degradation of steroid hormones by intestinal organisms, Bokkenheuser et al.[71] have hypothesized that the carriage of bacteria able to remove the C-21 hydroxyl group of C-21 steroids may be a marker for large bowel cancer. In particular tetrahydrodeoxycorticosterone, a potentially carcinogenic substrate, is dehydroxylated to noncarcinogenic pregnanolone by strains of *Eubacterium lentum* (Figure 5). In this limited study the carriage of these organisms was significantly lower in cancer subjects when compared to controls suggesting that the flora of at risk subjects was less able to degrade the potentially toxic tetrahydrodeoxycorticosterone.

In some breast cancer subjects despite ablation therapy estrogen dependent tumors cannot be supressed suggesting that additional sources of these steroids must be present. It has been hypothesized that in some of these subject the circulating levels of estrogens may be due to these substrates being synthesized in the intestine by the gut bacteria.[72]

Finally the intestinal flora may be associated with contraceptive failure in women taking steroid contraceptives and antibiotics simultaneously.[73] Effective contraception depends upon appropriate levels of serum estrogens being maintained and this may be suppressed in subjects taking antibiotics such as rifampicin, ampicillin, or tetracycline. Under normal conditions steroid hormones enter the bowel as conjugates where they are deconjugated and partially reabsorbed, the latter step contributing to maintaining the circulating levels of estrogens. In subjects under antibiotic therapy the intestinal flora may be suppressed, particularly the steroid deconjugating activity. Under these circumstances the reabsorption, and therefore the serum levels, of estrogens may be suppressed and the contraceptive effect compromised.

## REFERENCES


1. **Macdonald, I. A., Bokkenheuser, V. D., Winter, J., McLernon, A. M., and Mosbach, E. H.,** Degradation of steroids in the human gut, *J. Lipid Res.*, 24, 675, 1983.
2. **Snog-Kjaer, A., Prange, I., and Dam, H.,** Conversion of cholesterol into coprostanol by bacteria *in vitro, J. Gen. Microbiol.*, 14, 256, 1956.
3. **Wilkins, T. D. and Hackman, A. S.,** Two patterns of neutral steroid conversion in the feces of normal North Americans, *Cancer Res.*, 34, 2250, 1974.
4. **Korpela, J.,** Capillary gas-liquid chromatography of fecal free and esterified neutral sterols, Proc. Symp. Anal. Steroids, Eger, Hungary, 255, 1981.
5. **Eneroth, P. and Nyström, E.,** Quantification of cholesteryl sulfate and neutral steroid derivatives in human faeces after purification on liphophilic Sephadex gels, *Steroids*, 11, 187, 1968.
6. **Miettinen, T. A., Ahrens, E. H., and Grundy, S. M.,** Quantitative isolation and gas-liquid chromatographic analysis of total dietary and fecal neutral steroids, *J. Lipid Res.*, 6, 411, 1965.
7. **McNamara, D. J., Proia, A., and Miettinen, T. A.,** Thin-layer and gas-liquid chromatographic identification of neutral steroids in human and rat faeces, *J. Lipid Res.*, 22, 474, 1981.

8. **Reddy, B. S., Martin, C. W., and Wynder, E. L.,** Fecal bile acids and cholesterol metabolites in patients with ulcerative colitis, a high-risk group for development of colon cancer, *Cancer Res.*, 37, 1697, 1971.
9. **Huang, C. T. L., Rodriguez, J. T., Woodward, W. E., and Nichols, B. L.,** Comparison of patterns of fecal bile acid and neutral sterol between children and adults, *Am. J. Clin. Nutr.*, 29, 1196, 1976.
10. **Gustafsson, J. Å., and Werner, B.,** Fecal sterols in infants, *Acta Physiol. Scand.*, 73, 305, 1968.
11. **Kellog, T. F.,** Fecal bile acids and neutral sterols of gnotobiotic, antibiotic fed normal, and normal human children, in *Germfree Research*, Heneghan, J., Ed., Academic Press, New York, 1973, 79.
12. **Salyers, A. A., Sperry, J. F., Wilkins, T. D., Walker, A. R. P., and Richardson, N. J.,** Neutral steroid concentrations in the faeces of North American white and South African black populations at different risks for cancer of the colon, *S. Afr. Med. J.*, 51, 823, 1977.
13. **Korpela, J. T., Hämäläinen, E., and Adlercreutz, H.,** Effect of oxytetracycline on bacterial intestinal metabolism of neutral sterols and on serum lipids, *Scand. J. Gastroenterol.*, 19, 401, 1984.
14. **Grundy, S. M., Ahrens, E. H., and Salen, G.,** Dietary β-sitosterol as an internal standard to correct for cholesterol losses in sterol balance studies, *J. Lipid Res.*, 9, 374, 1968.
15. **Ullrich, I. M., Lai, H-Y., Vona, L., Reid, R. L., and Albrink, M. J.,** Alterations of faecal steroid composition induced by changes in dietary fiber consumption, *Am. J. Clin. Nutr.*, 34, 2054, 1981.
16. **Eneroth, P., Hellström, K., and Ryhage, R.,** Identification and quantification of neutral fecal steroids by gas-liquid chromatography and mass spectrometry: studies of human excretion during two dietary regimens, *J. Lipid Res.*, 5, 245, 1964.
17. **Denbesten, L., Connor, W. E., Kent, T. H., and Lin, D.,** Effect of cellulose in the diet on the recovery of dietary plant sterols from feces, *J. Lipid Res.*, 11, 341, 1970.
18. **Kritchevsky, D., Davidson, L. M., Mosbach, E. H., and Cohen, B. I.,** Identification of acidic steroids in feces of monkeys fed β-sitosterol, *Lipids*, 16, 77, 1981.
19. **Danielsson, H.,** Present status of research on catabolism and excretion of cholesterol, in *Advances in Lipid Research*, Paoletti, R. and Kritchevsky, D., Eds., Academic Press, New York, 1963, 335.
20. **Kellog, T. F. and Wostmann, B. S.,** Fecal neutral steroids and bile acids from germ free rats, *J. Lipid Res.*, 10, 495, 1969.
21. **Coleman, D. L. and Baumann, C. A.,** Intestinal sterols. III. Effects of age, sex and diet, *Arch. Biochem. Biophys.*, 66, 226, 1957.
22. **Aigueperse, J., Lutton, C., and Chevalier, F.,** Identification and origins of neutral fecal sterols in adult large sows: occurrence of eocternally secreted cholesterol, *Reprod. Nutr. Dev.*, 21, 545, 1981.
23. **Martin, W. J., Ravi Subbiah, M. T., Kottke, B. A., Birk, C. C., and Naylor, M. C.,** Nature of fecal sterols and intestinal bacterial flora, *Lipids*, 8, 208, 1973.
24. **Tint, G. S., Dayal, B., Batta, A. K., Shefer, S., Joanen, T., McNease, L., and Salen, G.,** The fecal bile acids and sterols of *Alligator mississippiensis*, *Gastroenterology*, 80, 114, 1981.
25. **Nomura, A. M. Y., Wilkins, T. D., Kamiyama, S., Heilbrun, L. K., Shimada, A., Stemmermann, G. N., and Mower, H. F.,** Fecal neutral steroids in two Japanese populations with different colon cancer risks, *Cancer Res.*, 43, 1910, 1983.
26. **Miettinen, T. A., Proia, A., and McNamara, D. J.,** Origins of fecal neutral sterols in rats, *J. Lipid Res.*, 22, 485, 1981.
27. **Eyssen, H., Piessens-Denef, M., and Parmentier, G.,** Role of the cecum in maintaining $\Delta^5$-steroid- and fatty acid-reducing activity of the rat intestinal microflora, *J. Nutr.*, 102, 1501, 1972.
28. **Eyssen, H. and Parmentier, G.,** Biohydrogenation of sterols and fatty acids by the intestinal microflora, *Am. J. Clin. Nutr.*, 27, 1329, 1974.
29. **Dam, H.,** The formation of coprostanol in the intestine. II. The action of intestinal bacteria on cholesterol, *Biochem. J.*, 28, 820, 1934.
30. **Rosenfeld, R. S. and Hellman, L.,** Reduction and esterification of cholesterol and sitosterol by homogenates of feces, *J. Lipid Res.*, 12, 192, 1971.
31. **Crowther, J. S., Drasar, B. S., Goddard, P., Hill, M. J., and Johnson, K.,** The effect of a chemically defined diet on the faecal flora and faecal steroid concentration, *Gut*, 14, 790, 1973.
32. **Sadziowski, M. R., Sperry, J. F., and Wilkins, T. D.,** Cholesterol-reducing bacterium from human faeces, *Appl. Environ. Microbiol.*, 34, 355, 1977.
33. **Brinkley, A. W., Gottesman, A. R., and Mott, G. E.,** Isolation and characterization of new strains of cholesterol-reducing bacteria from baboons, *Appl. Environ. Microbiol.*, 43, 86, 1982.
34. **Rosenfeld, R. S. and Gallagher, T. F.,** Further studies of the biotransformation of cholesterol (to coprostanol), *Steroids*, 4, 515, 1964.
35. **Parmentier, C. G. and Eyssen, H.,** Mechanism of biohydrogenation of cholesterol to coprostanol by *Eubacterium* ATCC 21408, *Biochem. Biophys. Acta*, 348, 279, 1974.
36. **Björkhem, I. and Gustafsson, J-Å.,** Mechanism of microbial transformation of cholesterol to coprostanol, *Eur. J. Biochem.*, 21, 428, 1971.

37. **Björkhem, I., Gustafsson, J-Å., and Wrange, Ö.,** Microbial transformation of cholesterol to coprostanol: properties of a 3-oxo-$\Delta^4$-steroid-5β-reductase, *Eur. J. Biochem.*, 37, 143, 1973.
38. **Goddard, P. and Hill, M. J.,** The *in vivo* metabolism of cholesterol by gut bacteria in the rat and the guinea pig, *J. Steroid Biochem.*, 5, 569, 1974.
39. **Owen, R. S., Mason, A. N., and Bilton, R. F.,** The degradation of cholesterol by *Pseudomonas sp.* NCIB 10590 under aerobic conditions, *J. Lipid Res.*, 24, 1500, 1983.
40. **Owen, R. W., Tenneson, M. E., Bilton, R. F., and Mason, A. N.,** The degradation of cholesterol by *Eschericia coli* isolated from human faeces, *Biochem. Soc. Trans.*, 6, 377, 1978.
41. **Zachariah, P. K., Slaga, T. J., Berry, D. L., Bracken, W. M., Buty, S. G., Martinsen, C. M., and Juchau, M. R.,** The ability of enteric bacteria to catalyze the covalent binding of bile acids and cholesterol to DNA and their inability to metabolize benzo/a/pyrene to a binding product and to known metabolites, *Cancer Lett.*, 3, 99, 1977.
42. **Eyssen, H. J., Parmentier, G. G., Compernolle, F. C., De Pauw, G., and Piessens-Denef, M.,** Biohydrogenation of sterols by *Eubacterium* ATCC 21, 408-*Nova* species, *Eur. J. Biochem.*, 36, 411, 1973.
43. **Gustafsson, J-Å., and Stenberg, Å.,** Steroids in meconium from male and female newborn infants, *Eur. J. Biochem.*, 22, 246, 1971.
44. **Taylor, W.,** The excretion of steroid hormones metabolites, *Vitam. Horm.*, 29, 201, 1971.
45. **Adlercreutz, H., Martin, F., Pulkkinen, M., Dencker, H., Rimer, U., Sjöberg, N-O., and Tikkanen, M. J.,** Intestinal metabolism of estogens, *J. Clin. Endoennd., Metab.*, 43, 497, 1976.
46. **Adlercreutz, H., Martin, F., Tikkanen, M. J., and Pulkinnen, M.,** Effect of ampicillin administration on the excretion of twelve oestrogens in pregnancy urine, *Acta Endocrinol. (Copenhagen)*, 80, 551, 1975.
47. **Casey, M. L. and MacDonald, P. C.,** Metabolism of deoxycorticosterone and deoxycorticosterone sulfate in men and women, *J. Clin. Inv.*, 70, 312, 1982.
48. **Eriksson, H. and Gustafsson, J-Å.,** Steroids in germfree and conventional rats: sulpho- and glucuroro-hydrolase activities in caecal contents from conventional rats, *Eur. J. Biochem.*, 13, 198, 1970.
49. **Eriksson, H., Gustafsson, J-Å., and Sjovall, J.,** Steroids in germfree and conventional rats: 21-dehydroxylation by intestinal organisms, *Eur. J. Biochem.*, 9, 550, 1969.
50. **Eriksson, H. and Gustafsson, J-Å.,** Steroids in germfree and conventional rats: distribution and excretion of labelleld pregnenolone and corticosterone in male and female rats, *Eur. J. Biochem.*, 15, 132, 1970.
51. **Graef, V., Furuya, E., and Nishikaze, O.,** Hydrolysis of steroid glucuronides with β-glucuronidase preparations from bovine liver, *Helix pomatia* and *E. coli, Clin. Chem.*, 23, 532, 1977.
52. **Huijghebaert, S. M., Sim, S. M., Back, D. J., and Eyssen, H. J.,** Distribution of estrone sulfatase activity in the intestine of germfree and conventional rats, *J. Steroid Biochem.*, 20, 1175, 1984.
53. **Back, D. J., Chapman, C. R., May, S. A., and Rowe, P. H.,** Adsorption of oestrone sulphate from the gastrointestinal tract of the rat, *J. Steroid Biochem.*, 14, 347, 1981.
54. **Bokkenheuser, V. D. and Winter, J.,** Biotransformation of steroids, in *Human Intestinal Microflora in Health and Disease*, Hentges, D., Ed., Academic Press, New York, 1983, 215.
55. **Cerone-McLernon, A. M., Winter, J., Mosbach, E. H., and Bokkenheuser, V. D.,** Side chain cleavage of cortisol by fecal flora, *Biochim. Biophys. Acta*, 666, 341, 1981.
56. **Winter, J., Bokkenheuser, V. D., and Ponticorvo, L.,** Bacterial metabolism of corticoids with particular reference to the 21-dehydroxylation, *J. Biol. Chem.*, 254, 2626, 1979.
57. **Kelly, W. G., de Leon, O., Winter, J., and Bokkenheuser, V. D.,** Exchange of hydrogen at C-21 during dehydroxylation of deoxycorticosterone by mixed cultures of human fecal flora, *J. Steroid Biochem.*, 8, 73, 1977.
58. **Winter, J., O'Rourke, S., Bokkenheuser, V. D., Hylemon, P. B., and Glass, T.,** 16 α-dehydration of corticoids by bacteria isolated from rat fecal flora, *J. Steroid Biochem.*, 16, 231, 1982.
59. **Schubert, K., Schlegel, J., and Hörhold, C.,** Der hydrierungsverlauf bei 17-ketosteroiden mit darmbakterien unter anaeroben bendingungen, *Z. Naturforsch.*, 17b, 84, 1962.
60. **Glass, T. L., Wheeler, L. A., Sutter, V. L., and Finegold, S. M.,** Transformation of 4-androsten-3,17-dione by growing cultures and cell extracts of *Clostridium paraputrificum, Biochim. Biophys. Acta*, 573, 332, 1979.
61. **Schubert, K., Schlegel, J., and Hörhold, C.,** Stereospecific hydrogenation of $\Delta^4$-, $\Delta^{1,4}$-, $\Delta^{4,6}$- and $\Delta^{1,4,6}$-3-ketosteroids by *Clostridium paraputrificum, Steroids*, Suppl. 1, 175, 1965.
62. **Lombardi, P., Goldin, B., Boutin, E., and Gorbach, S. L.,** Metabolism of androgens and estrogens by human faecal organisms, *J. Steroid Biochem.*, 9, 795, 1978.
63. **Aries, V. C., Goddard, P., and Hill, M. J.,** Degradation of steroids by intestinal bacteria. III. 3-oxo-5β-steroid-$\Delta'$-dehydrogenase and 3-oxo-5β-steroid-$\Delta^4$-dehydrogenase, *Biochim. Biophys. Acta*, 248, 482, 1971.
64. **Järvenpää, P., Kosunen, T., Totsis, T., and Adlercreutz, H.,** *In vitro* metabolism of estrogens by isolated intestinal microorganisms and by human faecal microflora, *J. Steroid Biochem.*, 13, 345, 1980.

65. **Smith, R. L.,** Biliary excretion and hepatoxicity of contraceptive steroids. Pharmacological models in contraceptive development, *Acta Endocrinol.*, Suppl. 185, 149, 1974.
66. **Huempel, M., Wendt, M., Jogs, G., Weiss, C., Reitz, S., and Speck, U.,** Intra individual comparison of pharmacokinetic parameters of d-norgestrel, lynestienol and cyproterone acetate in six women, *Contraception,* 16, 199, 1977.
67. **Bokkenheuser, V. D., Winter, J., Cohen, B. I., O'Rourke, S., and Mosbach, E. H.,** Inactivation of contraceptive steroid hormones by human intestinal clostridia, *J. Clin. Microbiol.*, 18, 1983, 500.
68. **Cruse, P., Lewin, M., and Clark, C. G.,** Dietary cholesterol is co-carcinogenic for human colon cancers, *Lancet,* i, 752, 1979.
69. **Hiramatsu, Y., Takada, H., Yamamura, M., Hioki, K., Saito, K., and Yamamoto, M.,** Effect of dietary cholesterol on azoxymethane-induced colon carcinogenesis in rats, *Carcinogenesis,* 4, 553, 1983.
70. **Thompson, M. H. and Hill, M. J.,** Some aetiological factors in large-bowel cancer, in *Host Factors in Human Carcinogenesis,* Vol. 39, Bartsch, H., Armstrong, B., and Davis, W., Eds., International Agency for Research on Cancer, Lyon, 1982, 241.
71. **Bokkenheuser, V. D., Winter, J., Mosenthal, A. C., Mosbach, E. H., McSherry, C. K., Ayengar, N. K. N., Andrews, A. W., Lebherz, W. B., Pienta, R. J., and Wallenstein, S.,** Fecal steroid 21-dehydroxylase, a potential marker for colorectal cancer, *Am. J. Gastroenterol.*, 78, 469, 1983.
72. **Hill, M. J., Goddard, P., and Williams, R. E. O.,** Gut bacteria and aetiology of cancer of the breast, *Lancet,* ii, 472, 1971.
73. **Gorbach, S. L.,** Estrogens, breast cancer, and intestinal flora, *Rev. Inf. Dis.*, 6 (Suppl.), 585, 1984.

Chapter 7

# METABOLISM OF AMMONIA, UREA, AND AMINO ACIDS, AND THEIR SIGNIFICANCE IN LIVER DISEASE

**Angela J. Vince**

## TABLE OF CONTENTS

## I. INTRODUCTION

Interest in ammonia formation in the human intestine is centered on its role as a toxin in the pathogenesis of portal-systemic encephalopathy. To this end it has been important to determine first the sources of intestinal ammonia and second the means of preventing its release. Evidence has been accumulating recently that proteins and their derivatives constitute a major source of ammonia formed in the intestine. This displaces the long-standing belief that urea is the single major source of intestinal ammonia. Other studies have shown how bacterial nitrogen metabolism is affected by the fermentation of various carbohydrates, particularly by increasing the uptake of ammonia and by decreasing the production of ammonia from amino acids. These effects of carbohydrate on ammonia formation could explain the beneficial effects of lactulose in portal-systemic encephalopathy, if indeed ammonia is toxic to patients with this condition.

## II. PROTEIN

It is estimated that from 0.5 to 4.0 g of fixed nitrogen enters the large intestine daily, of which at least 80% is probably contained in proteins. This suggests that from 3 to 20 g of protein are available for digestion by colonic bacteria. Man excretes approximately 1 g of fecal nitrogen daily, of which 60% is contained in bacterial bodies,[1] indicating that some of the protein nitrogen entering the colon has been metabolized and incorporated into bacterial protoplasm. The remainder, up to 3 g of fixed nitrogen, is metabolized by bacteria and then absorbed, probably as ammonia. A detailed appraisal of the possible sources and quantities of nitrogen entering the colon is given by Wrong et al.[2]

Proteins must be degraded to peptides and amino acids before bacteria can assimilate them. The bacterial proteases that have been well characterized are extracellular enzymes that hydrolyze both proteins and oligopeptides, and have very broad side-chain specificities.[3] While it is apparent that gram quantities of protein are metabolized by the intestinal flora, it is not obvious which component(s) of the flora is responsible. Proteolytic activity is reported to be absent or weak among the predominant human fecal organisms.[4] Proteases have been described in organisms that may be present as minor components of the fecal flora, e.g. *Escherichia coli, Proteus mirabilis, Streptococcus faecalis, Clostridium welchii, Bacteroides melaninogenus,* and *Bacillus* spp.[5-8] but none of these are generally considered to be strongly proteolytic, with the exception of some *Bacillus* spp. whose erratic recovery from the fecal flora may indicate that they are normally present as spores rather than as viable cells.

## III. PEPTIDES

Bacterial peptidases hydrolyze peptides to amino acids and peptide residues. It used to be thought that the peptidases of *Escherichia coli* were exclusively intracellular, but recently, a small fraction of the activity of dipeptidyl carboxypeptidase was demonstrated to be in the periplasmic space.[9] The hydrolysis of peptides in vitro is accompanied by a concomitant exodus of selected amino acid and peptide residues from the bacterial cell.[10,11] However, it is considered that this exodus may not occur in the mammalian gut, where the existence of a great variety of peptides may allow for a selective uptake that enables organisms to balance their nutritional needs.[10]

There has been no systematic study of peptide uptake by intestinal anaerobes, although there are several reports of bacteria, e.g., *Bacteroides fragilis, Bacteroides ruminicola, Bacteroides melaninogenicus,* and *Fusobacterium necrophorum* preferring peptides to amino acids as nitrogen sources,[12-15] despite the fact that peptides are rapidly hydrolyzed to supply amino acids once they are taken up.[16] In organisms in which it has been studied, this preference, with the exception of proline, is not due to an inability to take up amino acids,[17,14] but studies with other bacteria may reveal a lack of transport systems.

## IV. AMINO ACIDS

### A. Amino Acid Uptake

Bacteria take up amino acids by kinetically defined active transport systems for structurally related amino acids.[18] Cells maintain an amino acid pool which serves for the biosynthesis of proteins and other nitrogen containing metabolites, so that at any given time, the intracellular amino acid concentration is considerably higher, sometimes a thousandfold,[19] than that of the surrounding medium. Once taken up, amino acids may either be directly incorporated in bacterial proteins or they may be fermented to provide carbon skeletons and ammonia. Many organisms use ammonia as their preferred source of nitrogen, despite their ability to take up preformed amino acids. It has been suggested that this may be due to the conditions of low amino acid supply[20] and moderate ammonia concentration[21] that normally prevail in the gut.[22]

Individual amino acids, particularly those containing sulfur, e.g., cysteine and methionine appear to be essential for the growth of some anaerobes,[12,23] and many others, including aspartic acid, glutamic acid, histidine, lysine, glycine, alanine, γ-aminobutyrate, δ-aminovalerate, serine, threonine, and tyrosine can serve as major energy sources for selected species of anaerobes.[14,24,25]

### B. Amino Acid Synthesis

Most studies on amino acid biosynthesis have been performed on aerobic bacteria, but the preferential use of ammonia as a nitrogen source and the poor growth of organisms such as bacteroides on amino acid mixtures[22,26] suggests that synthesis of amino acids is fundamental to the metabolism of intestinal anaerobes.

In the rumen, because the energy for amino acid biosynthesis is mainly derived from glycolytic reactions the synthesis of microbial protein is dependent on the availability of carbohydrate. In addition, the products of carbohydrate fermentation, particularly carbon dioxide and acetate, but also propionate, butyrate and valerate are used for amino acid biosynthesis. Branched-chain fatty acids derived from the fermentation of amino acids, e.g., isobutyrate and isovalerate are also incorporated into fresh amino acids.[26,27] It is likely that fatty acids will also be used for amino acid synthesis by human intestinal bacteria, but the extent to which this occurs is unknown.

While it is likely that many of the pathways used by anaerobes are identical to those described for aerobes[18,28] some will certainly differ.[26] Several clostridia and *Bacteroides fragilis* use the α-isopropylmalate pathway for the synthesis of leucine,[29] a route used by many aerobes.[30] However, many ruminal anaerobes synthesize leucine by a pathway involving the reductive carboxylation of isovalerate to α-ketoisocaproate,[27] a reaction that is not known in aerobes.

### C. Amino Acid Degradation

#### *1. Deamination*

There are four direct pathways for deaminating single amino acids[31,32] and a fifth, the Stickland reaction in which pairs of acids are used. Ammonia is formed in all reactions. The four direct pathways are

1. Oxidation with the formation of an α-keto acid
2. Reduction, in which a saturated fatty acid is formed
3. Hydrolysis, in which an α-hydroxy fatty acid is formed
4. Removal of the elements of ammonia, giving an unsaturated fatty acid

The Stickland reaction is used by many *Clostridia*[24] that degrade single amino acids poorly. Certain pairs of amino acids are degraded in coupled oxidation-reduction reactions, with one amino acid being oxidatively deaminated to a keto acid and the other being reductively deaminated to form a saturated fatty acid. A consistent feature of the reaction is the reduction of proline, and generally arginine to δ-aminovaleric acid, coupled to the oxidation of serine, phenylalanine, and possibly other amino acids. The significance of this reaction in the colon is unknown, but it has been demonstrated for at least one intestinal organism, *Clostridium difficile*.[33] The reaction was not found in one survey of anaerobic cocci.[34]

An interesting extension to the occurrence of coupled deamination of amino acids has been described.[35] It has been suggested that in anaerobic digestion, the oxidation of alanine, valine, and leucine may occur in the absence of $H_2$-accepting amino acids such as glycine and proline, which are replaced by methanogenic bacteria acting as alternative extracellular $H_2$ acceptors.

The oxidative and particularly the reductive pathways are probably those most commonly used by the intestinal flora. The oxidation reactions of anaerobes (most reports are of *Clostridia*) are usually similar or identical to corresponding reactions catalyzed by aerobic organisms, except for limitations imposed by the absence of molecular oxygen or other high potential oxidant: oxidative transaminations and α-keto acid oxidations commonly occur in anaerobic amino acid degradation, but oxygenation reactions and fatty acid oxidations are excluded. A high proportion of intestinal anaerobes, e.g., *Bacteroides*, *Bifidobacteria* and *Clostridia*, are able to oxidatively deaminate the diamines cadaverine and putrescine.[31] This reaction is a step in the formation of the secondary cyclic amines piperidine and pyrrolidine from lysine and arginine respectively.

Reduction reactions comprise the major proportion of amino acid degradations in the intestine. The pathways followed by aerobic and anaerobic organisms may differ, with a greater variety of products being formed by anaerobic organisms. Over the last decade several new pathways and reactions, not previously observed in aerobic organisms, have been identified in anaerobes.[24] Electron acceptors used by anaerobes include amino acids, α- and β-keto acids, α, β unsaturated acids or their coenzyme A thiolesters and protons. Most of the work on degradation of amino acids by anaerobes has been done with *Clostridia*, in which it seems that the reduction of both aromatic and nonaromatic amino acids involves the sequence amino acid, 2-keto acid, 2-hydroxy acid, 2-3-unsaturated acid and finally, the saturated acid. All the component reactions have been demonstrated, except the conversion of the hydroxy acid to the unsaturated acid. It also remains to be demonstrated that the rates of the individual reactions are sufficient to account for the overall rate of reductive deamination.[24]

The ultimate reduction products include a variety of fatty acids, e.g., acetic, propionic, butyric, succinic and δ-amino valeric, ammonia, carbon dioxide, and molecular hydrogen.[24] The fermentation products from the aromatic amino acids include phenylacetic, phenylpropionic, and phenyllactic acids from phenyllalaine; hydroxyphenyllactic acid, hydroxyphenylpropionic acid, phenol and *p*-cresol from tyrosine; and indole, indoleacetic acid and indolepropionic acid from tryptophan. A relatively simple and rapid method for the qualitative identification of aromatic amino acids metabolized by anaerobic bacteria growing in a casein hydrolysate medium has been described.[36] Fumarate and succinate are formed from aspartate, and urocanic acid from histidine by the removal of elements of ammonia. Details of the fermentation pathways for individual amino acids may be found in several papers.[24,31,33,34,37-41] All groups of intestinal organisms contain deaminases, although the degree of activity varies considerably both within and between the different genera. Ammonia released by the deamination of amino acids and peptides is described later.

When considering the end products of fermentation it is essential to realize that the formation of identical products from the same substrate by different organisms does not necessarily mean that they have been formed by the same metabolic pathway, e.g., *Clostridium tetanomorphum* and *Peptococcus aerogenes* both ferment glutamate to acetate, butyrate, carbon dioxide, ammonia, and hydrogen, but they use entirely different metabolic pathways. *C. tetanomorphum* uses the methylaspartate pathway, in which the successive intermediates are 2S,2S-β-methylaspartate, mesaconate, S-citramalate, and finally pyruvate which is converted to the final products. *Peptococcus aerogenes* uses the hydroxyglutarate pathway, in which the successive intermediates are α-ketoglutarate, R-α-hydroxyglutarate and probably glutaconyl-CoA and crotonyl CoA, the latter presumably being converted to acetate, butyrate, and hydrogen. The use of different pathways in glutamate fermentation is not exclusive to different genera; some species of *Clostridia* use the methylaspartate pathway and others use the hydroxyglutarate pathway.[42]

*2. Decarboxylation*

Bacterial decarboxylases degrade amino acids to form an amine and carbon dioxide. They are inducible intracellular enzymes and generally (histidine is a notable exception) require pyridoxal phosphate as a co-factor. Many decarboxylases are highly specific for their substrate, acting only on a single amino acid, e.g., ornithine, glutamine, histidine, lysine, arginine, and tyrosine.[43] A less specific decarboxylase acting on leucine, isoleucine, valine, and α-amino butyric acid has also been described.[44]

While the activity of decarboxylases extends over the pH range 4.0 to 7.5, they are most active at acid pH.[45] There are several reports of amines, e.g. histamine and tyramine accumulating in ruminal fluid at low pH.[46,47] The release of the basic amines is thought to provide a mechanism for raising the pH towards a neutral value that is more favorable for the growth of most bacteria.[48] Once formed, amines may be further metabolized, probably by deamination, with the formation of ammonia.[49] In one study, a histaminase inhibitor was used to show that approximately one third of histamine produced at pH 6.5 is immediately degraded.[50]

Organisms with decarboxylase activity include *Enterobacteria, Enterococci, Lactobacilli, Clostridia, Bacteroides,* and *Bifidobacteria*. Activity is not distributed universally through any genus and individual species possess varying numbers of decarboxylases. In one comprehensive study of the decarboxylation of lysine, arginine, and ornithine, *Enterobacteria* and *Enterococci* were generally active but Gram-negative anaerobic bacteria were rarely so.[51]

The products of decarboxylation include cadavarine from lysine, putrescine from ornithine, histamine from histidine and tyramine from tyrosine. Cadavarine and putrescine are further metabolized by oxidative deamination and ring closure, to form the secondary amines piperidine and pyrrolidine, which are normally present in the colon. Their concentration is reduced to trace amounts in both feces and urine when oral neomycin is taken.[52] Tetracycline likewise reduced the concentration of several amines, e.g., agmatine (derived from arginine), putrescine, cadavarine, histamine, ethanolamine, tryptamine, and tyramine in the ileum of young pigs. Tyramine has also been demonstrated in fecal extracts from normal subjects and was formed when [$^{14}$C] tyrosine was incubated with colonic bacteria. The excretion of tyramine in urine is diminished, but not eliminated by oral neomycin and it is suggested that tyramine may also be formed in the body tissues.[53] Tyramine can be metabolized to form a weak neurotransmitter octopamine, through the action of a β-oxidase, either in the intestinal lumen, or in the tissues after absorption from the gut.[54]

The oral flora also has considerable decarboxylase activity, and amines from arginine, aspartic and glutamic acids, histidine, lysine and ornithine are formed in dental plaque material.[45] The physiological and pharmacological activities of these amines have been outlined elsewhere.[31]

## V. AMMONIA

### A. Uptake of Ammonia

Ammonia probably enters bacterial cells in two forms; (1) as free ammonia, $NH_3$, and (2) as the ammonium ion $NH_4^+$. Most biological membranes are permeable to the diffusion of $NH_3$ and it is suggested that even organisms which synthesize powerful $NH_4^+$ transport systems would repress them when growing in abundant $NH_4^+$ (generally greater than 10 m*M*) as under these conditions, diffusion of $NH_3$ across the membrane should be sufficient to support growth.[55] However, as the pK for the dissociation $NH_4^+ \rightleftharpoons NH_3 + H^+$ is 9.25 at physiological pH, 99% of the ammonia that normally surrounds bacteria will be $NH_4^+$, so the ability to transport ammonia in this ionized form would be advantageous, particularly when ammonia concentrations are limiting.

There appear to be two mechanisms by which ammonia enters cells; (1) nonspecific diffusion and (2) an energy consuming carrier system that can concentrate large ammonia gradients across a membrane, even in the absence of a pH gradient that could help by trapping $NH_4^+$ in the cells. Some bacteria, particularly those that fix nitrogen, have high intracellular ammonia concentrations, e.g. *Klebsiella pneumoniae* contains 3 m*M* ammonia.[56] Comparison of intra- and extracellular $NH_4^+$ levels in *K. pneumoniae* and *Clostridium pasteurianum* showed that up to a 100-fold $NH_4^+$ concentration existed across the membranes against which ammonia was transported.[55]

### B. Ammonia Assimilation

There is little information available on ammonia assimilation by human intestinal bacteria. Most will utilize ammonia, but the proportion that use it preferentially to other nitrogen sources, or as the sole source of nitrogen is unknown. *Ruminococcus bromii, Bacteroides vulgatus, Bacteroides diastonis, Bacteroides thetaiotamicron,* and *Lactobacillus bifidus* are among the intestinal anaerobes that use ammonia.[22,23,57]

Ammonia is the main nitrogen source for most ruminal bacteria, however the situation in the human gut may resemble that of the pig more closely than that of the rumen. In a study on the utilization of ammonia nitrogen by intestinal bacteria isolated from pigs, 17 of

24 *Bacteroides*, 8 of 20 *Eubacteria*, 8 of 13 *Peptococci*, 3 of 4 *Clostridia*, 19 of 20 *Enterobacteria*, and 1 of 8 *Streptococci* utilized ammonia nitrogen preferentially to proteose peptone nitrogen. *Bacteroides ruminicola* was studied in detail using $^{15}N$ from [$^{15}N$] ammonia. $^{15}N$ was incorporated into every amino acid, with the highest uptake in aspartic acid and alanine. The presence of single amino acids in the medium, particularly aspartic acid, glutamic acid, threonine, or alanine did decrease ammonia uptake, but ammonia was still utilized more effectively than any of the amino acids.[58]

*1. Mechanism of Ammonia Assimilation*

To date two main mechanisms for the assimilation of ammonia have been described and both result in the amination of α-oxoglutarate to glutamate.

The first mechanism involves the reductive fixation of ammonia under the action of glutamate dehydrogenase (GDH), which converts α-oxoglutarate directly to glutamate with NADPH or NADH as the electron donor. This reaction is freely reversible and appears to be the principal method of incorporating free ammonia into amino acids in most bacteria.[28] The Km of GDH for ammonia is high, indicating that this route for incorporating ammonia operates most efficiently at high concentrations of ammonia, although it may be the only route used by some organisms, e.g., *Streptococci*, even at low ammonia concentrations.[59] Glutamate dehydrogenase is found in both aerobes and anaerobes.[60]

The second and less important mechanism, involves two enzymes glutamine synthetase (GS) and glutamate synthase, which conjointly catalyze the amination of α-oxoglutarate to glutamate, with glutamine as an intermediate.[61] This reaction differs from that catalyzed by GDH,

1. In being irreversible
2. In requiring 1 mol of ATP per mole ammonia assimilated
3. In operating well at low ammonia concentrations, due to the high affinity of GS for ammonia

Glutamate synthase appears to be lacking in the ruminal anaerobes *Bacteroides amylophilus* and *Selenomonas ruminantum*,[60] and in some *Streptococci*,[59] which suggests that these organisms are dependent on the GDH pathway or others yet to be discovered for their uptake of ammonia.

The position of glutamine as an intermediate in the pathway for ammonia assimilation has important and far reaching consequences for the regulation of nitrogen metabolism in *Escherichia coli*, and presumably in other intestinal bacteria. Glutamine is a key compound, being able to transfer its amide nitrogen to a variety of compounds in the biosynthesis of AMP, CMP, tryptophan, histidine, glucosamine 6-phosphate, and carbamyl phosphate.[62] At high concentrations, ammonia can replace glutamine in most of these reactions. Thus, glutamine synthetase clearly occupies a central role in nitrogen metabolism; besides controlling uptake of ammonia, GS is also involved in the control of the synthesis of enzymes that are regulated by ammonia, e.g. urease.[60,63]

### C. Bacteria Producing Ammonia from Amino Acids and Peptides

There have been few studies on ammonia production from nonurea sources, e.g. peptides and amino acids by intestinal bacteria.[64-67] The most recent and detailed of these found that Gram-negative anaerobic rods, *Clostridia*, *Enterobacteria*, and aerobic spore-forming organisms were very active producers of ammonia; *Streptococci*, *Micrococci*, and Gram-positive nonsporing anaerobes formed moderate concentrations; *Lactobacilli* formed very little, as did two yeast isolates.[67] There was enormous variation between organisms of the same species in their ablity to form ammonia, e.g. six strains of *Bacteroides vulgatus* formed a mean of 36.7 mmol/ℓ ammonia with a range of 21.1 to 67.7 mmol/ℓ and 15 strains of *Escherichia coli* formed a mean of 19.4 mmol/ℓ with a range of 13.6 to 36.3 mmol/ℓ ammonia. Aerobic Gram-negative organisms showed greater variability than anaerobic Gram-negative organisms when ammonia produced in different media was compared. The mean concentrations for ammonia production by 32 isolates of Gram-negative anaerobes were 25.0 and 22.4 mmol/ℓ in brain-heart infusion broth and nutrient broth respectively, and the corresponding values for 32 Gram-negative aerobes were 29.6 and 19.8 mmol/ℓ, suggesting that the anaerobic organisms are affected less by, and are more adaptable to change in the supply of nutrients than are the aerobes.

### D. Factors Affecting the Production and Use of Ammonia

#### *1. pH*

Organisms that are mainly responsible for the production of ammonia from peptides and amino acids, e.g. the Gram-negative anaerobes and aerobes, and *Clostridia* all form considerably less ammonia at pH 5.0 than pH 7.0.[64,67] Less ammonia was also formed in fecal incubates when the pH was below 5.0,[68] although ammonia formed in incubates of pH 5.0 to 6.0 did not differ from that of the control regardless of the mechanism (HCl, lactic/acetic acid mixture) used to lower pH. The normal colonic pH is 5.5 to 7.0,[69] indicating that in vivo pH would not usually be a factor limiting the release of ammonia. However, under certain circumstances, e.g. when carbohydrates such as lactulose or hemicelluloses are fermented in the colon, the pH may fall below 5.0[69] and ammonia production would then be curtailed.[68]

#### *2. Ammonia Concentration*

It also seems unlikely that the ambient ammonia concentration is a factor limiting ammonia production in the colon. The mean concentration of fecal ammonia in normal subjects was found to be 11.2 mmol/ℓ with a 95% range of 2.9 to 44 mmol/ℓ.[21] These concentrations did not inhibit ammonia production by any of the major groups of intestinal bacteria.[67] Furthermore, strains of *Enterobacteria* tested in media containing 68 mmol ammonia/ℓ, which is several times greater than the mean concentration in the colon, produced amounts of ammonia similar to those formed in media containing 2 or 3 mmol/ℓ. Thus, even in uraemic subjects for whom there is evidence that the amounts of urea entering the colon and the amounts of ammonia formed (mean 17.9 mmol/ℓ, 95% range 4.3 to 75 mmol/ℓ) are both significantly greater than in normal subjects, the ambient ammonia concentration would probably not be a factor limiting release of ammonia.

#### *3. Viability*

Sterilization did not affect ammonia formed by fecal incubates, provided that the enzymes were not inactivated by the process. Feces that had been sterilized by $\gamma$-irradiation formed similar concentrations of ammonia to that formed by 25% fecal suspensions in which survival of bacteria was good. Autoclaved feces showed no increase in ammonia concentration.[70]

#### *4. Antibiotics*

The clinical improvement that often follows the administration of neomycin to patients with hepatic encephalopathy is attributed to suppression of the gut flora. However, as many of the nonsporing anaerobes that are very active ammonia producers are resistant to neomycin, e.g. *Bacteroides,* it is suggested that neomycin may inhibit ammonia production by some means other than its bactericidal effect. To test this hypothesis 56 strains of Gram-negative anaerobes and 55 strains of nonsporing Gram-positive anaerobes were grown in broths containing either 0, or 100 μg/mℓ neomycin; after 48 hr growth and ammonia concentration were measured. There was no difference between the ammonia concentration in the control and neomycin broths for either group of organisms, suggesting that even at this high concentration, neomycin does not inhibit bacterial deaminases.[71]

*5. Substrate Effect*

Ammonia cannot be efficiently assimilated unless there is carbohydrate present to provide both carbon skeletons and energy. Once all the carbohydrate has been used, nitrogen sources will be fermented for these purposes, with the inevitable result that excess ammonia will accumulate in the medium. There are two mechanisms affecting ammonia metabolism that operate in the presence of fermentable carbohydrate: (1) an increase in the uptake of ammonia into bacterial protoplasm, and (2) a decrease in the fermentation of nitrogenous substrates such as amino acids, due to inhibition in both the formation and the activity of enzymes which form ammonia.[72] The mechanisms operate simultaneously and their net effect is to reduce the ammonia concentration in the medium.

While the use of ammonia as a nitrogen source is taken for granted as fundamental to bacterial metabolism, the selective inhibition of enzymes to promote efficient use of available carbohydrate substrates is not fully appreciated. It has been known for many years that the presence of readily fermentable carbohydrate in the medium inhibits the formation and the activity of enzymes responsible for fermenting nitrogen sources such as urea and amino acids. Kendall and Farmer recognized in 1913 that the provision of a utilizable carbohydrate may arrest the proteolytic activities of bacteria,[73] and Gale and his co-workers showed that, as measured by ammonia formation, the formation of alanine and glutamic deaminases and aspartase was inhibited by glucose fermentation, and that the inhibition was independent of fall in pH.[74,75] This ability of glucose to inhibit the synthesis of inducible and constitutive enzymes has been termed catabolite repression.[76] In addition to inhibiting the synthesis of enzymes, carbohydrates also repress the activity of preformed enzymes, a process known as catabolite inhibition. These effects are not restricted to glucose, but may extend to other substances that can act as readily available sources of intermediary metabolites and energy.

There are many reports of carbohydrate fermentation inhibiting the amino acid deaminases of aerobic bacteria,[72] but there is little on the metabolism of intestinal anaerobes. In one study, pure cultures of various species of *Bacteroides* and other intestinal bacteria were incubated with single amino acids after prior growth in control media and media supplemented with either lactose or lactulose. Bacteria grown with these sugars showed a reduction in ammonia formation ranging from 0 to 99% from alanine, histidine, and methionine, while ammonia release from aspartate was stimulated rather than inhibited. Inhibition also occurred in a nutrient medium supplemented with lactose or glucose, although the effect was sometimes considerably less than that seen with washed cells and single amino acids.[67]

It is only comparatively recently that interest in the potential benefit for man of carbohydrate fermentation in the colon has been reawakened, with the recognition that undigested food residues reaching the colon may influence the metabolic activities of intestinal bacteria. People with hepatic encephalopathy, to whom ammonia may be toxic, are among those who may benefit from a decrease in colonic ammonia concentration. They often improve when treated with the synthetic disaccharide lactulose (1-4β-galactosido-fructose),[77] which is not metabolized by human disaccharidases, but passes unchanged to the ileum and colon, where it is metabolized by many intestinal bacteria.[67,78] Among the many theories postulated for the beneficial effect of lactulose are its potential to inhibit bacterial deaminases[67] and to increase bacterial uptake of ammonia by acting as a readily available source of carbon and energy.[68,79] Fermentation of lactulose not only prevented the normal accumulation of ammonia, but it actually resulted in a net decrease in ammonia concentration while it was being metabolized in fecal incubates. Once all the lactulose had been broken down, ammonia accumulated in the incubate as normal, provided that a pH of 5.5 or above was maintained[68,80] (see Figure 1). Glucose, mannitol, and sorbitol exhibited similar effects.[68]

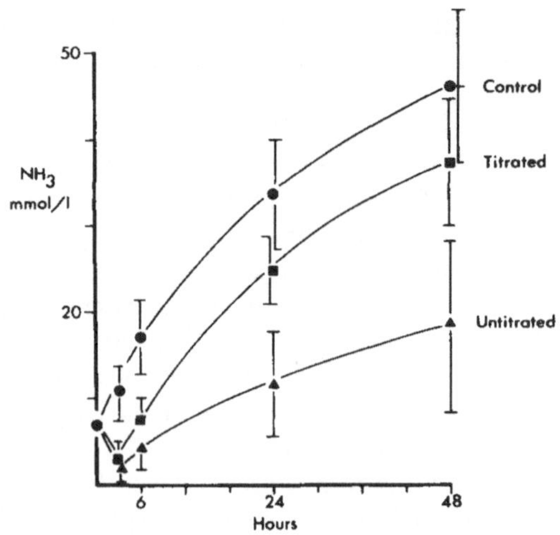


FIGURE 1. Effect of 0.03 *M* lactulose on ammonia concentration in a 25% fecal incubation system: ●, control; ■, lactulose titrated to keep pH at 5.5 or above; ▲, untitrated lactulose. Results are mean and standard deviation for six experiments on fecal specimens from four subjects. Reproduced by kind permission of Vince, A., *Gastroenterology*, 74, 544, 1978.

This effect is not limited to simple sugars; it also occurs with more complex carbohydrates, e.g. pectin and hemicelluloses, becoming more pronounced after increased quantities of the particular carbohydrate have been ingested.[81] Lactulose, pectin, and the hemicellulose arabinogalactan, but not cellulose, each lowered the ammonia concentration in fecal incubates in a study where ammonia production was measured before and after subjects had ingested increased quantitites (0.3 g/kg pectin and cellulose, and 0.8 g/kg lactulose and arabinogalactan) of each substance. Lactulose was the most effective substrate, but after subjects had ingested arabinogalactan, ammonia formation in 6 hr fecal incubates was reduced by 46%. Pectin also lowered ammonia concentration while it was being fermented. However, more ammonia was formed over 48 hr in incubates from people who had ingested extra pectin, than had been formed prior to its ingestion, presumably because pectin increased the excretion of fecal nitrogen by 32%. Other subjects taking 36 g pectin per day for 9 weeks also showed an increase in fecal nitrogen excretion of 47% at week 6 and 24% at week 9, despite the fact that pectin is totally digested in the gut.[82] No such increase in ammonia formation was seen in incubates from subjects who had ingested lactulose or arabinogalactan. Lactulose and arabinogalactan are obviously suitable candidates for lowering the concentration of colonic ammonia, while pectin would clearly have the reverse effect and could not be recommended for this purpose.

There have been no studies on the overall nutrient requirements of intestinal organisms from the human gut. However, in the rumen, ammonia is important for the nutrition of the fiber- and starch-digesting bacteria, as by the time the fiber-digesting bacteria have established themselves on the particles of digesta, any free amino acid has been utilized or absorbed, making the use of ammonia as a nitrogen source an attractive proposition[83] and a sensible survival mechanism. A similar situation may exist in the human colon, as it is known that considerable quantities of fiber and soluble starch may be available daily for fermentation,[84-86] and studies with fecal incubates have shown rapid uptake of ammonia while carbohydrate is fermented.[68,81]

## E. Sources of Intestinal Ammonia

### *1. Urea*

It has been generally assumed that the major source of fecal ammonia is urea, mainly because about 7 g of it are degraded daily in healthy subjects.[87] However, it is known that sufficient ammonia to account for normal fecal levels is generated when feces containing no urea are incubated in vitro,[70] suggesting that nitrogen sources other than urea are the major precursors of fecal ammonia. This has been confirmed in a study using $^{15}N$-labeled urea.[88]

Two healthy people were given $^{15}N$ urea as a constant intravenous infusion. Constant production of endogenous $^{14}N$ urea was ensured by giving identical 3-hourly meals throughout the study. After one complete intestinal transit, the enrichment of ammonia nitrogen was measured in samples of fecal dialysate and fecal ultracentrifugate. The $^{15}N$ labeling of the various samples of fecal ammonia averaged only 8.5% of the plasma $^{15}N$ urea, showing that endogenous urea is not the major source of fecal ammonia and that the circulating blood urea makes only a minor contribution to ammonia in the intestinal lumen. This does not necessarily mean that nonurea nitrogen sources contribute more ammonia than does urea for absorption from the intestine, but it does strengthen suggestions that urea hydrolysis is closely associated with the mucosa and that the ammonia released is rapidly reabsorbed into the portal vein, and does not diffuse far into the intestinal lumen.[87] It is not possible to proportion exactly the relative contributions of urea and nonurea sources to ammonia absorbed from the intestine, but it is apparent that the long standing assumption that urea is the single major source of intestinal ammonia is incorrect. Studies on portal blood using $^{15}N$ urea are needed to provide precise quantitation in man. In a study on the origins of portal blood ammonia in the dog, approximately equal amounts were derived from the small and large intestines. All the ammonia from the small intestine and 9% from the large intestine originated from glutamine in the perfusing blood. Urea accounted for only 42% of that derived from the large intestine.[89]

### *2. Nonurea*

The experiments with $^{15}N$ urea have shown that only 9% of fecal ammonia is derived from urea, indicating that over 90% of fecal ammonia is formed from other sources. The major sources are likely to be proteins, peptides, amino acids, and amines, which may originate from dietary residues, from the intestinal secretions and shed epithelial cells of the host, or from the digestion of bacterial cells. Small amounts will be formed from the breakdown of other nitrogen containing compounds such as creatinine and uric acid. Ammonia is also produced inside bacterial cells from the metabolism of urea and amino acids and this will be released when bacterial cells breakup.

Ammonia normally forms about 2% of the total fecal nitrogen. This fraction increases to 12% when 25% saline homogenates of feces are incubated outside the body, when the mean ammonia concentration rises to 33 mmol/ℓ (equivalent to 132 mmol/ℓ in indiluted feces), at 48 hr. This is much higher than the normal fecal ammonia level, suggesting that all fecal ammonia could be accounted for by bacterial activity. It is possible however, that particularly higher up in the intestine, some ammonia may be formed by nonbacterial mechanisms such as the digestion of glutaminc. Ccrtainly ammonia is present, albeit at much lower concentrations, in the intestines of germfree animals[90] and there is no reason why humans should differ in this respect.

## F. Fate of Intestinal Ammonia

Ammonia produced in the intestine may be either excreted, absorbed, or incorporated into bacterial protoplasm. Fecal nitrogen (60%) is contained in bacteria and a large proportion of this will have been derived from ammonia. The immediate precursor of most of this ammonia will be nonurea, with a small fraction coming ultimately from urea following preliminary incorporation into bacterial protoplasm.

About 400 mmol of ammonia are formed in the intestine each day, of which less than 2% is excreted as fecal ammonia, the remainder is either absorbed or utilized by bacteria. It is thought that absorption is mainly by passive nonionic diffusion through the mucosa into the portal vein.[91,92] Most of the ammonia is converted into urea in the liver, with a small proportion being used for the synthesis of nonessential amino acids. Essential amino acids may also be synthesized if their keto-acid analogues are available.[93] These analogues have been given to uremic subjects to stimulate the re-use of ammonia nitrogen and have many beneficial effects, i.e. they reduce plasma urea concentrations, increase plasma levels of essential amino acids, convert a negative into a positive external nitrogen balance and improve symptoms.[94,95] The mechanism of these effects is not completely clear. It is not due entirely to the transamination of the administered keto-acids, but may also be related to a decrease in the breakdown of the existing essential amino acids.[94] The recycling of ammonia nitrogen would theoretically be advantageous to patients with renal failure, allowing them to maintain their nitrogen balance on a low protein intake. Unfortunately any such benefit is likely to be small, as only 3% of the nitrogen in albumin synthesized by uraemic subjects on a low protein diet was derived from urea.[96]

## VI. UREA

### A. Bacteria Producing Ureases

Urea is hydrolyzed to ammonia and carbon dioxide by some 5 to 22% of intestinal organisms.[97-99] Urease activity is very widely distributed, occurring in strains of *Peptostreptococcus, Ruminococcus, Peptococcus, Coprococcus, Eubacterium, Bifidobacterium, Clostridium, Lactobacillus, Bacteroides, Fusobacterium, Enterobacterium,* and *Streptococcus.* and some enterobacteria[32,98,99] The most numerous urease-forming species in human feces are anaerobes, particularly *Peptostreptococcus productus,*[22,97,99] and *Eubacterium aerofaciens*[96] which both occur at counts of greater than $10^{10}$/g fecal dry weight.[4] While most strains of *P. productus* are thought to produce urease,[22,99] indicating that they are the predominant urease producing intestinal organism in most people,[22] the characteristic appears extremely variable in other species, e.g. only 4 of 15 strains of *E. aerofaciens,* 3 of 52 strains of *Lactobacillus fermentum* and 4 of 44 strains of *Bifidobacterium breve* were urease positive.[97,101] The specific activity of the enzyme also varies considerably, both between species of the same genus[100] and also between strains of the same species.[97]

### B. Properties and Control of Urease

Considerable variation also exists within other properties of bacterial ureases. The pH optimum for the urease of *Lactobacillus fermentum* was found to be 4.0, whereas that of *Bacteroides multiacidus* and *Proteus mirabilis* was 8.0,[97] with hydrolysis of urea by *P. mirabilis* being negligible below pH 5.3.[64] The ureases produced by different species of *Proteus* also appear to vary at a very basic level. Each of the four species has a characteristic and unique pattern of urease isoenzymes which suggests that they are different from each other and that they may have dissociated from genetically different enzymes.[101]

It seems that urease production may be associated with either the intact chromosome or with extrachromosomal elements. Attempts to demonstrate the presence of extrachromosomal elements in strongly ureolytic strains of bifidobacteria were unsuccessful.[100] However, the loss of an apparently plasmid-coded urease from *Streptococcus faecium* has been reported,[102] as has the unexplained disappearance of urease activity from individual strains of *Coprococcus catus, Fusobacterium prausnitzii* and *Streptococcus mitus* over 8 months storage.[99]

The control of urease formation clearly differs considerably between various groups of organisms, e.g. *Proteus vulgaris*[103] and certain bifidobacteria[100] continue to form urease under many growth conditions (high or low ammonia, high urea, high ammonia and low urea) whereas in others, e.g. *Peptostreptococcus productus* and *Selenomonas ruminantum* the enzyme is repressed by ammonia, complex nitrogen sources and even high levels of urea.[22,97,99] On the practical level, it seems likely that urease activity of many, but not all, anaerobes may be repressed, or a spuriously low specific activity be recorded if the enzyme is sought in media containing too much ammonia, urea or other nitrogenous nutrients such as yeast extract or *Trypticase*.[22,97,99,100] A complex medium and test has been developed for the rapid detection of urease in fermentative anaerobic bacteria,[99] and modifications of it have been used subsequently in two large studies.[97,100]

With such variety in the ability to form urease contained within the various members of the intestinal flora, it is predictable that some, if not all, of the urea entering the gut will be hydrolyzed under most of the conditions that normally prevail. In normal healthy people, about 7 g of urea,[87] or about 50% of the body pool,[2] is degraded daily. There are many reasons, reviewed extensively elsewhere,[2] for believing that all this urea is hydrolyzed by bacterial ureases in the intestine, the major sites of urea destruction coinciding with the maximum sites of bacterial colonization, i.e. the colon and terminal ileum, with small amounts degraded by the oral flora in the mouth.[105] Nearly all the urea diffusing into the gut is hydrolyzed under normal conditions and also when cathartics are given which increase the speed of intestinal transit, as fecal levels are negligible at 0.1 mmol/ℓ.[105] In uraemic subjects, in whom proportionately more urea (related to raised plasma levels) diffuses into the intestine, it was only detected in the feces of one third of the subjects studied,[21] suggesting that the intestinal flora can adapt and hydrolyze increased quantities of urea. This adaption is reported to be due to an increase in the proportion of bacteria that hydrolyze urea and not to any increase in the mean urease activity per strain.[98] However, with the diverse responses described previously, of different bacteria to changes in urea concentration, it seems probable that if the concentration of urea increased in the colon urease would be induced in some bacteria and repressed in others, with corresponding changes in their mean urease activities.

### C. Inhibition of Urea Hydrolysis

Because of the clinical interest in urea as the source of gram quantities, (approximately 3.5 g/day in subjects with normal plasma levels and possibly as much as 17 g in uraemic subjects),[87,106] of ammonia that may be toxic to patients with liver failure, considerable effort has been made both to reduce the amount of ammonia formed from urea each day and also to reduce the amount of ammonia available for the resynthesis of urea.

#### *1. Antibiotics*

Antibiotics have been shown to reduce urea hydrolysis, both in vivo and in vitro. When sufficient antibiotics are given to kill urea-splitting bacteria, urea appears in fecal dialysate at concentrations similar to those in plasma.[107] Neomycin therapy reduced the mean in vitro urease-specific activity in fecal incubates by 59%.[108] Because oxytetracycline and chloramphenicol inhibit plant ureases in vitro, it has been suggested that they might also inhibit bacterial ureases. This theory has not been supported by results from two studies: in the first urease from *Proteus mirabilis* was not inhibited by oxytetracycline, chloramphenicol, neomycin, kanamycin, or gentamicin at pH 7.2,[109] while the second reported it to be inhibited by oxytetracycline, trimethoprim, and cefamdole, but only at very high concentrations (not quoted) and only at acidic pH.[110]

Treatment with antibiotics although effective in the short-term is generally to be avoided because of the emergence of resistant strains, overgrowth with undesirable organisms, e.g. candida, *Staphylococci*, and *Clostridium difficile* and other side effects such as diarrhea and steatorrhea.

*2. Acetohydroxamic Acid*

Acetohydroxamic acid, an analogue of urea, is one of the most potent inhibitors of bacterial urease described so far.[109,110] It reduced the activity of urease from *Proteus mirabilis* by 98% and that of fecal incubates from 66 to 98%.[108] When given orally it lowered the concentration of ammonia excreted from the colon of normal subjects[111] and reduced the blood ammonia levels, but had no consistent beneficial effects in cirrhotic patients.[112] Unfortunately acetohydromaic acid is toxic and has teratogenic properties,[113] and so is of limited use clinically.

*3. Immunization*

Another approach has been to immunize subjects with urease isolated from the jack bean rather than a bacterial urease. Of the patients treated in this way, six out of eight formed antibodies and improved clinically.[114] If this form of therapy were to be used again, it would seem logical to immunize subjects with a bacterial urease, e.g. from *Peptostreptococcus productus* and not a plant urease, because we know that the two are not identical.[115,116]

*4. Limiting Ammonia Available to Form Urea*

More recently, another approach has been tried in uraemic subjects. The aim was to decrease the amount of ammonia that is available to form urea, the rationale for the therapy being the same as that for the use of lactulose in hepatic encephalopathy, and is given in detail in the section on factors affecting the production and use of ammonia.

There were 6-week courses of two different hemicelluloses, (arabinogalactan and ispaghula) given to uraemic subjects and they reduced the mean plasma urea by 11 and 19% respectively. Ispaghula stopped the plasma creatinine rising and increased fecal nitrogen excretion by 39% in one formal balance study.[117] It is suggested that such therapy might alleviate some of the symptoms of uraemia and postpone dialysis in patients with endstage renal disease.

## VII. LIVER DISEASE: ROLE OF BACTERIA IN PORTAL-SYSTEMIC ENCEPHALOPATHY (PSE)

In this syndrome comatose patients are suffering from cerebral intoxication by intestinal contents that have not been metabolized by the liver, a) because its cells are damaged, or b) because the blood supply has circumvented the liver. The reversible nature of the cerebral symptoms suggests a chemical origin of coma, and the major toxins are thought to be formed by bacteria from proteins and their derivatives.[118] The substances advanced as possible toxins include ammonia, methionine and its metabolites the mercaptans methanethiol and ethanethiol, tryptophan and its derivatives, short-chain fatty acids ($C_5$-$C_8$) and false neurotransmitters such as octopamine. A disturbance in the ratio of certain plasma amino acids has also been incriminated in the pathogenesis of coma. The evidence for and against the various toxins has been reviewed extensively[119-123] and is summarized briefly here, with reference to the likely effect of treatment on toxin formation.

### A. Toxins Incriminated in the Pathogenesis of PSE

*1. Ammonia*

Although it is certainly not the only toxin of importance in PSE, ammonia has remained the front runner for many reasons:

1. The syndrome is reproduced in some patients by oral or intravenous ammonium salts and is precipitated by other substances — urea, blood, or protein — that give rise to ammonia.

2. Blood ammonia concentrations are often elevated in patients with PSE; children with hyperammonia due to deficiencies of urea cycle enzymes have symptoms of encephalopathy.
3. There are well established mechanisms by which ammonia could interfere with cerebral metabolism. Combination of ammonia with α-ketoglutarate to form glutamic acid removes an essential intermediate in the Krebs cycle, which could reduce oxidative metabolism in the brain. This hypothesis is supported by an increase in the pyruvic acid concentration in the blood and cerebro-spinal fluid (CSF), and a decrease in oxygen consumption in the brain of patients with hepatic coma.[124] Further combination of ammonia with glutamic acid, to form glutamine would compound this deficiency. Raised glutamine in the CSF of patients a sensitive index of PSE.[125]
4. Ammonia may act in conjunction with mercaptans and fatty acids to produce encephalopathy.[126,127]

The development of coma in germfree dogs with porto-caval shunts[90] does not invalidate the case, as considerable quantities of ammonia are released in the small intestine of dogs by cellular metabolism of glutamine;[128] the kidney and muscle also release ammonia. There are however, some facts which suggest that ammonia intoxication is not always involved in PSE: (1) the brain in hepatic coma does not always remove ammonia from the blood, but may occasionally add to it,[119] and (2) blood ammonia levels are normal in 10% of cases (up to 1 μg/mℓ) regardless of the depth of coma.[129]

*2. Methionine*

When given orally to susceptible patients with liver disease methionine precipitates hepatic coma, with no increase in blood ammonia.[130] This effect could follow bacterial metabolism of methionine to form mercaptans, particularly methanethiol which is exceedingly toxic. Raised blood levels occur in some patients with PSE.[131] Intravenous methionine does not precipitate coma, indicating that methionine is not toxic per se.

*3. Tryptophan*

Ingestion of tryptophan may cause encephalopathy in susceptible patients and increased levels occur in the brain in experimental coma.[119] There are two mechanisms which have been suggested: (1) an alteration in the endogenous metabolism of tryptophan to serotonin, a substance that has profound effects on the central nervous system, and (2) indoles and skatoles, bacterial degradation products of tryptophan could inhibit brain respiration. There is little evidence to support the second hypothesis.

*4. Fatty Acids*

Various short- and medium-chain fatty acids (butyric, valeric, octanoic) have been implicated in the causation of hepatic coma. Patients with cirrhosis have raised blood and CSF levels. In children with isovaleric acidaemia, the onset of confusion and stupor is accompanied by a dramatic rise in this particular acid.[132] Octanoate is reported to potentiate the toxic effects of the mercaptans and ammonia.[126] Overall, no convincing association between fatty acids of any chain length and PSE has been established.

*5. False Neurochemical Transmitters*

The true sympathetic neurochemical transmitters are noradrenaline and dopamine. However, similar substances, e.g. phenylethanolamine and octopamine, are produced by the bacterial metabolism of phenylalanine and tyramine respectively. It is suggested that the bacterial products, particularly octopamine may accumulate and interfere with neurotransmission at the synapses of the brain.[133]

Tyramine is known to interfere with the formation of dopamine and adrenaline, and its metabolite octopamine is present at increased levels in the serum and urine in PSE.[134] However, proof is lacking that the true transmitters are actually depleted in PSE, and recent evidence suggests that they may not be, as patients did not improve on levodopa (a precursor of dopamine that, unlike dopamine, crosses the blood brain barrier) in a double blind controlled trial.[135]

*6. Amino Acids*

The plasma concentration of many amino acids is altered in PSE and it is claimed that these abnormalities are important.[121] In particular, a reduced ratio between valine + leucine + isoleucine, and tyrosine + phenylalanine has been linked to the development of PSE. However, in a large group of cirrhotic patients the ratio was reduced in both those with and without encephalopathy,[136] and treatment with the branched-chain amino acids, leucine, valine, and isoleucine which are reduced in the plasma of patients with PSE had no effect on encephalopathy as compared with placebo.[137]

In conclusion, it seems that the causes of PSE are multifactorial and that various toxins or combinations of toxins may be important in different subjects, or even in the same subject at different times. Raised blood levels of different substances may be nonspecific indicators of disturbances in liver and brain function, rather than important markers for the pathogenesis of PSE.

### B. Treatment

*1. Diet*

The major toxic substances in PSE are formed in the intestine by bacterial degradation of proteins and their derivatives.[118] Hence acute episodes of coma are treated by complete removal of dietary protein and chronic cases are often maintained on a low protein diet.[119] The quality of the protein appears important, as protein derived from vegetables may be tolerated better than animal protein,[138] and it is suggested that a vegetable protein diet rather than overall protein restriction should be considered in the management of PSE.[139] It is probable that the structural carbohydrates present in plant cell walls exert a substrate effect, identical to that described for lactulose and arabinogalactan,[68,81] and thus lower colonic ammonia concentrations. Fecal ammonia levels fall in human volunteers when they are fed fermentable carbohydrate.[86] Unfortunately part of this substrate effect will be negated by any extra protein that the undigested carbohydrate carries with it into the colon. Another plus for vegetable protein is that it contains lower concentrations of methionine and aromatic amino acids than animal protein.

*2. Antibiotics*

Neomycin, tetracycline, and metronidazole have all been used to treat PSE, and the evidence indicates that they act through their bactericidal/bacteristatic effects. Tetracycline is effective for short periods but its use is limited by the speed at which resistant organisms, particularly enterobacteria, emerge. Neomycin is traditionally regarded as the best antibiotic in PSE, but unfortunately its long term use may result in impaired hearing or deafness.[140] Neomycin has also been used in combination with lactulose, to obtain a larger reduction in both the blood ammonia concentration and the urea production rate than was observed when the drugs were used separately.[141,142] As some bacteria that metabolize lactulose are resistant to neomycin, e.g. *Bacteroides*, the effect of each drug can be mediated by a different sector of the intestinal flora when the combination is used. There are some patients who do not benefit from this combination.[143] Recently it has been found that metronidazole may be as effective as neomycin for treatment of PSE,[144] but further studies are needed to confirm this. Metronidazole is active against anaerobic organisms, e.g. *Bacteroides*, while neomycin is most active against aerobes, e.g. *Enterobacteria*. It would be interesting to see how ammonia production and uptake (± lactulose) is affected in fecal incubates from patients receiving each of these antibiotics, as this might indicate which fraction of the intestinal flora is most important in the genesis of ammonia from nonurea sources in vivo. Neomycin had no effect on ammonia production in fecal incubates from subjects who had not taken neomycin, because either the organisms that were killed contributed a negligible quantity of ammonia, or more likely, the ammonia producing enzymes were unaffected by the antibiotic and continued to produce ammonia despite loss of cell viability.[145]

*3. Lactulose*

This synthetic disaccharide was first used to treat PSE in 1966.[146] It is reported to provide effective control in from 70 to 85% of cases of chronic PSE,[147,148] improving neuropsychiatric symptoms and lowering blood ammonia.[143] It is not recommended for acute hepatic coma where nitrogenous intoxication is controlled best by neomycin.[119] Several mechanisms have been proposed for its beneficial effect; these have been reviewed extensively and are only briefly mentioned here.[120,147,149,150]

***a. Effect on Fecal Flora***

The original theory was that lactulose favors the growth of acidophilic bacteria, such as *Lactobacilli* and discourages growth of putrefactive ammonia producing species. In practice, it has been difficult to demonstrate any consistent effect of lactulose on the fecal flora,[151] and impossible to correlate clinical improvement with a reduction in the count of two groups of organisms, *Enterobacteria* and *Bacteroides*, that are considered to be active ammonia producers.[152]

***b. Cathartic Effect***

Lactulose acts as a carthartic, decreasing colonic transit time and hence the time available for toxins to be released from nitrogenous substrates. While there is no doubt that lactulose is an effective cathartic and is used as such in PSE, it is of greater benefit in PSE than other nonfermentable purgatives such as magnesium sulfate. This suggests that there is an additional factor at work other than its cathartic action. Sorbitol has also been used as a purgative in PSE, and may be of similar efficacy to lactulose. Like lactulose, it is largely unabsorbed in the small intestine, and passes on to be fermented in the terminal ileum and colon, where it exerts a substrate effect, which has been demonstrated in an in vitro fecal incubation system.[68]

***c. Trapping of $NH_4^+$ at Low pH***

Ammonia diffuses through biological membranes most readily in its unionized form and it was therefore suggested that when the colonic pH was lowered by lactulose fermentation, more ammonia would be ionized and hence trapped in the lumen. If this were the case, an increase in fecal ammonia concentration should occur in subjects taking lactulose. There were two studies which failed to show any such increase,[152,153] possibly because any ammonia that is trapped is incorporated into bacterial protoplasm. In any event, most of the colonic ammonia would normally be ionized (see Section V.A, Uptake of Ammonia) and it is doubtful whether a further shift in the proportion ionized would greatly influence its concentration in the colon.

***d. Substrate Effect***

Provision of a readily fermentable carbohydrate alters bacterial metabolism; (a) by increasing uptake of ammonia into bacterial protoplasm and (b) by inhibiting the breakdown of nitrogen sources that supply carbon and energy for bacteria in the absence of carbohydrate.

Each of these mechanisms will decrease the amount of ammonia available for absorption, which would normally be recycled as urea. They have been discussed earlier under the effect of substrate on ammonia formation.

Indirect evidence that the uptake of ammonia by bacteria is increased, is provided by the raised excretion of fecal nitrogen in subjects on lactulose.[154] Most of this increase in likely to be in bacterial protoplasm, as neither urea nor ammonia excretion in feces is increased.[153] Overall, lactulose causes a 25% decrease in the urea production rate, which probably follows a depletion in colonic ammonia reserves, due to its increased removal by bacteria and obviously also to its decreased production.[142] Although the excretion of fecal nitrogen is increased, lactulose does not alter the total nitrogen balance of the body, only the route by which nitrogen is excreted, as a compensatory decrease occurs in the urinary excretion of urea.

At present, in the absence of any other proposed mechanisms, this substrate effect seems likely to be responsible for a major portion of the beneficial effect of lactulose therapy. This effect is not restricted to lactulose. It has also been demonstrated in an in vitro fecal incubation system, for glucose, mannitol, sorbitol, and the hemicellulose arabinogalactan and ispaghula.[68,81,117]

#### *e. Other Effects of Lactulose*

Thus, there are ready explanations for the beneficial effect of lactulose in PSE if ammonia is the major toxin: mechanisms by which it may decrease the production of other toxins await demonstration. Inhibition in the fermentation of amino acids by lactulose or other substrates could have both beneficial and detrimental effects. Lactulose inhibits deamination of methionine,[67] but it is not known how it affects the production of methanethiol or ethanethiol. Glucose is known to inhibit tryptophan degradation[155] and other substrates may have similar effects; unfortunately, the accumulation of tryptophan could have deleterious effects in PSE providing extra substrate for the formation of serotonin. Reduction in the pH of the colon following lactulose fermentation could theoretically stimulate the decarboxylation of tyrosine, the precursor of the false neurotransmitter octopamine. In addition the acids produced from lactulose, particularly butyric acid, could have toxic effects,[122,78] although inhibition of amino acid fermentation could reduce the formation of other potentially toxic organic acids.

## ACKNOWLEDGMENTS

I would like to thank Professor O. M. Wrong for his help and encouragement during the time I have worked on nitrogen metabolism in his department. I am grateful to the Medical Research Council for financial support and to Mr. J. Wager for considerable help with the literature search.

## REFERENCES

1. **Stephen, A. M. and Cummings, J. H.,** The microbial contribution of human faecal mass, *J. Med. Microbiol.*, 13, 45, 1980.
2. **Wrong, O. M., Edmonds, C. H., and Chadwick, V. S.,** *The Large Intestine: Its Role in Mammalian Nutrition and Homeostasis*, MTP Press, Cambridge, 1981, chap. 10.
3. **Hagihara, B.,** Bacterial and mould proteases, in *the Enzymes*, Vol. 3, 2nd ed., Boyer, P. D., Ed., Academic Press, New York, 1960, 193.
4. **Moore, W. E. C. and Holdeman, L. V.,** Human fecal flora; the normal flora of 20 Japanese-Hawaiians, *Appl. Microbiol.*, 27, 961, 1974.

5. **Hausken, O. W.,** Serological studies on proteolytic enzymes of genus Clostridium, *Acta Pathol. Microbiol. Scand.*, 71B, 387, 1967.
6. **Bleivis, A. S. and Zimmerman, L. N.,** Properties of proteinase from streptococcus faecalis var. lique faciens, *J. Bacteriol.*, 88, 653, 1964.
7. **Matsubara, H. and Feder, J.,** Other bacterial, mould and yeast proteases, in *The Enzymes*, Vol. 3, 3rd ed., Boyer, P. D., Ed., Academic Press, New York, 1971, chap. 20.
8. **Hausmann, E. and Kaufman, E. J.,** Collagenase activity in a particulate fraction from *Bacteroides melaninogenicus, Biochim. Biophys. Acta*, 194, 612, 1969.
9. **Deutch, C. E. and Soffer, R. L.,** Escherichia coli mutants defective in dipeptidyl carboxypeptidase, *Proc. Natl. Acad. Sci. U.S.A.*, 75, 5998, 1978.
10. **Payne, J. W.,** Transport and utilisation of peptides by bacteria, in *Microorganisms and Nitrogen Sources*, Part 2, Payne, J. W., Ed., John Wiley & Sons, New York, 1980, 21.
11. **Payne, J. W. and Bell, G.,** Direct determination of the properties of peptide transport systems in *Escherichia coli* using a fluorescent labelling procedure, *J. Bacteriol.*, 137, 447, 1979.
12. **Pittman, K. A. and Bryant, M. P.,** Peptides and other nitrogen sources for growth of *Bacteroides ruminicola, J. Bacteriol.*, 88, 401, 1964.
13. **Wahren, A. and Gibbons, R. J.,** Amino acid fermentation by *Bacteroides melaninogenicus, J. Microbiol.*, 36, 149, 1970.
14. **Wahren, A. and Holme, T.,** Amino acid and peptide requirement of *Fusiformis necrophorus, J. Bacteriol.*, 116, 279, 1973.
15. **Lev, M.,** Casamino acids enhance growth of *Bacteroides melaninogenicus, J. Bacteriol.*, 129, 562, 1977.
16. **Pittman, K. A., Lakshmanan, S., and Bryant, M. P.,** Oligopeptide uptake by *Bacteroides ruminicola, J. Bacteriol.*, 93, 1499, 1967.
17. **Stevenson, R. M. W.,** Amino acid uptake systems in *Bacteroides ruminicola, Can. J. Microbiol.*, 25, 1161, 1979.
18. **Anraku, Y.,** Transport and utilisation of amino acids by bacteria, in *Microorganisms and Nitrogen Sources*, Part 1, Payne, J. W., Ed., John Wiley & Sons, New York, 1980, 1.
19. **Wilson, G. S. and Miles, A.,** in *Topley and Wilson's Principles of Bacteriology, Virology and Immunity*, Arnold, London, 1975, chap. 3.
20. **Owens, C. W. I. and Padovan, W.,** Limitations of ultracentrifugation and in vivo dialysis as methods of stool analysis, *Gut*, 17, 68, 1976.
21. **Wilson, D. R., Ing, T. S., Metcalf-Gibson, A., and Wrong, O. M.,** The chemical composition of faeces in uraemia, as revealed by *in vivo* faecal dialysis, *Clin. Sci.*, 35, 197, 1968.
22. **Varel, V. H. and Bryant, M. P.,** Nutritional features of *Bacteroides fragilis* subsp. *fragilis, Appl. Microbiol.*, 18, 251, 1974.
23. **Hassinen, J. B., Durbin, G. T., Tomarelli, R. M., and Bernhart, F. W.,** The minimal nutritional requirements of *Lactobacillus bifidus, J. Bacteriol.*, 62, 771, 1951.
24. **Barker, H. A.,** Amino acid degradation by anaerobic bacteria, *Ann. Rev. Biochem.*, 50, 23, 1981.
25. **Rogosa, M.,** *Acidaminococcus* gen.n., *Acidaminococcus ferment*, sp.n., Anaerobic gram-negative diplococci using amino acids as sole energy source for growth, *J. Bacteriol.*, 98, 756, 1969.
26. **Allison, M. J.,** Biosynthesis of amino acids by ruminal microorganisms, *J. Anim. Sci.*, 29, 797, 1969.
27. **Allison, M. J. and Bryant, M. P.,** Biosynthesis of branched-chain amino acids from branched-chain fatty acids by rumen bacteria, *Arch. Biochem.*, 101, 269, 1963.
28. **Dawes, E. A. and Large, P. J.,** Class II reactions: synthesis of small molecules, in *Biochemistry of Bacterial Growth*, Mandelstam, J. and McQuillen, K., Eds., Blackwell Scientific, Oxford, 1968, chap. 4.
29. **Wiegel, J.,** α-Isopropylmalate synthase as a marker for the leucine biosynthetic pathway in several clostridia and in *Bacteroides fragilis, Arch. Microbiol.*, 130, 385, 1981.
30. **Umbarger, H. E.,** Amino acid biosynthesis and its regulation, *Ann. Rev. Biochem.*, 47, 533, 1978.
31. **Drasar, B. S. and Hill, M. J.,** *Human Intestinal Flora*, Academic Press, New York, 1974, chaps. 6 and 16.
32. **Sabbaj, J., Sutter, V. L., and Finegold, S. M.,** Urease and deaminase activities of fecal bacteria in hepatic coma, *Antimicrobiol. Agents Chemother.*, 181, 1970.
33. **Mead, G. C.,** The amino acid-fermenting clostridia, *J. Gen. Microbiol.*, 67, 47, 1971.
34. **Whiteley, H. R.,** Fermentation of amino acids by *Micrococcus aerogenes, J. Bacteriol.*, 74, 324, 1957.
35. **Nagase, M. and Matsuo, T.,** Interactions between amino-acid-degrading bacteria and methanogenic bacteria in anaerobic digestion, *Biotechnol. Bioeng.*, 24, 2227, 1982.
36. **Elsden, S. R. and Hilton, M. G.,** Amino acid utilisation patterns in clostridial taxonomy, *Arch. Microbiol.*, 123, 137, 1979.
37. **Elsden, S. R., Hilton, M. G., and Waller, J. M.,** The end products of the metabolism of aromatic amino acids by clostridia, *Arch. Microbiol.*, 107, 283, 1976.
38. **Wong, J. C., Dyer, J. K., and Tribble, J. L.,** Fermentation of L-aspartate by a saccharolytic strain of *Bacteroides melaninogenicus, Appl. Microbiol.*, 33, 69, 1977.

39. **Barker, H. A.,** Fermentation of nitrogenous organic compounds, in *The Bacteria,* Vol. 2, Gunsalus, I. C. and Stanier, R. Y., Eds., Academic Press, New York, 1961, 151.
40. **Prins, R. A.,** Biochemical activities of gut micro-organisms, in *Microbial Ecology of the Gut,* Clarke, R. T. J. and Bauchop, T., Eds., Academic Press, New York, 1977, chap. 3.
41. **Stadman, T. C.,** Lysine metabolism by clostridia, *Adv. Enzymol.,* 38, 413, 1973.
42. **Buckel, W.,** Analysis of the fermentation pathways of clostridia using double labelled glutamate, *Arch. Microbiol.,* 127, 167, 1980.
43. **Gale, E. F.,** The bacterial amino acid decarboxylases, *Adv. Enzymol.,* 6, 1, 1946.
44. **Haughton, B. G. and King, H. K.,** Induced formation of Leucine decarboxylase in *Proteus vulgaris, Biochem. J.,* 80, 268, 1961.
45. **Hayes, M. L. and Hyatt, A. T.,** The decarboxylation of amino acids by bacteria derived from human dental plaque, *Arch. Oral Biol.,* 19, 361, 1974.
46. **Sanford, J.,** Formation of histamine in ruminal fluid, *Nature (London),* 199, 829, 1963.
47. **Dunlop, R. H. and Hammond, P. B.,** D-lactic acidosis of ruminants, *Ann. N.Y. Acad. Sci.,* 119, 1109, 1965.
48. **Morris, D. R. and Fillingame, R. H.,** Regulation of amino acid decarboxylation, *Ann. Rev. Biochem.,* 43, 303, 1974.
49. **Michel, M. C.,** Metabolism de la flore intestinale du porc degradation des formes L et D des acides amines, *Ann. Biol. Anim.,* 6, 33, 1966.
50. **Wicki, A. and Schatzmann, H. J.,** Histamine formation by ruminal fluid from cattle in vitro, *Experientia,* 33, 422, 1977.
51. **Johnson, K. A.,** The production of secondary amines by the human gut bacteria and its possible relevance to carcinogenesis, *Med. Lab. Sci.,* 34, 131, 1977.
52. **Asatoor, A. M., Chamberlain, M. J., Emmerson, B. T., Johnson, J. R., Levi, A. J., and Milne, M. D.,** Metabolic effects of oral neomycin, *Clin. Sci.,* 33, 111, 1967.
53. **Asatoor, A. M.,** The origin of urinary tyramine. Formation in tissues and by intestinal microorganisms., *Clin. Chim. Acta,* 22, 223, 1968.
54. **Perry, T. L., Hestrin, M., MacDougall, L., and Hanson, S.,** Urinary amines of intestinal bacterial origin, *Clin. Chim. Acta,* 14, 116, 1966.
55. **Kleiner, D.,** The transport of $NH_3$ and $NH_4^+$ across biological membranes, *Biochim. Biophys. Acta,* 639, 41, 1981.
56. **Yoch, D. C. and Pengra, R. M.,** Effect of amino acids on the nitrogenase system of *Klebsiella pneumoniae, J. Bacteriol.,* 92, 618, 1966.
57. **Bryant, M. P.,** Nutritional features and ecology of predominant anaerobic bacteria of the intestinal tract, *Am. J. Clin. Nutr.,* 27, 1313, 1974.
58. **Takahashi, M., Benno, Y., and Mitsuoka, T.,** Utilisation of ammonia nitrogen by intestinal bacteria isolated from pigs, *Appl. Environ. Microbiol.,* 39, 30, 1980.
59. **Griffith, C. J. and Carlsson, J.,** Mechanism of ammonia assimilation in streptococci, *J. Gen. Microbiol.,* 82, 253, 1974.
60. **Smith, C. J. and Bryant, M. P.,** Introduction to metabolic activities of intestinal bacteria, *Am. J. Clin. Nutr.,* 32, 149, 1979.
61. **Meers, J. L., Tempest, D. W., and Brown, C. M.,** Glutamine (amide): 2-oxo, glutarate amino transferase oxidoreductase (NADP), an enzyme involved in the synthesis of glutamate by some bacteria, *J. Gen. Microbiol.,* 64, 187, 1970.
62. **Ginsburg, A. and Stadtman, E. R.,** Regulation of glutamine synthetase in *Escherichia coli,* in *The Enzymes of Glutamine Metabolism,* Prusiner, S. and Stadtman, E. R., Eds., Academic Press, New York, 1973, 9.
63. **Friedrich, B. and Magasanik, B.,** Urease of *Klebsiella aerogenes:* control of its synthesis by glutamine synthetase, *J. Bacteriol.,* 131, 446, 1977.
64. **Vince, A., Dawson, A. M., Park, N., and O'Grady, F. W.,** Ammonia production by intestinal bacteria, *Gut,* 14, 171, 1973.
65. **Phear, E. A. and Ruebner, B.,** The in vitro production of ammonium and amines by intestinal bacteria in relation to nitrogen toxicity as a factor in hepatic coma. *Br. J. Exp. Pathol.,* 37, 253, 1956.
66. **O'Grady, F. W.,** Differences in ammonia production by faecal bacteria of patients with hepatic encephalopathy, *Proc. R. Soc. Med.,* 59, 1246, 1966.
67. **Vince, A. J. and Burridge, S. M.,** Ammonia production by intestinal bacteria, the effects of lactose, lactulose and glucose, *J. Med. Microbiol.,* 13, 177, 1980.
68. **Vince, A., Killingley, M., and Wrong, O. M.,** Effect of lactulose on ammonia production in a fecal incubation system, *Gastroenterology,* 74, 544, 1978.
69. **Bown, R. L., Gibson, J. A., Sladen, G. E., Hicks, B., and Dawson, A. M.,** Effects of lactulose and other laxatives on ileal and colonic pH as measured by a radiotelemetry device, *Gut,* 15, 999, 1974.

70. **Vince, A., Down, P. F., Murison, J., Twigg, F. J., and Wrong, O.**, Generation of ammonia from non-urea sources in a faecal incubation system, *Clin. Sci. Mol. Med.*, 51, 313, 1976.
71. **Vince, A.**, Unpublished data, 1975.
72. **Paigen, K. and Williams, B.**, Catabolite repression and other control mechanisms in carbohydrate utilization, *Adv. Microb. Physiol.*, 4, 251, 1970.
73. **Kendall, A. I. and Farmer, C. J.**, Studies in bacterial metabolism. VII, *J. Biol. Chem.*, 13, 63, 1912.
74. **Stephenson, M. and Gale, E. F.**, Factors influencing bacterial deamination. I. The deamination of glycine, dl-alanine and l-glutamic acid by *Bacterium coli*, *Biochem. J.*, 31, 1316, 1937.
75. **Epps, H. M. R. and Gale, E. F.**, The influence of the presence of glucose during growth on the enzymic activities of *Escherichia coli:* comparison of the effect with that produced by fermentation acids, *Biochem. J.*, 36, 619, 1942.
76. **Magasanik, B.**, Catabolite repression, *Cold Spring Harbor Symp. Quant. Biol.*, 26, 249, 1961.
77. **Bircher, J., Muller, J., Guggenheim, P., and Hammerli, U. P.**, Treatment of chronic portal-systemic encephalopathy with lactulose, *Lancet*, i, 890, 1966.
78. **Sahota, S. S., Bramley, P. M., and Menzies, I. S.**, The fermentation of lactulose by colonic bacteria. *J. Gen. Microbiol.*, 128, 319, 1982.
79. **Mason, V. C.**, An explanation of the effect of lactulose in the treatment of hepatic encephalopathy, *Gastroenterology*, 66, 1271, 1974.
80. **Lewis, H., Rojas, S., Uribe, M., Galvan, E., and Gil, S.**, Similar effect of lactose and lactulose on in vitro fecal ammonia generation and bacterial flora, *Gastroenterology*, 78, 1206, 1980.
81. **Vince, A., Wrong, O. M., and Wager, J.**, in press, 1986.
82. **Cummings, J. H., Southgate, D. A. T., Branch, W. J., Wiggins, H. S., Houston, H., Jenkins, D. J. A., Jivraj, T., and Hill, M. J.**, The digestion of pectin in the human gut and its effect on calcium absorption and large bowel function, *Br. J. Nutr.*, 41, 477, 1979.
83. **Hungate, R. E.**, *The Rumen and it Microbes*, Academic Press, New York, 1966, chap. 7.
84. **Cummings, J. H.**, Dietary fiber, *Br. Med. Bull.*, 37, 65, 1981.
85. **Anderson, I. H., Levine, A. S., and Levitt, M. D.**, Incomplete absorption of the carbohydrate in all-purpose wheat flour, *New Eng. J. Med.*, 304, 891, 1981.
86. **Cummings, J. H.**, Fermentation in the human large intestine: evidence and implications for health, *Lancet*, i, 1206, 1983.
87. **Wolpert, E., Phillips, S. F., and Summerskill, W. H. J.**, Transport of urea and ammonia production in the human colon, *Lancet*, 2, 1387, 1971.
88. **Wrong, O. M., Vince, A. J., and Waterlow, J. C.**, The origins and bacterial metabolism of faecal ammonia, in *Colon and Nutrition*, Kasper, H. and Goebell, H., Eds., MTP Press, London, 1982, chap. 4.
89. **Weber, F. L. and Veach, G. H.**, The importance of the small intestine in gut ammonium production in the fasting dog, *Gastroenterology*, 77, 235, 1979.
90. **Nance, F. C. and Kline, D. G.**, Eck's fistula encephalopathy in germfree dogs, *Ann. Surg.*, 174, 856, 1971.
91. **Down, P. F., Agostini, L., Murisen, J., and Wrong, O. M.**, The interrelations of faecal ammonia, pH and bicarbonate: evidence of colonic absorption of ammonia by non-ionic diffusion, *Clin. Sci.*, 43, 101, 1972.
92. **Castel, D. O. and Moore, E. W.**, Ammonia absorption from the human colon, *Gastroenterology*, 60, 33, 1971.
93. **Richards, P., Metcalfe-Gibson, A., Ward, E. E., Wrong, O. M., and Houghton, B. J.**, Utilisation of ammonia nitrogen for protein synthesis in man, and the effect of protein restriction and uraemia, *Lancet*, 2, 845, 1967.
94. **Richards, P.**, The metabolism and clinical relevance of the keto acid analogues of essential amino acids, *Clin. Sci. Mol. Med.*, 54, 589, 1978.
95. **Walser, M., Coulter, A. W., Dighe, S., and Crantz, F. R.**, The effect of keto-analogues of essential amino acids in severe chronic uremia, *J. Clin. Invest.*, 52, 678, 1973.
96. **Varcoe, R., Halliday, D., Carson, E. R., Richards, P., and Tavill, A. S.**, Efficiency of utilisation of urea nitrogen for albumin synthesis by chronically uraemic and normal man, *Clin. Sci. Mol. Med.*, 48, 379, 1975.
97. **Suzuki, K., Benno, Y., Mitsuoka, T., Takebe, S., Kobashi, K., and Hase, J.**, Urease-producing species of intestinal anaerobes and their activities, *Appl. Microbiol.*, 37, 379, 1979.
98. **Brown, C. L., Hill, M. J., and Richards, P.**, Bacterial ureases in uraemic men, *Lancet*, 2, 406, 1971.
99. **Wozny, M. A., Bryant, M. P., Holdeman, L. V., and Moore, W. E. C.**, Urease assay and urease producing species of anaerobes in the bovine rumen and human feces, *Appl. Microbiol.*, 33, 1097, 1977.
100. **Crociani, F. and Matteuzzi, D.**, Urease activity in the Genus Bifidobacterium, *Ann. Microbiol.*, 133A, 417, 1982.

101. **Senior, B. W., Bradford, N. C., and Simpson, D. S.,** Ureases of proteus strains in relation to virulence for the urinary tract, *J. Med. Microbiol.*, 13, 507, 1980.
102. **Cook, A. R.,** The elimination of urease activity in *Streptococcus faecium* as evidence for plasmid-coded urease. *J. Gen. Microbiol.*, 92, 49, 1976.
103. **Kaltwasser, H., Kramer, J., and Conger, W. R.,** Control of urease formation in certain anaerobic bacteria, *Arch. Mikrobiol.*, 81, 178, 1972.
104. **Kopstein, J. and Wrong, O. M.,** The origin and fate of salivary urea and ammonia in man, *Clin. Sci. Mol. Med.*, 52, 9, 1977.
105. **Owens, C. W. I. and Padovan, W.,** Quantitative method for estimating fecal amino acids, *Clin. Chem.*, 21, 1437, 1975.
106. **Jones, E. A., Smallwood, R. A., Craigie, A., and Rosenoer, V. M.,** The enterohepatic circulation of urea nitrogen, *Clin. Sci.*, 37, 825, 1969.
107. **Wilson, D. R., Ing, T. S., Metcalf-Gibson, A., and Wrong, O. M.,** In vivo dialysis of faeces as a method of stool analysis. III. The effect of intestinal antibiotics, *Clin. Sci.*, 34, 211, 1968.
108. **Aoyagi, T. and Summerskill, W. H. J.,** Inhibition by acetohydroxamic acid of human mucosal and fecal urease specific activity, *Lancet*, 1, 296, 1966.
109. **Vince, A. J. and Chadwick, V. S.,** The effect of antibiotics and acetohydroxamic acid on bacterial hydrolysis of urea, *J. Antimicrobiol. Chemother.*, 4, 189, 1978.
110. **Suh, B., Strockbine, N. A., Chang, P. S., and Kunin, C. M.,** Inhibition of bacterial urease by substrate analogues, *Clin. Res.*, 29, 730A, 1981.
111. **Wolpert, E., Phillips, S. F., and Summerskill, W. H. J.,** Ammonia production in the human colon. Effects of cleansing, neomycin and acetohydroxamic acid, *New Engl. J. Med.*, 283, 159, 1970.
112. **Summerskill, W. H. J., Thorsell, F., Feinberg, J. H., and Aldrete, J. S.,** Effects of urease inhibition in hyperammonia: clinical and experimental studies with acetohydroxamic acid, *Gastroenterology*, 54, 20, 1968.
113. **Fishbein, W. N. and Daly, J. E.,** Urease inhibitors for hepatic coma. II. Comparative efficiency of four lower hydroxamate homologs *in vitro* and *in vivo*, *Proc. Soc. Exp. Biol. Med.*, 134, 1083, 1970.
114. **Visek, W. J.,** Urease immunity in liver disease. A new approach, *Gastroenterology*, 46, 329, 1969.
115. **Reithal, F. J.,** Ureases, in *The Enzymes*, Vol. 4, 3rd ed., Boyer, P. D., Ed., Academic Press, New York, 1971, chap. 1.
116. **Hase, J. and Kobashi, K.,** Inhibition of *Proteus vulgaris* urease by hydroxamic acids, *J. Biochem.*, 62, 293, 1967.
117. **Rampton, D. S., Cohen, S. L., Crammond, V., Gibbons, J., Lilburn, M., Rabet, J. Y., Vince, A. J., Wager, J. D., and Wrong, O. M.,** Treatment of chronic renal failure with dietary fibre, *Clin. Nephrol.*, 21, 159, 1984.
118. **Sherlock, S., Summerskill, W. H. J., and Dawson, A. M.,** The treatment and prognosis of hepatic coma, *Lancet*, ii, 689, 1956.
119. **Sherlock, S.,** Hepatic enceohalopathy, in *Diseases of the Liver and Biliary System*, 6th ed., Blackwell Scientific, Oxford, 1981, chap 8.
120. **Weber, F. L.,** The use of lactulose in the treatment of portal-systemic encephalopathy, in *Nutritional Pharmacology*, Spiller, G. A., Ed., Alan R. Liss, New York, 1981, chap 8.
121. **Fischer, J. E.,** The etiology of hepatic encephalopathy — nutritional implications, *Acta Chir. Scand.*, S507, 50, 1981.
122. **Schenker, S., Breen, K. J., and Hoyumpa, A. M.,** Hepatic encephalopathy: current status, *Gastroenterology*, 66, 121, 1974.
123. **Hoyumpa, A. M., Desmond, P. V., Avant, G. R., Roberts, R. K., and Schenker, S.,** Hepatic encephalopathy, *Gastroenterology*, 76, 184, 1979.
124. **Dawson, A. M., De Groote, J., Rosenthal, W. A., and Sherlock, S.,** Blood pyruvic-acid and alpha-keto glutaric-acid levels in liver disease and hepatic coma, *Lancet*, 1, 392, 1957.
125. **Hourani, B. T., Hamlin, E. M., and Reynolds, T. B.,** Cerebrospinal fluid glutamine as a measure of hepatic encephalopathy, *Arch. Int. Med.*, 127, 1033, 1971.
126. **Zieve, L., Doizaki, W. M., and Zieve, F. J.,** Synergism between mercaptans and ammonia or fatty acids in the production of coma: a possible role for mercaptans in the pathogenesis of hepatic coma, *J. Lab. Clin. Med.*, 83, 16, 1974.
127. **Derr, R. F. and Zieve, L.,** Effect of fatty acids on the disposition of ammonia, *J. Pharm. Exp. Therap.*, 197, 675, 1976.
128. **Weber, F. L., Maddrey, W. C., and Walser, M.,** Amino acid metabolism of dog jejunum before and during absorption of keto analogues, *Am. J. Physiol.*, 232, 263, 1977.
129. **Phear, E. A., Sherlock, S., and Summerskill, W. H. J.,** Blood ammonium levels in liver disease and hepatic coma, *Lancet*, 1, 836, 1955.
130. **Phear, E. A., Ruebner, B., Sherlock, S., and Summerskill, W. H. J.,** Methionine toxicity in liver disease and its prevention by chlortetracycline, *Clin. Sci.*, 15, 93, 1956.

131. **McLain, C. J., Zieve, L., Doizaki, W. M., Gilberstadt, S., and Anstad, G. R.,** Blood methanethiol in alcoholic liver disease with and without hepatic encephalopathy, *Gut*, 21, 318, 1980.
132. **Tanaka, K., Budd, M. A., Efan, M. L., and Isselbacher, K. J.,** Isovaleric acidemia: a new genetic defect of leucine metabolism, *Proc. Natl. Acad. Sci.*, 56, 236, 1976.
133. **Fischer, J. E. and Baldessarini, R. J.,** False neurotransmitters and hepatic failure, *Lancet*, 2, 75, 1971.
134. **Livingstone, A. S., Potvin, M., Goresky, C. A., Finlayson, M. H., and Hinchey, E. J.,** Changes in the blood brain barrier in the hepatic coma following hepatectomy, *Gastroenterology*, 93, 697, 1977.
135. **Michel, H., Solere, M., Grainer, P., Cauvet, G., Bali, J. P., Pons, F., and Bellet-Hermann, H.,** Treatment of cirrhotic hepatic encephalopathy with L-dopa. A controlled trial, *Gastroenterology*, 79, 207, 1980.
136. **Morgan, M. Y., Milsom, J. P., and Sherlock, S.,** Plasma ratio of valine, leucine and isoleucine to phenylalanine and tyrosine in liver disease, *Gut*, 19, 1068, 1978.
137. **Eriksson, L. S., Persson, A., and Wahren, J.,** Branched chain amino acids in treatment of chronic hepatic encephalopathy, *Gut*, 23, 801, 1982.
138. **Greenberger, N. J., Carley, J., Schenker, S., Bettinger, I., Stammes, C., and Beyer, P.,** Effect of vegetable and animal protein diets in chronic hepatic encephalopathy, *Am. J. Dig. Dis.*, 22, 845, 1977.
139. **De Bruijn, K. M., Blendis, L. M., Zilm, D. H., Carlen, P. L., and Anderson, G. H.,** Effect of dietary protein manipulations in subclinical portalsystemic encephalopathy, *Gut*, 24, 53, 1983.
140. **Dawson, A. M., McLaren, J., and Sherlock, S.,** Neomycin in the treatment of hepatic coma, *Lancet*, 2, 1263, 1957.
141. **Pirotte, J., Guffens, J. M., and Devos, J.,** Comparative study of basal arterial ammonemia and of orally-induced hyperammonemia in chronic portal systemic encephalopathy treated with neomycin, lactulose and an association of neomycin and lactulose, *Digestion*, 10, 435, 1974.
142. **Weber, F. L., Fresard, K. M., and Lally, B. R.,** Effects of lactulose and neomycin on urea metabolism in cirrhotic subjects, *Gastroenterology*, 82, 213, 1982.
143. **Conn, H. O. and Lieberthall, M. M.,** *The Hepatic Coma Syndromes and Lactulose*, Williams & Wilkins, Baltimore, 1979, chap 15.
144. **Morgan, M. H., Read, A. E., and Speller, D. C. E.,** Treatment of hepatic encephalopathy with metronidazole, *Gut*, 23, 1, 1982.
145. **Vince, A. J.,** Unpublished data, 1976.
146. **Bircher, J., Muller, J., Guggenheim, P., and Hammerli, U. P.,** Treatment of chronic portal-systemic encephalopathy with lactulose, *Lancet*, 1, 890, 1966.
147. **Conn, H. O.,** Lactulose: a drug in search of a modus operandi, *Gastroenterology*, 74, 624, 1978.
148. **Bircher, J., Hammerli, U. P., Scollo-Lavizzari, G., and Hoffmann, K.,** Treatment of chronic portal-systemic encephalopathy with lactulose, *Am. J. Med.*, 51, 148, 1971.
149. **Bircher, J.,** The many effects of lactulose. A rational approach to its therapeutic use, *Drugs*, 4, 1, 1972.
150. **Avery, F. S., Davies, E. F., and Brogden, R. M.,** Lactulose: a review of its therapeutic and pharmacological properties with particular reference to ammonia metabolism and its mode of action in portal systemic encephalopathy, *Drugs*, 4, 7, 1972.
151. **Vince, A., Zeegen, R., Drinkwater, J. E., O'Grady, F. W., and Dawson, A. M.,** The effect of lactulose on the faecal flora of patients with hepatic encephalopathy, *J. Med. Microbiol.*, 7, 163, 1974.
152. **Zeegen, R., Drinkwater, J. E., Fenton, J. C. B., Vince, A., and Dawson, A. M.,** Some observations on the effects of treatment with lactulose on patients with chronic hepatic encephalopathy, *Q. J. Med.*, 39, 245, 1970.
153. **Agostini, L., Down, P. F., Murison, J., and Wrong, O. M.,** Faecal ammonia and pH during lactulose administration in man: comparison with other cathartics, *Gut*, 13, 859, 1972.
154. **Weber, F. L.,** The effect of lactulose on urea metabolism and nitrogen excretion in cirrhotic patients, *Gastroenterology*, 77, 518, 1979.
155. **Woods, D. P.,** Indole information by *Bacterium coli*. II. The action of washed suspensions of *Bacterium coli* on indol derivations, *Biochem. J.*, 29, 649, 1935.

Chapter 8

# GUT BACTERIA AND THE METABOLISM OF AROMATIC AMINO ACIDS

**A. G. Renwick**

## TABLE OF CONTENTS

## I. INTRODUCTION

The metabolism of aromatic amino acids by the gut microflora is complex and can involve both the synthesis and degradation of these essential compounds. In common laboratory animals and man, catabolism and the formation of toxic metabolites is of greater importance than the potential for synthesis. In ruminants the microflora of the rumen possess a greater capacity for synthetic reactions and the introduction of radio-labeled or $^{15}N$ precursors can be shown to be incorporated into various amino acids, including those which are the subject of this chapter. However, it appears that the contribution this may make to the host animal is minimal and thus this chapter will concentrate on degradative reactions and the evidence for their toxicological significance in vivo.

The pathways of metabolism of aromatic amino acids are given in Figure 1. Many of the products are weak organic acids and after absorption from the gut may be subjected to further metabolism by the animal tissues. Thus, the products finally eliminated by the host may have undergone further oxidation of the side-chain or conjugation with glucuronic acid, sulfate, or amino acids. In the following discussion information on metabolism by pure strains of bacteria and mixed cultures of intestinal bacteria is presented and its relationship to the pattern of urinary metabolites considered.

## II. THE METABOLISM OF TRYPTOPHAN

### A. Studies in Vitro

Incubation of tryptophan with cultures of ruminal microorganisms has demonstrated the formation of indole, indoleacetic acid,[1] skatole,[2] and tryptamine.[3] Studies using L-[U-benzene ring-$^{14}C$] tryptophan[4] showed that the major metabolites were skatole (39% of the added $^{14}C$), indole (7%), and indoleacetic acid (4%). However, increasing the glucose concentration in the medium dramatically reduced the amounts of skatole and indole formed such that at 110 m*M* glucose only 1.5% of the $^{14}C$ was present as skatole, 0.1% as indole, 6.5% as indoleacetic acid, and the remainder as the unchanged amino acid or polar unextracted metabolites. Although trace amounts of tryptamine were detected in the medium, there was negligible $^{14}C$ activity associated with this material. In the same series of experiments [$^{14}C$] skatole was the only metabolite detected on incubation of [2-$^{14}C$] indoleacetic acid, with ruminal organisms, while both skatole and indole were formed from indolepyruvate, indolelactate, and indolealdehyde. Indoleacetic acid and tryptophol were formed from indoleacetaldehyde (see Figure 2). Prolonged incubation times were necessary to produce a significant amount of metabolism with any of the substrates. The addition of various antibiotics to the incubations suppressed skatole formation and usually increased the amount of indoleacetic acid present. L-5-Hydroxytryptophan was degraded to 5-hydroxyindole and 5-hydroxyskatole, but other analogues including D-tryptophan, indolepropionate, indolebutyrate, indoleacrylate, tryptamine, 5-hydroxytryptamine, and tryptophol were not degraded.[4] The major difference between the scheme for tryptophan metabolism by ruminal microorganisms (Figure 2) and that proposed generally for aromatic amino acids (Figure 1) is the absence of indolepropionic acid, possibly due to the lack of formation of indoleacrylic acid. Indolepropionic acid has been detected in rat cecal contents and feces (see later).

Tryptophan transaminase which catalyzes the formation of indolepyruvate and glutamate from tryptophan and α-ketoglutarate has been detected in various organisms including *E. coli*[5] and *Clostridium sporogenes*.[6] *Escherichia coli* cell extracts contained 2 enzymes; transaminase A which utilized phenylalanine, tyrosine, tryptophan, and aspartate with glutamate as substrates and transaminase B which acted on isoleucine, valine, leucine, norleucine, and norvaline with glutamate. Transamination could occur also between the members of each group, but there was little or no transamination with those of the other group.

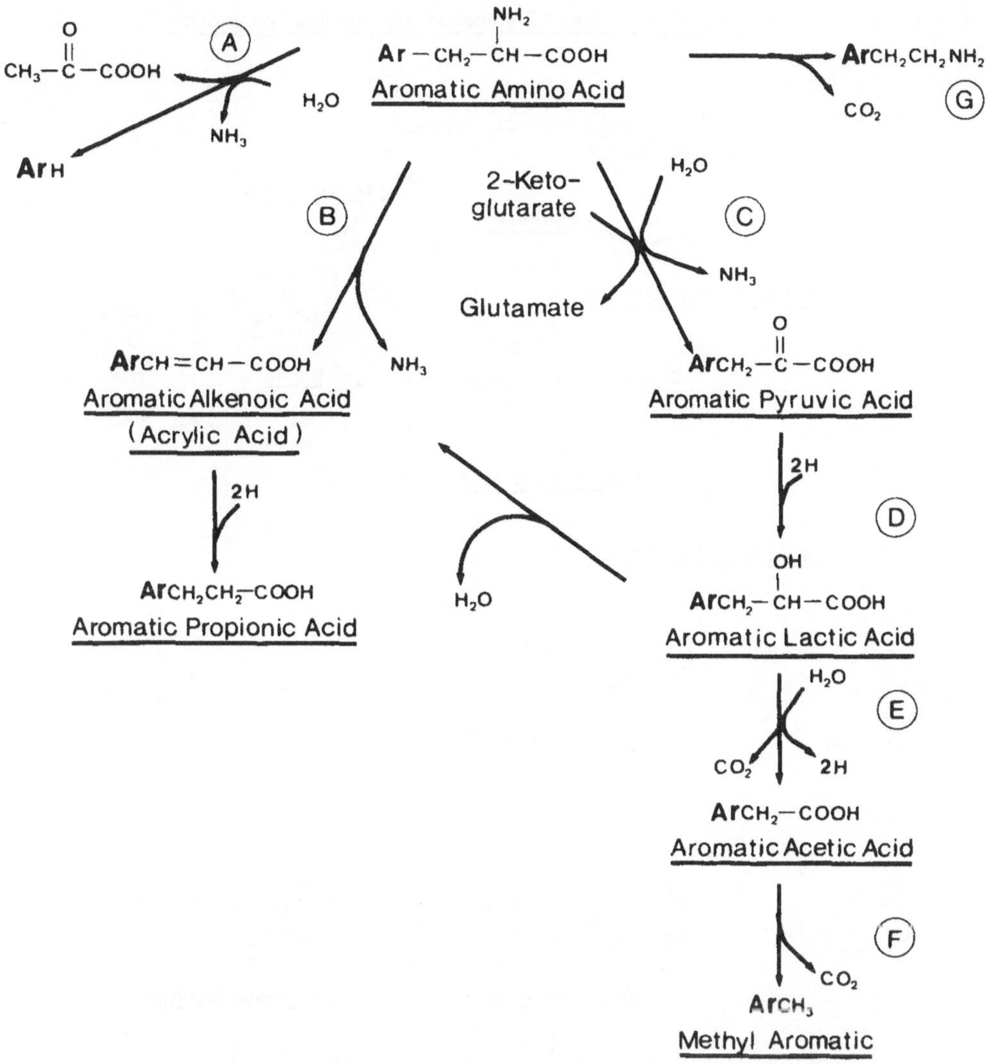


FIGURE 1. General scheme for the microbial metabolism of aromatic amino acids. (A) Formation of the unsubstituted atomatic nucleus, e.g., tryptophan is metabolized to indole, tyrosine to phenol but phenylalanine is not converted to benzene. (B) This reaction has not been demonstrated in bacteria but can be performed in plants via the enzyme amino acid ammonia lyase. Thus, in bacteria the unsaturated acid probably arises from the corresponding lactic acid. (C) Transamination to the arylpyruvic acid. (D) Reduction to the aryllactic acid. (E) Decarboxylation to the arylacetic acid. (F) Decarboxylation to the methyl aromatic compound, e.g., tryptophan to skatole and tyrosine to cresol, but phenylalanine does not give rise to toluene. (G) Decarboxylation of the amino acid to an amine.

The enzyme in the *Clostridium* species was purified 200-fold and shown to require pyridoxal phosphate for maximal activity. During purification the tryptophan transaminase was separated from a phenylalanine transaminase, but the final product could still perform transamination reactions with both phenylalanine and tyrosine as substrates, suggesting a low substrate specificity.

The reduction of indolepyruvic acid is performed by indolelactate dehydrogenase and is a reversible reaction. The enzyme has been detected in cell-free extracts of *Clostridium sporogenes*[7] and has a low specificity since it also possesses a high activity with phenyl lactate and *p*-hydroxyphenyllactate. Indolelactate has been detected as an important metab-

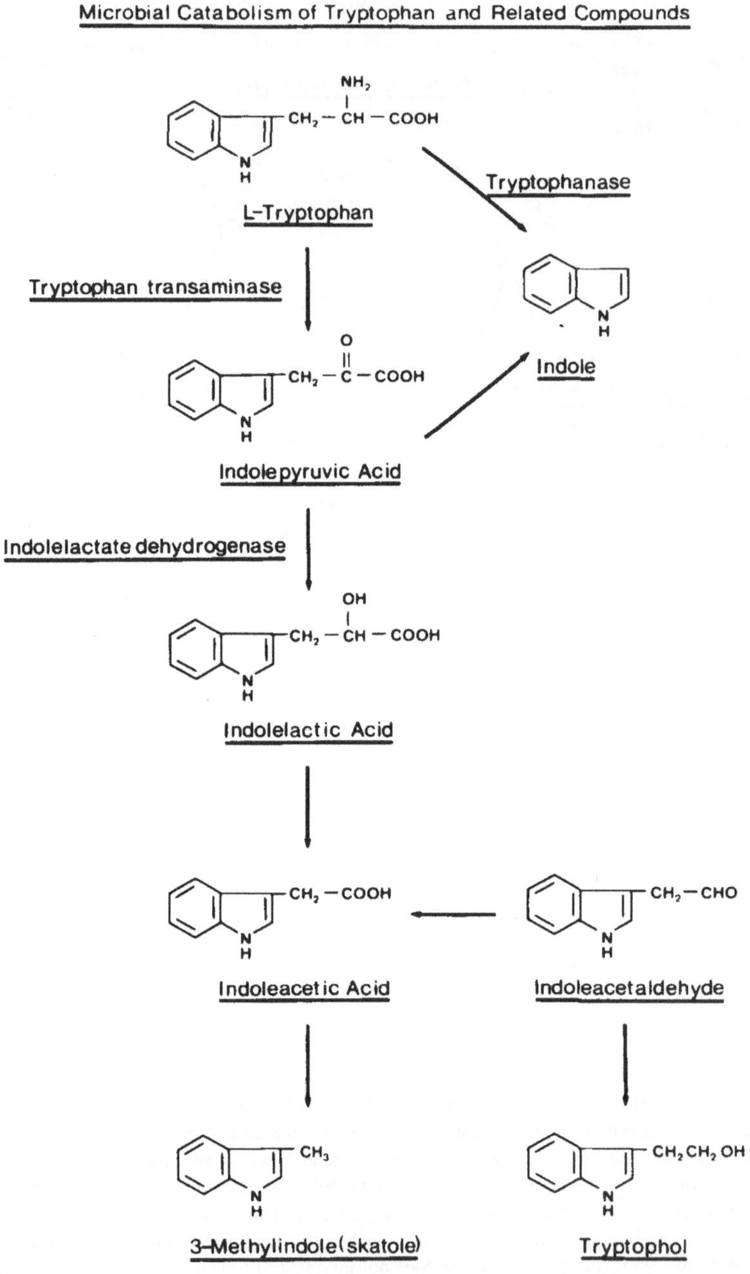


FIGURE 2. The metabolism of tryptophan and related compounds by ruminal microorganisms. The scheme is derived from incubation studies with [$^{14}$C] tryptophan and rumen microorganisms. Other metabolites (tryptamine and indolepropionic acid) are detected in rat cecal contents and feces (Table 1).

olite of tryptophan in strains of *Bifidobacterium*.[8] All 51 strains investigated were found to form indolelactic acid from tryptophan in vitro, although at different rates.

Indoleacetic acid is formed from tryptophan by human fecal bacteria,[9] and 3 strains of intestinal anaerobes out of 23 tested showed this ability in vitro.[10] However, tryptamine was not converted to indoleacetic acid by pure or mixed fecal cultures[9,10] which is consistent with the route described above.

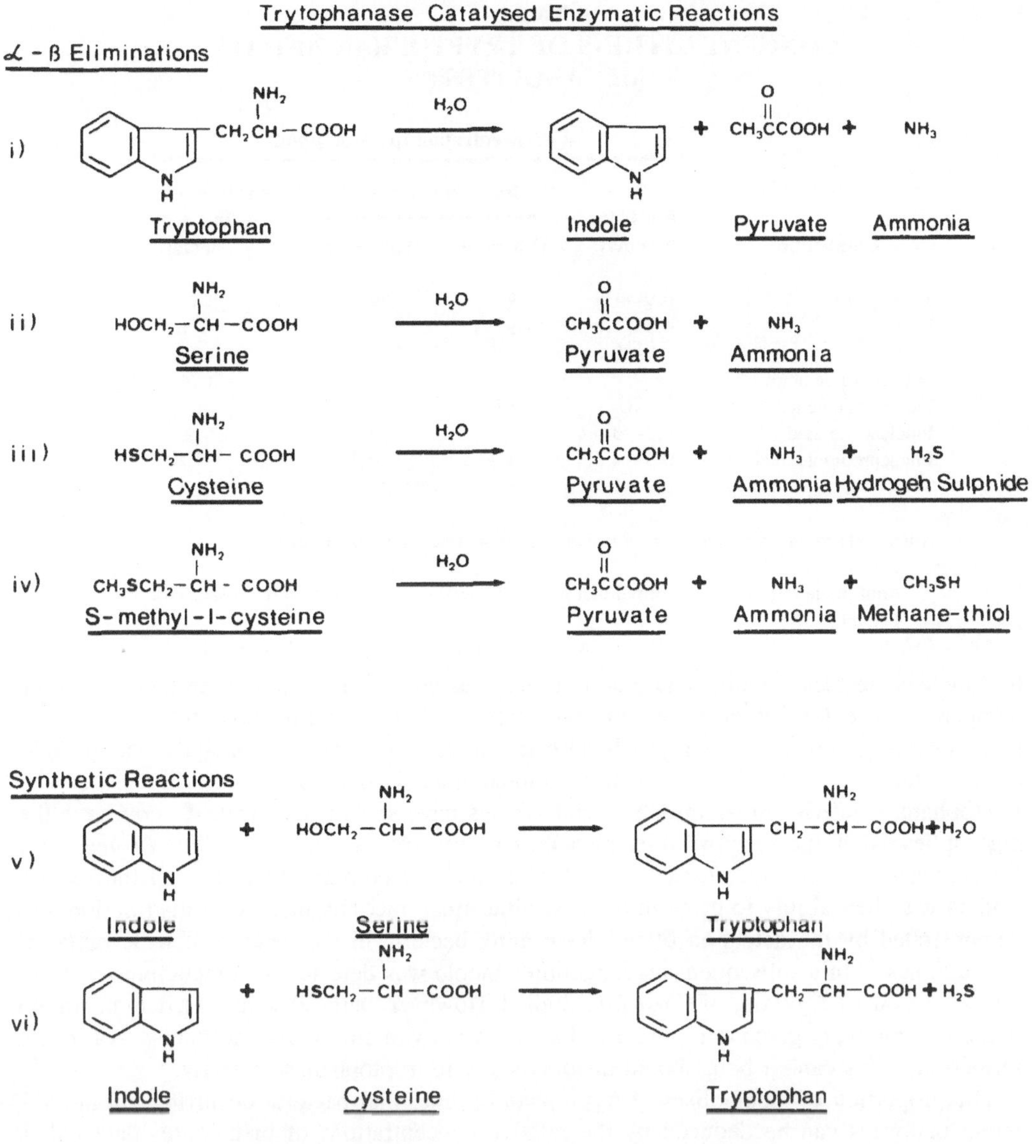


FIGURE 3. Tryptophanase catalyzed enzymatic reactions.

The formation of indole from tryptophan is catalyzed by the enzyme tryptophanase with liberation of ammonia and pyruvate. Initial studies[11] utilized *E. coli* as a source of the enzyme and demonstrated that tryptophanase would perform a range of enzymatic reactions which were either α, β-eliminations, or in the presence of indole, were synthetic in nature (Figure 3). The presence of excess indole inhibits indirectly those reactions leading to pyruvate formation, by increasing the affinity of the enzyme for inhibitory amino acids such as alanine and glycine.[12] However, indole is highly lipid soluble and absorbed rapidly from the gastrointestinal tract,[13] and thus accumulation of indole and end product inhibition appear unlikely in vivo. The presence of excess tryptophan results in induction of the enzyme, which can represent up to 7.5% of the cellular protein.[14] The importance of the *E. coli* enzyme in the gastrointestinal tract was investigated using germfree, monocontaminated, and conventional mice.[15] The feces and cecal contents were analyzed for tryptophan, tryptophanase, and indole in these animals and in mice fed normal diet or a diet supplemented with 0.025 g L-tryptophan/100 g. The tryptophan supplement was associated with a two-

**Table 1**
**CONCENTRATIONS OF TRYPTOPHAN AND ITS METABOLITES[20-22]**

| Compound | Concentration (μg/g or μg/mℓ) | | | |
|---|---|---|---|---|
| | Rat feces (normal) | Rat cecal contents[a] (Normal) | Rat cecal contents[a] (Saccharin) | Rat fecal culture (in vitro) (normal) |
| Tryptophan | 0.9—1.8 | 9 | 59 | ? |
| Indole | 4.5—7.8 | 9 | 8 | 131 |
| Skatole | 0—0.8 | ? | ? | 9 |
| Indolepyruvic acid | ND | ? | ? | 74 |
| Indolelactic acid | ND | ND | 3.0 | tr |
| Indoleacetic acid | 0.5—3.5 | 4.0 | 0.3 | 5 |
| Indolepropionic acid | 0.3—4.5 | 3.3 | 0.5 | ND |
| Tryptamine | 0.2—1.7 | 8 | 2 | 3 |

*Note:* ND = not detected, ? = not reported, tr = trace; not quantitated.

[a] Adult male rats were fed normal diet or a diet containing 7.5% w/w sodium saccharin for 1 month.

fold increase in fecal tryptophanase activity but was not accompanied by an increase in the numbers of *E. coli* or other indole-positive aerobes. Comparison of the numbers of *E. coli* in the various groups with the tryptophanase activity led the authors to conclude that <0.02% of the total activity in the gut of conventional mice is due to *E. coli*. Studies on the tryptophanase activity of various bacterial species have shown that while *E. coli* have the highest levels of the enzyme it is also present in other species including *Proteus* sp., *Corynebacterium* sp., and *Bacteroides*.[16,17] An attribute common to nearly all the species studied was their ability to grow in the gastrointestinal tract and their wide distribution was demonstrated by the detection of indole-forming bacteria in the feces of a wide range of host animals.[16] In a subsequent investigation[18] indole was detected as a metabolite of tryptophan in 8 out of 23 strains of *Clostridia* studied. However, in the absence of a demonstration of the stoichiometry given in Figure 3 and the possibility of alternative mechanisms of indole formation,[19] this cannot be assigned unequivocally to tryptophanase activity.

The proportion of metabolism of tryptophan by intestinal bacteria occurring via the different pathways can be deduced by the relative concentrations of metabolites detected in vitro with mixed cultures, or ex vivo by the concentrations present in the contents of the cecum or in the feces. Clearly, neither is ideal, since the former approach suffers from the possibility of manipulation by the choice of medium (see above), while the latter may give spurious conclusions due to differences in absorption of the various metabolites by the colon. The results of various studies are summarized in Table 1 and suggest that the pathways proposed in Figure 1 adequately describe the fate of tryptophan.

### B. Studies In Vivo

In man, infusion into the ascending colon of either L-tryptophan (2 g) or casein (240 g; equivalent to 2 g of tryptophan) resulted in a large increase in indican excretion and a small and more variable increase in indoleacetic acid.[13] Analysis of the urine samples by paper chromatography suggested that these were the only indolic metabolites to increase due to tryptophan infusion. Indican (indoxyl-sulfate) is the main metabolite in urine after administration of indole to man.[23] Chronic oral administration of various dietary levels of tryptophan to man showed a highly significant correlation between the amount of tryptophan ingested and the amount of indican excreted in the urine.[24] A relatively constant proportion (3%) of the tryptophan was converted to indican which is similar to the amount produced after a single oral dose of either L-tryptophan or *N*-acetyl-L-tryptophan.[23]

The extent of microbial catabolism of tryptophan is, therefore, dependent on the amount of substrate which is presented to the gut microflora. Abnormalities of the gastrointestinal tract can alter the amount of substrate available and provide valuable additional information. For example, elevated urinary excretion of 5-hydroxyindoleacetic acid, indoleacetic acid, and indican were detected in coeliac patients, either untreated or given gluten-free diet, who exhibited steatorrhoea.[25] In two of these patients oral tetracycline administration abolished indican excretion, reduced indoleacetic acid but had negligible effect on 5-hydroxyindole-acetic acid elimination. This agrees with the conclusion of Fordtran et al.[13] that the gut flora are not an important source of 5-hydroxyindoleacetic acid.

Although the gut microflora are the principal and probably the sole source of indole in the cecum[26] and of indican in the urine, it is likely that some of the indoleacetic acid excreted in normal subjects arises from tissue metabolism.[9]

Indoleacetic acid excretion is elevated in patients with various malabsorption conditions[27] especially in pancreatogenous steatorrhoea, where the reduced intestinal protease activity would allow a greater delivery of protein to the gut microflora. The non-nutritive sweetener, saccharin, is a weak inhibitor of proteolytic enzymes[21,28] and administration of high dietary levels to rats results in an increase in the amounts of protein and tryptophan in the cecum (which is the principal site of tryptophanase activity) and in the daily excretion of indican.[21]

Patients with Hartnup disease have a defective uptake of L- and D-tryptophan by the cells of the jejunum and thus, excrete increased amounts of bacterial metabolites. In such individuals a single oral dose of tryptophan is followed by a sustained excretion of elevated levels of indican and indoleacetic acid, compared with the much smaller transient rise seen with normal subjects.[29] Of greater interest was the detection of indolylacrylylglycine and bis-indolylindoxyl in the urine of these patients, which reached peak levels 12 hr after the dose and were proposed as bacterial metabolites. The former compound is the major metabolite of indoleacrylic acid, which in the rat can also undergo glucuronic acid conjugation.[30] Detection of this compound is thus consistent with the scheme proposed in Figure 1. Dimerized indoxyl has been detected as a bluish pigment in the urine of infants with tryptophan malabsorption[31] and in rats fed extremely high dietary levels (approximately 15%) of salt.[32] However, it is probable that the dimerization occurs as a chemical reaction in the urine ex vivo, following β-glucuronidase hydrolysis of indoxylglucuronide.[32]

Bacterial colonization of the upper intestine would be expected to lead to increased microbial tryptophan catabolism and increased indican excretion has been reported in such patients.[33] However, there was a poor correlation between indican excretion and quantitative jejunal cultures[34] and attempts to utilize indican excretion as a diagnostic indicator would yield false positive results.[35] Thus, it appears that indican excretion is an index of tryptophan metabolism in the hind gut but not in the proximal intestine even in the presence of bacterial contamination.

## III. METABOLISM OF PHENYLALANINE AND TYROSINE

### A. Studies In Vitro

Phenylalanine and tyrosine are metabolized extensively by the gut microflora to yield a range of products (see Figure 1 for general scheme). Major metabolites produced in vitro with pure cultures of organisms include the arylacetic and arylpropionic acids, phenol and cresol.

Mayrand and Bourgeau[36] analyzed the culture medium from 388 strains of anaerobes from 17 genera for the presence of phenylacetic acid as part of a taxonomic survey. The metabolite was found only in *Bacteroides* and *Clostridium* spp., but not in all species, with a 500-fold variation in the amount detected in the positive species. Incubation of 23 strains of *Clostridia* showed that the eight strains which accumulated phenylacetic acid from phenylalanine in most cases also formed 4-hydroxyphenylacetic acid from tyrosine (six strains) and indole-acetic acid from tryptophan (five strains).[18] The corresponding propionic acids were produced as the only detectable metabolites by four different strains, while others could form in addition phenyllactic and 4-hydroxyphenyllactic acids. The formation of 3-phenylpropionic acid by *C. sporogenes* was preceded by transient accumulation of the unsaturated cinnamic acid but neither this, nor phenyllactic acid was detected in the final incubation which is consistent with the pathway proposed in Figure 1.

Both phenol and *p*-cresol (4-hydroxytoluene) are also formed by isolated strains of bacteria. Of the 23 strains of *Clostridia* studied by Elsden et al.[18] six formed phenol but only one produced *p*-cresol *(C-difficile)*. Growth of pure isolated gut bacteria in a medium containing 0.01% tyrosine showed that facultative anaerobes, such as *E. coli, S. faecalis,* and *Proteus* sp., produced only phenol, while strict anaerobes, such as *Bacteroides fragilis, Bifidobacterium* spp., and strains of *Clostridia* produced only cresol.[37] The formation of phenol from tyrosine probably involves the enzyme tyrosine phenol lyase, which is analogous to tryptophanase in that it converts tyrosine to phenol, pyruvate, and ammonia, and required pyridoxal as a co-factor.[38] The enzyme purified from *C. tetanomorphum* was highly specific for L-tyrosine and did not act on the D-isomer, phenylalanine, or tryptophan.

Studies utilizing mixed cultures prepared from various sources such as sheep rumen contents, rat cecal contents and human feces, are consistent with the above studies, but also demonstrate further reactions of these metabolites.

The rumen contents of sheep contained 3-phenylpropionic acid as a major aromatic acid component with smaller amounts of phenylacetic and benzoic acids.[39] [$^{14}$C] Phenylalanine was converted in vitro and in vivo to [$^{14}$C] phenylacetic acid, while [$^{14}$C] tyrosine was metabolized largely to [$^{14}$C]3-phenylpropionic acid via dehydroxylation of [$^{14}$C] (4-hydroxyphenyl) propionic acid. The main metabolites, phenylacetic and 3-phenylpropionic acid were well absorbed from the rumen and did not undergo measurable interconversion. [$^{14}$C] Phenylalanine introduced into the rumen in vivo, was eliminated largely in the urine as [$^{14}$C] phenylacetic acid, while [$^{14}$C] tyrosine was excreted mainly as [$^{14}$C] benzoic acid which was produced readily from [$^{14}$C] 3-phenylpropionic by kidney tissue slices.

The contents of the cecum of the rat incubated with nutrient medium plus tyrosine produced 4-hydroxyphenylpropionic and 4-hydroxyphenylacetic acids plus phenol and *p*-cresol. The same metabolites were also formed on incubation with the 4-hydroxypyruvic acid, but surprisingly in lower yield (see Figure 4).[40] 4-Hydroxybenzoic and 4-hydroxyphenylacetic acids were decarboxylated to phenol and *p*-cresol respectively, while 4-hydroxycinnamic acid (which had not been detected as a metabolite of tyrosine) formed 4-vinylphenol, 4-ethylphenol, and 4-hydroxypropionic acid.[40,41] The decarboxylation reactions by these mixed cultures occurred only in substrates which had a 4-hydroxy substitution since 3- and 2-isomers were not decarboxylated.[42] The recently described novel rearrangement of 4-hydroxyphenylpropionic acid to 3-hydroxyphenylpropionic acid by human fecal organisms[43] via a movement of the side chain is of considerable interest. Studies using deuterated substrate with human feces showed that tyrosine was metabolized to 4-hydroxyphenyllactic acid, 4-hydroxyphenylpropionic acid and 4-hydroxyphenylacetic acid; 4-hydroxyphenyllactic acid was converted to 4-hydroxy- and 3-hydroxyphenylpropionic acids, 4-hydroxyphenylacetic acid and phenylpropionic acid; 4-hydroxyphenylpropionic acid was metabolized to the 3-hydroxyisomer (see above) and phenylpropionic acid; 4-hydroxyphenylacetic acid formed cresol and 4-hydroxybenzoic acid formed phenol.[44]

Based on these various largely in vitro studies the metabolism of tyrosine and phenylalanine can be represented by the scheme given in Figure 4.

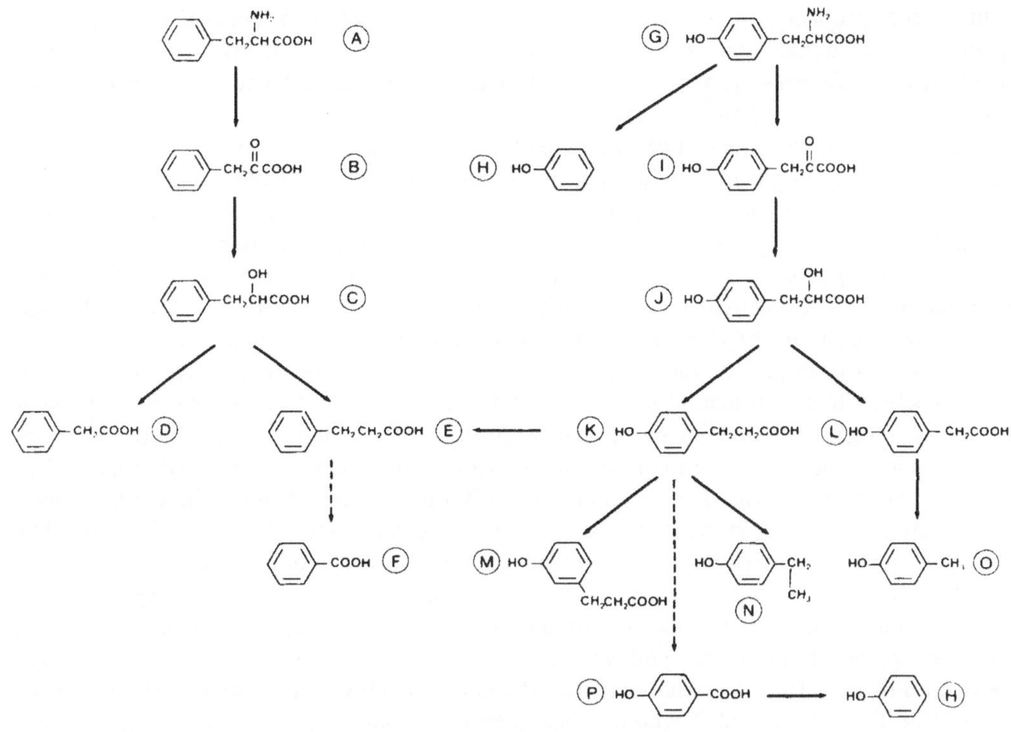


FIGURE 4. The microbial metabolism of phenylalanine and tyrosine. (A) Phenylalanine, (B) phenylpyruvic acid, (C) phenyllactic acid, (D) phenylacetic acid, (E) phenylpropionic acid, (F) benzoic acid, (G) tyrosine, (H) phenol, (I) 4-hydroxyphenylpyruvic acid, (J) 4-hydroxyphenyllactic acid, (K) 4-hydroxyphenylpropionic acid, (L) 4-hydroxyphenylacetic acid, (M) 3-hydroxyphenylpropionic acid, (N) ethylphenol, (O) *p*-cresol, (P) 4-hydroxybenzoic acid. (→) Reactions performed by the gut microflora. (--→) Reactions performed by the tissues.

## B. Studies In Vivo

Evidence for the role of the gut microflora in the metabolism of phenylalanine and tyrosine in vivo arises largely from studies in germfree and antibiotic treated animals, feeding studies in ruminants and pathophysiological conditions in man.

The principal phenols present in the urine of rats fed a purified casein-sucrose diet were phenol and cresol, with trace amounts of 4-ethylphenol.[45] When the food also contained plant material, increased levels of 4-ethylphenol and other phenols were detected. The excretion of phenol and *p*-cresol increased in rats when the dietary concentration of casein was increased from 0 up to 20% but there was no additional elimination when the protein represented 60% of the diet.[46] The excretion of 4-ethylphenol was not related to the amount of protein in the diet. A subsequent study[47] showed a 96% reduction in the excretion of phenol, *p*-cresol and 4-ethylphenol by rats treated with neomycin demonstrating that these three urinary phenols are of bacterological origin. This was confirmed by the virtual absence of these and other phenols from the urines of germfree rats.[48] In rats fed a casein-sucrose diet supplemented with tyrosine, the intracecal administration of neomycin produced a reduction in the excretion of both phenol (80% decrease) and *p*-cresol (96% decrease). These simple phenols accounted for 34% of the radioactivity eliminated in the urine after intracecal administration of [$^3$H] tyrosine, but only 0.5% after intragastric administration.[49] Germfree status in mice is associated with elevated levels of various amino acids in the cecum and colon,[26] including tyrosine and phenylalanine. This may have been due in part to the absence of microbial catabolism,[26] however, the cecal contents of feces of germfree rabbits contained lower levels of amino acids than did conventional animals.[50] The amino acid concentrations in the cecum probably reflect the sum of various factors such as proteolytic activity in the upper intestine, transit time, the extent of intestinal sloughings, bacterial amino acid degradation, and synthesis from precursors, the extent of degradation in the cecum of host proteolytic enzymes and also the extent of coprophagy.

Studies in ruminants are more complex since ruminal organisms in vitro can both synthesize[51,52] and degrade[53] a wide range of amino acids. Phenylalanine and tyrosine may be synthesized by rumen microorganisms in vitro from phenol, phenylacetic or 4-hydroxyphenylacetic acid but these routes are quantitatively of little importance compared with synthesis from glucose.[54] [$^{14}$C] Phenylacetic acid was incorporated mostly into [$^{14}$C] phenylalanine while [$^{14}$C] 4-hydroxyphenylacetic acid formed [$^{14}$C] tyrosine. The study also showed incorporation of $^{14}$C activity from [1-$^{14}$C] 4-hydroxyphenylacetic acid into bacterial tryptophan and from [1-$^{14}$C] indole-3-acetic acid into bacterial phenylalanine and tyrosine, which suggests that degradation to simple precursors may have preceded some of the synthesis. In a series of studies in ruminants, Martin[55-57] concluded that the formation of phenylpropionic acid by the rumen microorganisms is the major source of benzoic acid in the urine, the excretion of which is about 10 to 100 times greater than that in nonruminants. The phenylpropionic acid was formed from tyrosine and other dietary precursors. The principal phenol in sheep urine was *p*-cresol which was formed from 4-hydroxyphenylacetic acid and which accounted for the metabolism of nearly one half of the tyrosine content of casein, when this was infused into the rumen. Thus, it would appear that while the ruminal organisms have the ability to synthesize amino acids such as phenylalanine and tyrosine, the overall balance for the animal is that of degradation. This is supported by findings such as those of Yokoyama et al.[58] who demonstrated that weanling pigs fed antibiotics showed a greater growth rate and final body weight than controls and Elliott and Little[59] who showed that although [$^{35}$S] sulfate would be converted to [$^{35}$S] cysteine in the gut this was of negligible nutritional importance.

Various clinical conditions are associated with abnormal patterns of urinary phenolic amino acid metabolites, which may originate either in tissue metabolism, as in phenylketonuria or in metabolism within the lumen of the gut. Many of the biochemical consequences of abnormalities of gut structure and function have been investigated by determination of the excretion of volatile phenols, i.e., phenol and cresol. Bone et al.[37] studied a variety of conditions and found

1. That ileostomy, which is accompanied by a loss of the anaerobic flora of the distal gut was accompanied by decreased *p*-cresol but not phenol excretion
2. That diverticular disease was associated with elevated phenol elimination in the single patient studied
3. That patients with colostomy, megacolon, large bowel cancer, and familial polyposis did not excrete grossly abnormal quantities of these phenols

Increased phenols and acidic metabolites have been found in patients with coeliac disease, intestinal resection, cystic fibrosis of the pancreas, and unclassified enteritis.[60-63] The principal phenylalanine metabolites detected were benzoic and phenylacetic acids.[60-62] It has been suggested[63] that the latter may have been formed via decarboxylation of phenylalanine followed by oxidation of the β-phenylethylamine produced. However, the data obtained were also consistent with the pathway presented in Figure 4. 4-Hydroxybenzoic and 4-hydroxyphenylacetic acids were the principal metabolites of tyrosine detected in the urine while the feces also contained 4-hydroxyphenylpropionic and 4-hydroxyphenyllactic acids. It is interesting to note that patients with severely impaired amino acid absorption excreted small amounts of 4-hydroxyphenylacrylic and 4-hydroxyphenylpropionic acid in the urine, providing further evidence of the presence of the α, β unsaturated acid as part of the microbial metabolism of aromatic acids.[60] The principal simple phenols excreted by these patient groups were phenol and cresol, which would arise from an increase in protein in the hind gut.

## IV. BACTERIAL AROMATIC AMINO ACID METABOLITES AND TOXICITY

### A. Amines

The bacterial decarboxylation of aromatic amino acids produces amines, such as tyramine and tryptamine,[9] which possess indirectly acting sympathomimetic or pressor activity. The tyrosine decarboxylase of *Streptococcus faecalis* has a relatively broad substrate specificity that may be limited by the permeability of the bacterial cell wall to potential substrates.[64] In normal subjects such bacterial metabolites would produce negligible systemic effects since they are absorbed and transported by the hepatic portal vein to the liver. The high levels of monoamine oxidase activity in both the gut wall and liver would result in extensive first pass inactivation. However, in patients with cirrhosis there is a reduced first pass elimination and hypertyraminemia can result. The role of the gut flora as the principal source of tyramine in such patients has been shown by the decrease in plasma tyramine levels, which follows treatment with berberine, an inhibitor of bacterial decarboxylation.[65]

### B. 3-Methylindole (Skatole) as a Toxic Metabolite

Pulmonary edema and emphysema can be induced in cattle by the intraruminal, but not intravenous or intraperitoneal administration of tryptophan.[66] A similar lesion could be produced by parenteral administration of 3-methylindole but not indole[67] suggesting that this was the active microbial metabolite involved.[68] The administration of 3-methylindole to goats at a dose of 0.04 g/kg by intravenous injection produced rapid changes in the ultrastructure of the capillary endothelial cells and alveolar cells.[69] Enteral administration of tryptophan at 0.35 g/kg was associated with signs of pulmonary edema in three and death in one out of six treated cows, but this was prevented by the administration of 2.5 g chlortetracycline 12-hourly for 4 days prior to the dose of tryptophan. The concentration of 3-methylindole in the rumen contents of normal animals was 15 $\mu g/m\ell$ while that in antibiotic treated cows was $<1$ $\mu g/m\ell$, which is consistent with the proposed role of 3-methylindole formation by the gut microflora in this toxic effect.[70] The toxicity is associated with covalent binding of radio-labeled 3-methylindole to lung tissue, which is probably due to the formation of a reactive metabolite by the P450 system of the endoplasmic reticulum.[71] The metabolism of 3-methylindole is complex with at least ten different metabolites in the urine, which is the principal (87 to 92% of dose) route of elimination.[72] The nature of the toxic metabolite has not been elucidated, but its covalent binding results in depletion of glutathione, especially in the lung. A degree of protection can be afforded by cysteine administration which causes decreased covalent binding, a decreased severity of lung injury and an increased survival time.[73] The similarity of the above lesion to the natural acute bovine emphysema, which occurs in cattle soon after transfer from a poor to a lush green forage, suggests that both may share a common underlying mechanism.[67] If this is so, then the severity of the natural lesion would reflect the balance between the extent of 3-methylindole formation from tryptophan and the extent for formation of sulfur amino acids from other S sources.[74,75]

### C. Indole as a Toxic Metabolite

The microbial formation of indole may have important implications with respect to the development of tumors of the urinary bladder and colon. Indole and related microbial tryptophan metabolites are not mutagenic[22] and thus are unlikely to act as initiators of carcinogenesis. However, tryptophan has been shown to act in vivo as a co-carcinogen[76] and a promoter[77] of carcinogenesis; that is, it increases the tumorigenic potential of cells exposed to subcarcinogenic doses of known bladder carcinogens.

Most studies on tryptophan metabolism in relation to bladder carcinogenesis have concentrated on the formation of aromatic amino metabolites, such as 3-hydroxyanthranilate, by the tissues since various aromatic amines are bladder carcinogens. However the microbial formation of indole may also be important since indole like tryptophan is a co-carcinogen for the urinary bladder.[78,79] Since indole is of greater potency than the parent amino acid, it represents an active metabolite produced by the gut flora. Indole formation has received little attention in relation to human bladder cancer and an observation of increased indole in a single patient with bladder cancer[80] was not substantiated by a larger study.[81] However, an adequate case-control study on indican excretion has not been performed.

It has been reported recently that the effects of saccharin on the urinary bladder of the rat may be mediated in part by increased indole formation.[21] In addition to acting as an inhibitor of protease enzymes discussed above, saccharin alters the intestinal microflora of the rat.[21,82] This resulted in a different pattern of tryptophan metabolites in the cecum with increases in both indole and indolelactic acid.[21] These effects were seen in a single generation at the dietary levels at which saccharin, like tryptophan, is a promoter[77] and also throughout a 2-generation feeding study,[90] the protocol under which saccharin increases the incidence of bladder tumors without exposure to a known initiator.

The possibility of tryptophan metabolism being related to cancer of the colon arises from studies on the relationship of diet with the incidence of this disease. Populations with a high incidence ingest a diet which is high in protein and fat and low in fiber, that is a typical Western diet.[83,84] Since a high meat diet increases the levels of tryptophan and tryptophanase activity in the feces of experimental animals,[17] a human population receiving a high meat diet should have a greater formation of indole within the colon.

### D. Phenol and *p*-Cresol as Toxic Metabolites

Phenol and cresol are weak acids of moderate toxicity since the oral $LD_{50}$ in rats is about 0.4 g/kg.[46] The amounts of these phenols formed in rats fed normal diets (1 to 2 mg/day[45,46]) are thus of minimal importance and are altered little by large changes in the protein content of the diet.[46] However, feeding a diet containing 10% tyrosine resulted in the daily excretion of 0.2 to 0.6 g of total phenols, a value similar to the acute oral $LD_{50}$. It has been suggested that simple phenol formation may contribute to the toxic manifestations, such as growth inhibition, which accompany the feeding of high dietary levels of tyrosine. Oral administration of antibiotics increases the biological value of dietary protein in rats and roosters[85] while the administration of chlortetracycline, sulfamethazine or penicillin to weanling pigs increased the growth rate and final body weight compared with animals given lincomycin or no antibiotics.[86] There was a marked inverse correlation between growth rate and the excretion of *p*-cresol which was the major phenolic metabolite in the urine. The authors suggested that *p*-cresol may have acted as a growth depressant; however, it is possible that the two parameters may have shared a common cause, rather than demonstrating a cause-effect relationship.

Bone et al.[37] investigated the possibility of a link between the formation of simple phenols in man and the incidence of colo-rectal cancer. Both phenol and *p*-cresol promote the development of skin cancer following the application of a single initiating dose of a polycyclic aromatic hydrocarbon.[87,88] Since colo-rectal cancer is more prevalent in the Western world, where there is a higher intake of protein, it was postulated that the formation of simple phenols within the bowel might result in an increased development of tumors at this site. However, comparison of phenol and cresol excretion in ten normal subjects with the data for six patients with large bowel cancer showed no difference. This suggests that if phenols are involved in the development of human colon cancer, then it is only to a very minor degree.

## V. CONCLUSIONS

The gut flora convert aromatic amino acids to unique metabolites. These are frequently of high lipid solubility, are well absorbed from the gut and undergo further metabolism by the host tissues prior to elimination in the urine. Individual intestinal microorganisms produce different end products of amino acid metabolism and it is possible that an end product for one organism may be metabolized further by a different organism within the bowel. A number of specific enzymes involved in the pathways of metabolism of aromatic amino acids described in this chapter have been characterized. (The possible importance of the Stickland reaction[89] in the conversion of amino acids to ammonia and their acetic acid analogues is considered in Chapter 7.)

The products of microbial metabolism of aromatic amino acids are of potential toxicological significance to the host. Although any toxicity is minimal under normal conditions, abnormalities of the structure or function of the gut and major alterations to the composition of the diet can result in increased exposure of the host to such metabolites.

## REFERENCES


1. **Lacoste, A. M.,** Degradation du tryptophane par les bacteries de la panse de ruminants., *C.R. Acad. Sci. Paris,* 252, 1233, 1961.
2. **Lewis, T. R. and Emery, R. S.,** Intermediate products in the catabolism of amino acids by rumen microorganisms, *J. Dairy Sci.,* 45, 1363, 1962.
3. **Schatzmann, H. J. and Gerber, H.,** Production of tryptamine from tryptophan by ruminal fluid *in vitro, Zentralbl. Veterinaermed. Riehe A,* 19, 482, 1972.
4. **Yokoyama, M. T. and Carlson, J. R.,** Dissimilation of tryptophan and related indolic compounds by ruminal microorganisms *in vitro, Appl. Microbiol.,* 27, 540, 1974.
5. **Rudman, D. and Meister, A.,** Transamination in *Escherichia coli, J. Biol. Chem.,* 200, 291, 1953.
6. **O'Neil, S. R. and Demoss, R. D.,** Tryptophan transaminase from *Clostridium sporogenes, Arch. Biochem. Biophys.,* 127, 361, 1968.
7. **Jean, M. and Demoss, R. D.,** Indolelactate dehydrogenase from Clostridium sporogenes, *Can. J. Microbiol.,* 14, 429, 1968.
8. **Aragozzini, F., Ferrari, A., Pacini, N., and Gualandris, R.,** Indole-3-lactic acid as a tryptophan metabolite produced by *Bifidobacterium* spp., *Appl. Environ. Microbiol.,* 38, 544, 1979.
9. **Weissbach, H., King, W., Sjoerdsma, A., and Udenfrend, S.,** Formation of indole-3-acetic acid and tryptamine in animals, *J. Biol. Chem.,* 234, 81, 1959.
10. **Chung, K., Anderson, G. M., and Fulk, G. E.,** Formation of indoleacetic acid by intestinal anaerobes, *J. Bacteriol.,* 124, 573, 1975.
11. **Newton, W. A. and Snell, E. E.,** Catalytic properties of tryptophanase, a multi-functional pyridoxal phosphate enzyme, *Proc. Natl. Acad. Sci. U.S.A.,* 51, 382, 1964.
12. **Kazarinoff, M. N. and Snell, E. E.,** Some effects of indole on the interaction of amino acids with tryptophanase, *J. Biol. Chem.,* 255, 6228, 1980.
13. **Fordtran, J. S., Scroggie, W. B., and Polter, D. E.,** Colonic absorption of tryptophan metabolites in man, *J. Lab. Clin. Med.,* 64, 125, 1964.
14. **Burns, R. O. and Demoss, R. D.,** Properties of tryptophanase from *Escherichia coli, Biochim. Biophys. Acta,* 65, 233, 1962.
15. **Botsford, J. L. and Demoss, R. D.,** *Escherichia coli* tryptophanase in the enteric environment, *J. Bacteriol.,* 109, 74, 1972.
16. **Demoss, R. D. and Moser, K.,** Tryptophanase in diverse bacterial species, *J. Bacteriol.,* 98, 167, 1969.
17. **Chung, K. T., Fulk, G. E., and Slein, M. W.,** Tryptophanase of fecal flora as a possible factor in the etiology of colon cancer, *J. Natl. Cancer Inst.,* 54, 1073, 1975.
18. **Elsden, S. R., Hilton, M. G., and Waller, J. M.,** The end products of the metabolism of aromatic amino acids by clostridia, *Arch. Microbiol.,* 107, 283, 1976.
19. **Roth, C. W., Hoch, J. A., and Demoss, R. D.,** Physiological studies of biosynthetic indole excretion in *Bacillus alvei, J. Bacteriol.,* 106, 97, 1971.

20. **Anderson, G. M.,** Quantitation of tryptophan metabolites in rat feces by thin-layer chromatography, *J. Chromatogr.*, 105, 323, 1975.
21. **Sims, J. and Renwick, A. G.,** The effects of saccharin on the metabolism of dietary tryptophan to indole, a known cocarcinogen for the urinary bladder of the rat, *Toxicol. Appl. Pharmacol.*, 67, 132, 1983.
22. **Bowden, J. P., Chung, K., and Andrews, A. N.,** Mutagenic activity of tryptophan metabolites produced by rat intestinal microflora, *J. Natl. Cancer Inst.*, 57, 921, 1976.
23. **Bryan, G. T.,** Quantitative studies on the urinary excretion of indoxy sulphate (indican) in man following administration of L-tryptophan and acetyl-L-tryptophan, *Am. J. Clin. Nutr.*, 19, 105, 1966.
24. **Bryan, G. T.,** Urinary excretion of indoxyl sulphate (indican) by human subjects ingesting semisynthetic diets containing variable quantities of L-tryptophan, *Am. J. Clin. Nutr.*, 19, 113, 1966.
25. **Kowlessar, O. D., Haeffner, L. J., and Benson, G. D.,** Abnormal tryptophan metabolism in patients with adult celiac disease, with evidence of deficiency of vitamin B6, *J. Clin. Invest.*, 43, 894, 1964.
26. **Whitt, D. D. and Demoss, R. D.,** Effect of microflora on the free amino acid distribution in various regions of the mouse gastrointestinal tract, *App. Microbiol.*, 30, 609, 1975.
27. **Haverback, B. J., Dyce, B., and Thomas, H. V.,** Indole metabolism in the malabsorption syndrome, *N. Eng. J. Med.*, 262, 754, 1960.
28. **Lok, E., Iverson, F., and Clayson, D. B.,** The inhibition of urease and proteases by sodium saccharin, *Cancer Lett.*, 16, 163, 1982.
29. **Milne, M. D., Crawford, M. A., Girao, C. B., and Loughridge, L.,** The metabolic abnormality of Hartnup disease, *Biochem. J.*, 72, 30P, 1959.
30. **Smith, H. G., Smith, W. R. D., Jepson, J. D., and Sorenson, K.,** The metabolism and excretion of indolylacrylic acid in the rat, *Biochem. Pharmacol.*, 19, 1689, 1970.
31. **Drummond, K. N., Michael, A. F., Ulstrom, R. A., and Good, R. A.,** The blue-diaper syndrome: familial hypercalcemia with nephrocalcinosis and indicanuria, *Am. J. Med.*, 37, 928, 1964.
32. **Sapira, J. D., Somani, S., Shapiro, A. P., Sheib, E. T., and Reihl, W.,** Some observations concerning mammalian indoxyl metabolism and its relationship to the formation of urinary indigo pigments, *Metabolism*, 20, 474, 1971.
33. **Greenberger, N. J., Saegh, S., and Ruppert, R. D.,** Urine indican excretion in malabsorption disorders, *Gastroenterology*, 55, 204, 1968.
34. **Mayer, P. J. and Beeken, W. L.,** The role of urinary indican as a predictor of bacterial colonisation in the human jejunum, *Am. J. Dig. Dis.*, 20, 1003, 1975.
35. **Aarbakke, J. and Schjonsby, H.,** Value of urinary simple phenol and indican determinations in the diagnosis of the stagnant loop syndrome, *Scand. J. Gastroenterol.*, 11, 409, 1976.
36. **Mayrand, D. and Bougeau, G.,** Production of phenylacetic acid by anaerobes, *J. Clin. Microbiol.*, 16, 747, 1982.
37. **Bone, E., Tamm, A., and Hill, M.,** The production of urinary phenols by gut bacteria and their possible role in the causation of large bowel cancer, *Am. J. Clin. Nutr.*, 29, 1448, 1976.
38. **Brot, N., Smit, Z., and Weissbach, H.,** Conversion of L-tyrosine to phenol by *Clostridium tetanomorphum*, *Arch. Biochem.*, 112, 1, 1965.
39. **Scott, T. W., Ward, P. F. V., and Dawson, R. M. C.,** The formation and metabolism of phenyl-substituted fatty acids in the ruminant, *Biochem. J.*, 90, 12, 1964.
40. **Bakke, O. M.,** Studies on the degradation of tyrosine by rat caecal contents, *Scand. J. Gastroenterol.*, 4, 603, 1969.
41. **Scheline, R. R.,** Metabolism of phenolic acids by the rat intestinal microflora, *Acta Pharmacol. Toxicol.*, 26, 189, 1968.
42. **Scheline, R. R.,** Decarboxylation and demethylation of some phenolic benzoic acid derivatives by rat caecal contents, *J. Pharm. Pharmacol.*, 18, 664, 1966.
43. **Fuchs-Mettler, M., Curtius, H. C., Baerlocher, K., and Ettlinger, L.,** A new rearrangement reaction in tyrosine metabolism, *Eur. J. Biochem.*, 108, 527, 1980.
44. **Curtius, H. C., Mettler, M., and Ettlinger, L.,** Study of the intestinal tyrosine metabolism using stable isotopes and gas chromatography-mass spectrometry, *J. Chromatogr.*, 126, 569, 1976.
45. **Bakke, O. M.,** Urinary simple phenols in rats fed purified and non-purified diets, *J. Nutr.*, 98, 209, 1969.
46. **Bakke, O. M.,** Urinary simple phenols in rats fed diets containing different amounts of caseine and 10% tyrosine, *J. Nutr.*, 98, 217, 1969.
47. **Bakke, O. M.,** The effect of neomycin and experimental coprostasis on the excretion of simple phenols in the rat, *Scand. J. Gastroenterol.*, 4, 419, 1969.
48. **Bakke, O. M. and Midtvedt, T.,** Influence of germ-free status on the excretion of simple phenols of possible significance in tumour promotion, *Experientia*, 26, 519, 1970.
49. **Aarbakke, J. and Bakke, O. M.,** Localisation of microbial L-tyrosine degradation in the intestine of coprophagy-prevented rats, *Scand. J. Gastroenterol.*, 7, 417, 1972.
50. **Yoshida, T., Pleasants, J. R., Reddy, B. S., and Wostmann, B. S.,** Amino acid composition of cecal contents and feces in germfree and conventional rabbits, *J. Nutr.*, 101, 1423, 1971.

51. **Sauer, F. D., Erfle, J. D., and Mahadevan, S.**, Amino acid biosynthesis in mixed rumen cultures, *Biochem. J.*, 150, 357, 1975.
52. **Stevenson, I. L.**, The production of extracellular amino acids by rumen bacteria, *Can. J. Microbiol.*, 24, 1236, 1978.
53. **Chalupa, W.**, Degradation of amino acids by the mixed rumen microbial population, *J. Anim. Sci.*, 43, 828, 1976.
54. **Kristensen, S.**, Ruminal biosynthesis of aromatic amino acids from arylacetic acids, glucose, shikimic acid and phenol, *Br. J. Nutr.*, 31, 357, 1974.
55. **Martin, A. K.**, The origin of urinary aromatic compounds excreted by ruminants. I. The metabolism of quinic, cyclohexanecarboxylic and non-phenolic aromatic acids to benzoic acid, *Br. J. Nutr.*, 47, 139, 1982.
56. **Martin, A. K.**, The origin of urinary aromatic compounds excreted by ruminants. II. The metabolism of phenolic cinnamic acids to benzoic acid, *Br. J. Nutr.*, 47, 155, 1982.
57. **Martin, A. K.**, The origin of urinary aromatic compounds excreted by ruminants. III. The metabolism of phenolic compounds to simple phenols, *Br. J. Nutr.*, 48, 497, 1982.
58. **Yokoyama, M. T., Tabori, C., Miller, E. R., and Hogberg, M. G.**, The effects of antibiotics in the weaning pig diet on growth and the excretion of volatile phenolic and aromatic bacterial metabolites, *Am. J. Clin. Nutr.*, 35, 1417, 1982.
59. **Elliott, R. and Little, D. A.**, Fate of cyst(e)ine synthesised by microbial activity in the ruminant caecum, *Aust. J. Biol. Sci.*, 30, 203, 1977.
60. **Van der Heiden, C., Waalman, S. K., Ketting, D., and De Bree, P. K.**, Urinary and faecal excretion of metabolites of tyrosine and phenylalanine in a patient with cystic fibrosis and severely impaired amino acid absorption, *Clin. Chim. Acta*, 31, 133, 1970.
61. **Van der Heiden, C., Wauters, E. A. K., Kotting, D., Duran, M., and Wadman, S. K.**, Gas chromatographic analysis of urinary tyrosine and phenylalanine metabolites in patients with gastro-intestinal disorders, *Clin. Chim. Acta*, 34, 289, 1971.
62. **Duran, M., Ketting, D., De Bree, P. K., Van der Heiden, C., and Wadman, S. K.**, Gas chromatographic analysis of urinary volatile phenols in patients with gastro-intestinal disorders and normals, *Clin. Chim. Acta*, 45, 341, 1973.
63. **Seakins, J. W. T.**, The determination of urinary phenylacetyl-glutamine as phenylacetic acid. Studies on its origin in normal subjects and children with cystic fibrosis, *Clin. Chim. Acta*, 35, 121, 1971.
64. **Blaschko, H. and Longlands, M. G.**, Some observations on the bacterial decarboxylation of metatyrosine, *Naunyn-Schmiedebergs Arch. Pharmacol.*, 273, 430, 1972.
65. **Watanabe, A., Obata, T., and Nagashima, H.**, Berberine therapy of hypertyraminemia in patients with cirrhosis, *Acta Med. Okayama*, 36, 277, 1982.
66. **Carlson, J. R., Dyer, I. A., and Johnson, R. J.**, Tryptophan-induced interstitial pulmonary emphysema in cattle, *Am. J. Vet. Res.*, 29, 1983, 1968.
67. **Yokoyama, M. T. and Carlson, J. R.**, Microbial metabolites of tryptophan in the intestinal tract with special reference to skatole, *Am. J. Clin. Nutr.*, 32, 173, 1979.
68. **Carlson, J. R., Yokoyama, M. T., and Dickinson, E. O.**, Induction of pulmonary edema and emphysema in cattle and goats with 3-methylindole, *Science*, 176, 298, 1972.
69. **Bradley, B. J. and Carlson, J. R.**, Ultrastructural pulmonary changes induced by intravenously administered 3-methylindole in goats, *Am. J. Pathol.*, 99, 551, 1980.
70. **Hammond, A. C., Bray, T. M., Cummins, K. A., Carlson, J. R., and Bradley, B. J.**, Reduction of ruminal 3-methylindole production and prevention of tryptophan induced acute bovine pulmonary edema and emphysema, *Am. J. Vet. Res.*, 39, 1404, 1978.
71. **Hanafy, M. S. M. and Bogan, J. A.**, The covalent binding of 3-methylindole metabolites to bovine tissue, *Life Sci.*, 27, 1225, 1980.
72. **Hammond, A. C., Carlson, J. R., and Willett, J. D.**, The metabolism and disposition of 3-methylindole in goats, *Life Sci.*, 25, 1301, 1979.
73. **Nocerini, M. R., Carlson, J. R., and Breeze, R. G.**, Effect of glutathione status on covalent binding and pneumotoxicity of 3-methylindole in goats, *Life Sci.*, 32, 449, 1983.
74. **Bird, P. R. and Hume, I. D.**, Sulphur metabolism and excretion studies in ruminants. IV. Cystine and sulphate effects upon the flow of sulphur from the rumen and upon sulphur excretion by sheep, *Aust. J. Agric. Res.*, 22, 443, 1971.
75. **McMemiman, N. P., Ben-Ghedalia, D., and Elliott, R.**, Sulphur and cystine incorporation into rumen microbial protein, *Br. J. Nutr.*, 36, 571, 1976.
76. **Dunning, W. T., Curtis, M. R., and Maun, M. E.**, The effect of added dietary tryptophane on the occurrence of 2-acetylaminofluorene-induced liver and bladder cancer in rats, *Cancer Res.*, 10, 454, 1950.
77. **Cohen, S. M., Arai, M., Jacobs, J. B., and Friedell, G. H.**, Promoting effects of saccharin and DL-tryptophan in urinary bladder carcinogenesis, *Cancer Res.*, 39, 1207, 1979.

78. **Dunning, W. F. and Curtis, M. R.,** The role of indole in incidence of 2-acetylaminofluorene-induced bladder cancer in rats, *Proc. Soc. Exp. Biol. Med.*, 99, 91, 1958.
79. **Oyasu, R., Kitajima, T., Hopp, M. L., and Sumie, H.,** Enhancement of urinary bladder tumorigenesis in hamsters by co-administration of 2-acetylaminofluorene and indole, *Cancer Res.*, 32, 2027, 1972.
80. **Krstulovic, A. M. and Matzura, C.,** Rapid analysis of tryptophan metabolites using reverse-phase high-performance liquid chromatography with fluorimetric detection, *J. Chromatogr.*, 163, 72, 1979.
81. **Wolf, H. and Brown, R. R.,** Tryptophan metabolism in bladder cancer patients from Copenhagen, *Scand. J. Urol. Nephrol.*, 13, 143, 1979.
82. **Anderson, R. L. and Kirkland, J. J.,** The effect of sodium saccharin in the diet on cecal microflora, *Food Cosmet. Toxicol.*, 18, 353, 1980.
83. **Hill, M. J., Drasar, B. S., Aries, V., Crowther, J. S., Hawksworth, G. M., and Williams, R. E. O.,** Bacteria and aetiology of cancer of large bowel, *Lancet*, 1, 95, 1971.
84. **Wynder, E. L. and Reddy, B. S.,** Metabolic epidemiology of colorectal cancer, *Cancer*, 34, 801, 1974.
85. **Campbell, L. D., Eggum, B. O., and Wolstrap, J.,** Influence of dietary antibiotics on nitrogen, amino acid and energy digestion by rats and adult roosters, *Can. J. Anim. Sci.*, 62, 537, 1982.
86. **Yokoyama, M. T., Tabori, C., Miller, E. R., and Hogberg, M. G.,** The effects of antibiotics in the weaning pig diet on growth and the excretion of volatile phenolic and aromatic bacterial metabolites, *Am. J. Clin. Nutr.*, 35, 1417, 1982.
87. **Boutwell, R. K. and Bosch, D. K.,** The tumour promoting action of phenol and related compounds for mouse skin, *Cancer Res.*, 19, 413, 1959.
88. **Wynder, E. L. and Hoffmann, D.,** Experimental tobacco carcinogenesis, *Science*, 162, 862, 1968.
89. **Stickland, L. H.,** Studies in the metabolism of the strict anaerobes (genus *Clostridium*). I. The chemical reactions by which Cl. sporogenes obtains its energy, *Biochem. J.*, 28, 1746, 1934.
90. **Sims, J. and Renwick, A. J.,** unpublished data.

Chapter 9

# NITRATE METABOLISM

**M. J. Hill**

## TABLE OF CONTENTS

# I. INTRODUCTION

Nitrate is a major source of dietary nonprotein nitrogen, the normal intake being approximately 100 mg/day. It is readily metabolized by bacterial enzymes to nitrite (acting as an electron acceptor in the respiratory chain during anaerobic metabolism) and may be metabolized further to nitrogen (to remove nitrite, which is highly toxic to bacteria) or to ammonia (when it can be utilized in bacterial biosynthesis).

Nitrite reductases have been used by bacteriologists in bacterial taxonomy and by bacterial physiologists because of their importance in bacterial respiration and as structural components of the bacterial cytoplasmic membrane. The products of nitrate metabolism may also be of great importance clinically since nitrite may be an important intermediate in the formation of *N*-nitroso compounds.

In this chapter the in vitro metabolism of nitrate will be described, followed by a discussion of the in vivo nitrate metabolism and its possible importance in human disease.

# II. ENZYMES IN NITRATE METABOLISM

## A. Nitrate Reductase

Nitrate is reduced to nitrite as part of the anaerobic respiratory system in a wide range of organisms and as part of the nitrate assimilation system in a more restricted group of bacteria (but commonly in yeasts, fungi, and algae). Nitrate respiration, which is linked to formate reductase and cytochrome b, has been widely studied both in itself and as a taxonomic tool. Zobell[1] investigated the factors affecting the reduction of nitrate to nitrite in semi-solid medium and further work by Cohn[2] and by Cook[3] established the test as one which was reproducible and of taxonimic use. Table 1 summarizes the distribution of nitrate reductase among the major species of gastrointestinal bacteria. In general the lactic acid bacteria *(Lactobacilli, Bifidobacteria,* and *Streptococci* other than fecal strains) are inactive while *Enterobacteria, Propionic* bacteria, Viellonellae, *Staphylococci* and *Micrococci, Clostridia* and some species of *Bacteroides*, and *Eubacteria* are nitrate reducers. Although assay of nitrate reductase is a widely used taxomic test there are few instances where it is crucial in the identification of an organism.

The enzyme is normally assayed under anaerobic conditions because it has been demonstrated that it is only synthesized in the absence of oxygen,[4,5] once synthesized, the activity of the enzyme is not affected by oxygen concentration. It is an inducible enzyme and is normally particle-bound, being an integral structure in protoplast membrane together with formate dehydrogenase and cytochrome b. It has a pH-optimum of 7 to 7.5 and its synthesis may be affected by pH as well as by substrate concentration and oxygen tension; Rogosa[6] showed that the nitrate reductase of *Lactobacillus fermenti* was not produced when the pH was below 5.5. Enoch and Lester[7] extracted nitrate reductase from *E. coli* cells grown anaerobically in the presence of nitrate. They showed that the enzyme is an essential structural component of the cytoplasmic membrane of induced cultures and is in association with formate dehydrogenase and cytochrome b; it accounts for 52% of the membrane protein. Early studies had shown it to be a molybdoprotein of 4 to 500,000 mol wt;[8] Enoch and Lester[7] and Chaudry and MacGregor[9] showed that it contains three polypetide subunits A, B, and C with 145000, 60000, and 20000 mol wt respectively. Subunit C was shown to be cytochrome b, subunit A was shown to contain the active site of the enzyme[10] and to be the site of the molybdenum co-factor. Subunit B is thought to be associated with membrane attachment.[11] The complete nitrate reductase contains 2A + 2B + 4C units (total of 490,000 mol wt, which is very similar to the 498,000 mol wt observed by Enoch and Lester[7]).

Under conditions of nitrate respiration nitrate acts as the ultimate acceptor of electrons from formate, the electron chain being[7]

**Table 1**
**PRODUCTION OF NITRATE REDUCTASE BY INTESTINAL BACTERIA**

| Generally producers | Generally nonproducers |
|---|---|
| *Haemophilus* spp. | *Lactobacillus* spp. |
| *Propionibacterium* spp. | *Streptococcus* spp. |
| *Clostridium* spp. | *Bifidobacterium* spp. |
| *Peptococcus* spp. | *Bacteriodes fragilis* |
| *Eubacterium lentum* | *Peptostrepticoccus* spp. |
| *Arachnia* spp. | *Fusobacterium* spp. |
| *Actinomyces* spp. | *Neisseria* spp. |
| *Veillonella* spp. | *Gemella* spp. |
| *Bacteriodes corrodens* | *Corynebacterium* spp. |

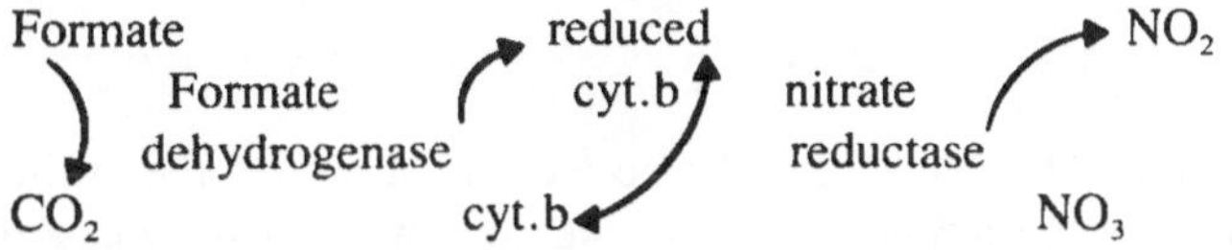


However, quinones are also known to be important components of the electron chain. Quinone-deficient strains of *Staph aureus* are also nitrate reductase deficient[8] and the activity can be restored by the addition of ubiquinone-6, 8, or 10 or menaquinone.

There are two classes of nitrate reductase, namely type A and B.[10] Type A is membrane bound, is only produced under anaerobic conditions, can reduce chlorate as well as nitrate, is very sensitive to azide and is widely distributed among those species in which it has been sought. It appears to be responsible for nitrate respiration. The type B enzyme in contrast, is soluble, can be produced aerobically, does not reduce chlorate (and is in fact, inhibited by chlorate) is fairly resistant to azide, is less widely distributed in the bacterial kingdom, and appears to be involved in nitrate assimilation.

There have been no good recent reviews of nitrate reductases, but the distribution of the enzyme was reviewed by Hall[13] and its structure by MacGregor[14] an older but more thorough review was that by Payne.[8]

### B. Nitrite Reductase

In nitrate assimilation nitrate is reduced to nitrite then in a series of steps through hydroxylamine to ammonia, which is then utilized in amino acid synthesis. Bacteriologists testing for nitrate reductase have known for a long time that in many organisms the end point of nitrate reduction is not nitrite and that a nitrite reductase exists. This has only been the subject of limited study in bacteria but has been more extensively investigated in the yeasts, algae, and fungi where nitrate assimilation is more widely distributed. The nitrite reductase of *E. coli* was studied by Kemp and Atkinson[15] who demonstrated the existence of an enzyme complex. One component was a soluble nitrite reductase with a requirement for reduced NADP while another soluble enzyme had a requirement for reduced FMN. A third enzyme activity could be induced by anaerobic growth in the presence of nitrate or nitrite and required NADH to reduce nitrite or hydroxylamine to ammonia; this enzyme functioned via cytochrome $C_{552}$ with the release of $CO_2$, but had limited activity in the absence of cyt.c. It was suggested by Payne[8] that the nitrite reductase had as its principal function the removal of toxic nitrite rather than nitrite assimilation. Cole and Wimpenny[16] showed that nitrite reductase synthesis was inhibited under anerobic conditions by high nitrate concentrations and was associated with cytochrome c activity. The pH-optimum of

**Table 2**
**THE REDUCTION OF NITRATE TO NITRITE IN THE SALIVA OF VARIOUS INDIVIDUALS[20]**

| Person | Size of nitrate challenge dose (mg) | Peak concentration of nitrate and nitrite (mg/mℓ) | Percentage as nitrite |
|---|---|---|---|
| A | 25 | 38 | 45 |
| | 50 | 68 | 47 |
| | 100 | 118 | 47 |
| | 170 | 207 | 47 |
| B | 170 | 215 | 56 |
| C | 100 | 180 | 78 |
| | 170 | 365 | 78 |
| D | 100 | 111 | 46 |
| E | 25 | 35 | 0 |
| | 50 | 56 | 11 |
| | 100 | 118 | 7 |

the enzyme was 7 to 7.5. Rogosa[6] reported that at pH 7.5 the nitrite reductase of *S. aureus* was as active as the nitrate reductase and so continuously removed nitrite as rapidly as it was synthesized; in these organisms the end product of nitrate metabolism was nitrogen. Nitrite reductase is part of a nitrate assimilation complex of enzymes coproduced by some species of *Clostridia, Escherichia, Klebsiella,* and *Micrococcus,* but a simple nitrite reductase functioning as a scavenger to detoxify nitrite is produced by many other species (although it has not been searched for systematically). Nitrite reductases produced by bacteria have been reviewed by Coleman et al.[17]

## III. NITRATE METABOLISM IN VIVO

The pharmacology of dietary nitrate has been studied extensively by Bartholomew et al.[18-20] In summary dietary nitrate is absorbed from the upper small intestine, secreted in saliva, sweat, tears, gastric juice, and probably in a range of other secretions and is ultimately excreted in the urine. Absorption from the upper small intestine must be very rapid since the nitrate concentration in saliva increases rapidly after a nitrate dose to reach a maximum value 30 to 60 min later and then to return slowly to reach the baseline value 4 to 6 hr later. Detailed studies by Walters and Smith[21] suggest that the decrease is not linear but indicates cycles of secretion and absorption. Ductal saliva contains no nitrite but the saliva present in the oral cavity contains a rich bacterial flora (see Chapter 1) which reduces nitrate to nitrite; the extent of this reduction varies widely between persons (Table 2) and presumably reflects the composition of the oral bacterial flora, although no attempt has been made to study this relationship systematically. Secretion in sweat follows the same temporal pattern as that in saliva, reaching a maximum value within 1 hr and then returning to the baseline concentration 3 to 4 hr later. However, the nitrate in sweat does not undergo reduction to nitrite even though the organisms found on the skin (which include *Staphylococci, Micrococci, Coryneforms,* and *Propionibacteria*) are potent producers of nitrate reductase. This presumably indicates that other conditions (salt concentration, oxygen tension, etc.) are not conducive to production of the enzyme. Secretion of nitrate in gastric juice was demonstrated by Ruddell et al.[22] and the secretion of nitrate into the vagina can be inferred from the studies of Allsobrook et al.[23] Nitrate is ultimately excreted in the urine, the urinary concentration reaching a maximum value 3 to 6 hr after nitrate ingestion and returning to the control value within 18 to 24 hr (depending on the dosage used). Approximately 65 to 70% of a challenge dose is excreted in the urine during the 24 hr following challenge;[20,24,25] the

**Table 3**
**THE RELATION BETWEEN THE GASTRIC BACTERIAL FLORA AND pH (SEE CHAPTER 1)[29]**

| pH | Type of flora found |
|---|---|
| Less than 4 | Only salivary organisms which had still to be killed by the acid conditions |
| 4 to 5 | Salivary organisms, particularly the acid-tolerant organisms able to proliferate at this pH (e.g. *Lactobacilli*, Veillonellae, etc.) |
| Above 5 | A resident flora including, in addition to the salivary organisms, some bacteria more characteristic of feces (e.g. *Strepfaecalis*, bile resistant *Bacteroides* spp., coliforms occasionally) |

**Table 4**
**OBSERVATIONS ON THE RELATION BETWEEN GASTRIC JUICE pH AND GASTRICE JUICE NITRITE CONCENTRATION**

| Observation | Ref. |
|---|---|
| Early studies with single samples | |
| Nitrite present in very small amounts in the normal acid stomach and in large amounts in the achlorhydric stomach | 22, 27, 29, 30, 32 |
| Later studies with continuous sampling | |
| Nitrite concentration in gastric juice is well correlated with gastric juice pH between pH 4 and 7 | 31, 33, 35 |

risk of the challenge can be accounted for as losses in sweat or as metabolized in the mouth and stomach. Although endogenous synthesis of nitrate in the human body has been suggested[26] there was no evidence from the studies of Bartholomew et al.[18-20] that this was a major (or even a significant) contribution to the total nitrate load.

Nitrate metabolism by bacteria has been studied in four sites in the body, namely the oral cavity, the stomach, the infected urinary bladder, and the colon. In the oral cavity, it has been demonstrated that saliva is a rich source of nitrate reductase, but salivary flow is such that in general less than 40% of nitrate has undergone reduction to nitrite and in some persons the amount of reduction product is less than 5% of the total; over a period of 8 hr the proportion of nitrate reduced in the saliva of an individual remains constant and is independent of the nitrate concentration. The major salivary organisms which produce nitrate reductase are the *Micrococci*, *Staphylococci*, Veillonellae, *Arachnia*, and *Bacteroides*.

Nitrate reduction in the stomach has been investigated more thoroughly following the suggested link between gastric bacteria, nitrite, and gastric carcinogenesis[27] which is discussed more fully in Chapter 16. The normal human stomach has a pH of 2 and contains few recoverable bacteria (see also Chapter 1).[28] However, when acid secretion is decreased as a result of natural aging, surgery, dietary stress etc., it has a decreased ability to sterilize the gastric lumen; an empirical observation[29] is that if the gastric juice pH does not fall below four the swallowed salivary organism can survive while if it fails to fall below five a resident gastric flora establishes which contains fecal-type organisms as well as those of salivary origin (Table 3). In the acid stomach the nitrite concentration is very low[22,30-32] but in persons with decreased acidity the nitrite concentration increases with the pH (Table 4)

**Table 5**
**THE EVIDENCE FOR *N*-NITROSATION BY GUT BACTERIA**

| Observation | Ref. |
|---|---|
| *E. coli* catalyzes the *N*-nitrosation reaction at pH 7 (confirmed by many others) | 38 |
| *N*-Nitrosation catalyzed by autoclaved bacteria, so the effect is not due to an enzyme, (a similar conclusion was reached by Reference 40) | 39 |
| *N*-Nitrosation can be catalyzed by resting bacterial cells and the kinetics indicate that the catalysis is enzymic | 41 |
| *N*-Nitrosation by resting cells of *E. coli* is enzymic | 43 |

*Note:* See also Table 6.

in an approximately linear relation.[33-36] The nitrate reducing organisms isolated include fecal *Streptococci, Enterobacteria,* and in some cases *Clostridia* in addition to those organisms commonly found in saliva. In some studies nitrite reductase has also been assessed (e.g. Dolby et al.);[37] in this case in addition to pH the immune status of the patient appeared to be an important factor in determining enzyme activity since the nitrite reductase activity was much higher in patients with pernicious anemia and hypogammaglobulinemia than in those with pernicious anemia alone.

Nitrate reduction to nitrite has been studied in the urinary bladder by a number of groups. Normal human urine is sterile and, although often a rich source of nitrate, contains no nitrite. Urinary tract infections are common however, and are mostly due to organisms which produce nitrate reductase (e.g. *E. coli, Klebsiella* spp., *Proteus* spp., or *Staphylococcus* spp., which between them account for more than 95% of urinary tract infections in Western persons). Thus, in populations in which the intake of nitrate is always sufficient the presence of nitrite in urine has proved to be a reliable and very rapid means of detecting bacterial infection; it has proved less reliable in areas where the drinking water contains little nitrate because then the urine often contains nitrate concentrations which are too low to yield detectable amounts of nitrite. In a recent study of urine samples from persons living in Norfolk almost 5% of samples contained nitrite and therefore urinary tract infection which was presumably assymptomatic.[49]

## IV. *N*-NITROSO COMPOUND FORMATION

Although *N*-nitroso compounds cannot be formed directly from nitrate they may be produced from the metabolites of nitrate (nitrite and the oxides of nitrogen). *N*-nitroso compounds may be formed by the action of nitrous acid on a suitable secondary nitrogen compound; secondary amines give *N*-nitrosamines, amides give *N*-nitrosamides, urea analogues give *N*-nitrosoureas etc. The reaction proceeds at a maximum rate at acid pH values; the pH optimum depends on the nature of the parent amino compound, being around two for secondary amines, three to four for peptides, less than two for ureas etc. This created no interest amongst bacteriologists until 1968 when Sander[38] showed that bacteria could catalyse the *N*-nitrosation reaction and since the organisms studied also produced nitrate reductase, they were able to utilize nitrate as well as nitrite as the substrate for the reaction.

### A. Bacterial *N*-Nitroso Synthetase

Sander[38] first demonstrated that diphenylamine was *N*-nitrosated in the presence of *Esch. coli* and this observation has since been confirmed by many groups (Table 5). However,

**Table 6**
**THE PRELIMINARY OBSERVATIONS BY LEACH ET AL. ON THE CATALYSIS OF *N*-NITROSATION BY *E. COLI*[43]**

| Characteristic | Observation |
|---|---|
| Comparison of strains | Very wide variation between strains in rate of catalysis and so unlikely to be simply an effect caused by a bacterial metabolite |
| pH optimum | The catalyzed reaction proceeded maximally at about pH 7; below pH 6 the acid-catalyzed reaction was dominant |
| Variation in amine | Different amines are *N*-nitrosated at different rates and the pH- optimum of the nitrosation is related to the pK of the amine |
| Addition of nutrients | When nutrients which allow bacterial respiration were added, the rate of *N*-nitrosation increased |

the mechanism has been and still remains, the subject of much controversy. Some are of the opinion that the role of bacteria in the formation of *N*-nitroso compounds is to reduce nitrate to nitrite (so providing an essential reactant) and to generate an acid pH via fermentation of sugars (so providing the $H^+$ ion catalyst for the reaction). Since many groups have demonstrated catalysis of nitrosation using nitrite the first of these suggestions, although coincidentally important, cannot be the whole story. Similarly, many groups (see Table 2) have shown *N*-nitrosation catalyzed by autoclaved bacteria which demonstrates that generation of $H^+$ ions is not essential in *N*-nitrosation. The demonstration by some groups of *N*-nitrosation catalyzed by autoclaved cultures[39-41] suggests that the catalyst is heat resistant and is likely to be a cation component of the bacterial cell wall or some other surface factor. However, this, too, cannot be the whole story since catalysis by resting cultures of bacteria in buffer solutions containing nitrosatable amine and nitrite has been demonstrated, and shown to be enzymic.[42]

Recent careful studies by Leach et al.[43] designed to resolve the question of bacterial involvement in *N*-nitrosation have supported the theory that a bacterial *N*-nitrososynthetase exists, with a pH optimum close to seven and varying with the nature of the nitrosatable amine. The early conclusions of Leach et al. are summarized in Table 6.

### B. Bacterial *N*-Nitrosation In Vivo

If bacteria can catalyze the N-nitrosation reaction in vivo then *N*-nitrosation should be demonstrable when ever bacteria, nitrosatable amino compound, and nitrate or nitrite coexist in the body. Table 7 lists the sites where this might be expected to occur and *N*-nitroso compounds have been demonstrated at all of the sites. The relevance of such *N*-nitroso compounds to human carcinogenesis is discussed in more detail in Chapter 16.

## V. NITRATE TOXICITY

Nitrate is totally nontoxic in humans and other animals but its metabolites, particularly nitrite, cause major toxicity problems particularly in children and in young adults but also possibly in adults.

### A. Infant Methemoglobinemia

Methemoglobinemia, also known as the blue baby syndrome is caused by the blocking of the oxygen-binding sites on hemoglobin (to form methemoglobin) and so reducing the ability of the blood to transport oxygen to the tissues. It was associated with ingestion of

**Table 7**
**SUMMARIZED DATA ON THE RELATION BETWEEN NITRATE INTAKE AND GASTRIC CANCER (GC) INCIDENCE**

| Population studied | Source of nitrate | Observation |
|---|---|---|
| Narino, Colombia | Food and water | GC increases with nitrate intake |
| Chile | Food | — |
| Japan | Food | — |
| China | Food | — |
| Hungary | Drinking water | — |
| Italy | Drinking water | — |
| Denmark | Drinking water | — |
| Israel | Drinking water | — |
| Worksop, England | Drinking water | — |

nitrate in the late 1940s[44,45] and the mechanism elucidated shortly after. In the newborn infant the gastric pH is high and the gastrointestinal tract is therefore colonized throughout. When nitrate is ingested it is reduced by the gastric and small bowel flora and the nitrite released is then absorbed and combines with hemoglobin to form methemoglobin. The nitrite is bound strongly and is not displaced by oxygen; erythrocytes contain only low levels of methemoglobin reductase in young infants. Later gastric acid secretion begins to reach bactericidal levels and cleanses the upper gastrointestinal tract, so decreasing the amount of nitrite produced from ingested nitrate. In addition the erythrocyte methemoglobin reductase activity increases to adult levels and so is able to regenerate hemoglobin. This methemoglobinemia is only a clinical problem in young infants and results from the nitrate in drinking water used to make up bottle feeds or from that of vegetable origin in the solid feeds given to weanlings. Very young infants who are breast fed are not at risk of infant methemoglobinemia regardless of the nitrate content of the drinking water, indicating that nitrate is not secreted at high concentrations in breast milk. In babies to be given formula feeds WHO recommends that where the drinking water contains more than 50 $mgL^{-1}$ nitrate a supply of low nitrate water be supplied specifically for reconstituting the formula feeds. Very close control is kept on the nitrate content of all solid foods to be given to young infants.

### B. Slowed Reflexes in School Children

Although methemoglobinemia is a widely recognized clinical disease only in young infants it has been claimed that it may also cause disease in school children. Studies of school children showed that those drinking high-nitrate water had slower reflexes and were unable to perform simple calculations as rapidly as children drinking low-nitrate water.[46] The authors relate these slowed reflexes to the level of circulating methemoglobin and suggest a causal relationship, but such studies have not yet been reported from elsewhere.

### C. Neural Abnormalities in Newborn Infants

It has been reported[47] that in areas where the drinking water nitrate concentration is relatively high the incidence of neural abnormalities in newborn infants is also high. It has been suggested that the mechanism of this relationship might be the transplacental passage of endogenously formed *N*-nitroso compounds causing a teratogenic effect in the fetus. There has been no time yet for these findings to be supported or refuted.

### D. Gastric Cancer

The relationship between nitroso compounds and gastric cancer is discussed in more detail in Chapter 16. At this point it is relevent, however, to note that there is a large body of evidence relating nitrate exposure to gastric cancer (Table 7) and this evidence is independent of that relating *N*-nitroso compounds to the disease.

The epidemiological evidence relating nitrate to gastric carcinogenesis is far from overwhelming; all of the studies can in retrospect be seen to leave much to be desired and the criticisms have been expressed well by Fraser and Chilvers.[48] However, the overall impression left by the studies *in toto* is that although the relationship is not proven it would be an extremely rash person who chose to dismiss it.

## VI. CONCLUSIONS

Of the various sources of nitrogen in food, by far the greatest is protein (10 to 15 g $day^{-1}$) and in comparison nitrate (at about 100 mg/day) is a minor contributer. Nevertheless it is clear that (1) as a result of farming and horticultural practice the daily intake of nitrate is rising in Western populations, and (2) there may be hazards associated with a large rise in nitrate intake. These hazards are due to the microbial metabolism of nitrate to nitrite (which is a toxic compound) and then further (either chemically or microbially) to *N*-nitroso compounds.

This is a field which is likely to grow in importance in the coming years rather than diminish.

## REFERENCES


1. **Zobell, C. W.,** Factors influencing the reduction of nitrates and nitrites by bacteria in semi-solid media, *J. Bacteriol.*, 24, 273, 1932.
2. **Cohn, H. J.,** On the detection of nitrate reduction, *J. Bacteriol.*, 31, 225, 1936.
3. **Cook, G. T.,** A plate test for nitrate reduction, *J. Clin. Pathol.*, 3, 359, 1950.
4. **Wimpenny, J. W. and Cole, J. A.,** The regulation of metabolism in facultative bacteria. III. The effect of nitrate, *Biochim. Biophys. Acta*, 148, 233, 1967.
5. **Lester, R. L. and De Moss, J. A.,** Effect of molydclate and selenite on formate and nitrate metabolism in *Escherichia coli*, *J. Bacteriol.*, 105, 1006, 1971.
6. **Rogosa, M.,** Experimental conditions for nitrate reduction by certain strains of the genus *Lactobacillus*, *J. Gen. Microbiol.*, 24, 401, 1961.
7. **Enoch, H. G. and Lester, R. L.,** The purification and properties of formate dehydrogenase and nitrate reductase from *Escherichia coli*, *J. Biol. Chem.*, 250, 6693, 1975.
8. **Payne, W. J.,** Reduction of nitrogenous oxides and microorganisms, *Bacteriol. Rev.*, 37, 409, 1973.
9. **Chaundhry, G. R. and MacGregor, C. H.,** *Escherichia coli* nitrate reductase subunit A: its role as the catalytic site and evidence for its modification, *J. Bacteriol.*, 254, 387, 1983.
10. **Chaundhry, G. R. and MacGregor, C. H.,** Cytochrome b from *Escherichia coli* nitrate reductase, *J. Biol. Chem.*, 258, 5819, 1983.
11. **De Moss, J. A.,** Limited proteolysis of nitrate reductase purified from membranes of *Escherichia coli*, *J. Biol. Chem.*, 252, 1696, 1977.
12. **Pichinoty, F. and Piechaud, M.,** Recherches des nitrate reductases bacteriennes A et B: methodes, *Ann. Inst. Pasteur Paris*, 114, 77, 1968.
13. **Hall, J. B.,** Nitrate reducing bacteria, in *Microbiology*. Schlessinger, D., Ed., American Society for Microbiology, Washington, D.C., 1978, 296.
14. **MacGregor, C. H.,** Regulation and assembly of nitrate reductase from *Escherichia coli*, in *Microbiology*, Schlessinger, D., Ed., American Society for Microbiology, Washington, D.C., 1978, 330.
15. **Kemp, A. D. and Atkinson, D. E.,** Nitrite reductase of *Esch. coli* specific for reduced nicotinamide aderine dinecleotide, *J. Bacteriol.*, 92, 628, 1966.
16. **Cole, J. A. and Wimpenny, J. W.,** Metabolic pathways for nitrate reduction in *Escherichia coli*, *Biochim. Biophys. Acta*, 162, 39, 1968.
17. **Coleman, K. J., Newman, B. M., Cornish-Bowden, A. J., and Cole, J. A.,** Nitrite reduction by bacteria, in *Microbiology*, Schlessinger, D., Ed., American Society for Microbiology, Washington, D.C., 1978, 334.

18. **Bartholomew, B. A., Butt, A., Caygill, C., Darbar, R., and Hill, M. J.,** The origin of urinary nitrate, *Proc. Nutr. Soc.*, 38, 124, 1979.
19. **Bartholomew, B. A., Caygill, C., and Hill, M. J.,** Investigations into the sources of urinary nitrate, *Biochem. Soc. Trans.*, 9, 333, 1981.
20. **Bartholomew, B. A. and Hill, M. J.,** The pharmacology of dietary nitrate and the origin of urinary nitrate, *Food Chem. Toxicol.*, 22, in press, 1984.
21. **Walters, C. L. and Smith, P. L. R.,** The effect of water-borne nitrate on salivary nitrite, *Food Cosmet. Toxicol.*, 19, 297, 1981.
22. **Ruddell, W. S. J., Bone, E. S., Hill, M. J., Blendis, L. M., and Walters, C. L.,** Gastric juice nitrite: a risk factor for cancer in the hypochlorhydric stomach?, *Lancet*, 2, 1037, 1976.
23. **Allsobrook, A. J., Duplessis, L. S., Harrington, J. S., Nunn, A. J., and Nunn, J. R.,** Nitrosamines in the human vaginal vault, in *N-Nitroso Compounds in the Environment*, Bogovski, P. and Walker, E. A., Eds., International Agency for Research on Cancer, Lyon, 1974, 197.
24. **Wagner, D. A., Shultz, D. S., Dean, W. M., Young, V. R., and Tannenbaum, S. R.,** Metabolic fate of an oral dose of $^{15}$N-lebeled nitrate in humans: effect of diet supplementation with ascorbic acid, *Cancer Res.*, 43, 1921, 1983.
25. **Ishiwata, H. and Tanimua, A.,** Ingestion and excretion of nitrate in humans and experimental animals, in *Environmental Mutagens and Carcinogens*, Sugimura, T., Kondo, S., and Takebe, H., Eds., Alan Liss, New York, 1982, 571.
26. **Tannenbaum, S. R., Fett, D., Young, V. R., Land, P. D., and Bruce, W. R.,** Nitrate and nitrite are formed by endogenous synthesis in the human intestine, *Science*, 200, 1487, 1978.
27. **Correa, P., Haenszel, W., Cuello, C., Tannenbaum, S., and Archer, M.,** A model for gastric cancer epidemiology, *Lancet*, 2, 58, 1974.
28. **Drasar, B. S., Shiner, M., and MacLeod, G. M.,** The bacterial flora of the gastrointestinal tract in healthy and achlorhydric persons, *Gastroenterology*, 56, 71, 1969.
29. **Bartholomew, B. A., Hill, M. J., Hudson, M. J., Ruddell, W. S. J., and Walters, C. L.,** Gastric bacteria, nitrate, nitrite and nitrosamines in patients with pernicious anema, and in patients treated with cimetidine, in *N-nitroso Compounds: Analysis, Formation and Occurrence*, Walker, E., Griciute, L., Gastagnero, M., and Borszonyi, M., Eds., International Agency Research on Cancer, Lyon, 1980, 595.
30. **Ruddell, W. S. J., Bone, E. S., Hill, M. J., and Walters, C. L.,** Pathogenesis of gastric cancer in pernicious anemia, *Lancet*, 1, 521, 1978.
31. **Milton-Thompson, G. J., Lightfoot, N. F., Ahmet, Z. et al.,** Intragastric acidity, bacteria, nitrite and N-nitrosocompounds before, during and after cimetidine treatment, *Lancet*, 1, 1091, 1982.
32. **Reed, P. I., Smith, P. L. R., Haines, K., House, F. R., and Walters, C. L.,** Gastric juice N-nitrosamines in health and gastroduodenal disease, *Lancet*, 2, 550, 1981.
33. **Hall, C. N., Cook, A., Darkin, D. et al.,** Evaulation of the nitrosamine hypothesis of gastric carcinogenesis in man, *Clin. Sci.*, 66, 34P, 1984.
34. **Watt, P. C. H., Sloan, J. M., Donaldson, J. D., Patterson, C. C., and Kennedy, T. L.,** Relationship between histology and gastric juice pH and nitrite in the stomach after operation for duodenal ulcer, *Gut*, 25, 246, 1984.
35. **Keighley, M. R. B., Youngs, D., Poxon, V. et al.,** Intragastric N-nitrosation is unlikely to be responsible for gastric carcinoma developing after operations for duodenal ulcer, *Gut*, 25, 238, 1984.
36. **Schlag, P., Bockler, R., Ulrich, H. et al.,** Are nitrite and N-nitrosocompounds in gastric juice risk factors for carcinoma in the operated stomach? *Lancet*, 1, 727, 1989.
37. **Dolby, J., Webster, A. D., Borriello, S. P. et al.,** Bacterial colonisation and nitrite concentration in the achlorhydric stomachs of patients with primary hypogammaglobulinemia or classical pernicious anemia, *Scand. J. Gastroenterol.*, 19, 105, 1984.
38. **Sander, J.,** Nitrosaminsynthese durch Baketerien, *Z. Physiol. Chem.*, 349, 429, 1968.
39. **Klubes, P., Cerna, I., Rabinowitz, A. D. et al.,** Factors affecting the dimethylinitrosamine formation from simple precursers by rat intestinal bacteria, *Food Cosmet. Toxicol.*, 10, 757, 1972.
40. **Collins-Thompson, D. L., Senn, P., Aris, B., and Schwinghamer, L.,** Non-enzymic in vitro formation of nitrosamines by bacteria isolated from meat products, *Can J. Microbiol.*, 18, 1968, 1972.
41. **Kunisaki, N. and Hayashi, M.,** Formation of N-nitrosamines from secondary amines and nitrite by resting cells of *Escherichia coli* B, *Appl. Env. Microbiol.*, 37, 279, 1979.
42. **Archer, M. C., Saul, R. L., Lee, L.-J., and Bruce, W. R.,** Analysis of nitriate, nitrite and nitrosamines in human feces, in *Gastrointestinal Cancer: Endogenous Factors*, Bruce, W. R. et al., Eds., Cold Spring Harbor Laboratory, New York, 1981, 321.
43. **Leach, S., Challis, B., Cook, A. R., and Hill, M. J.,** The bacterial catalysis of the N-nitrosation of secondary amines, *Biochem. Soc. Trans.*, in press.
44. **Comly, H. H.,** Cyanosis in infants caused by nitrates in well water, *JAMA*, 129, 112, 1945.
45. **Cornblath, M. and Hartman, A. F.,** Methaemoglobinemia in young infants, *J. Pediatr.*, 33, 421, 1948.

46. **Petukhov, N. I. and Ivanov, A. G.,** Investigation of certain physiocophysiological reactions in children suffering from methaemoglobinemia due to nitrates in water, *Gig. Sanit.*, 35, 26, 1979.
47. **Dorsch, M. M., Scrag, R. K., McMichael, A. J., Baghurst, P. A., and Dyer, K. F.,** Congenital malformation and maternal drinking water supply: a case-control study, *Am. J. Epidemiol.*, 119, 473, 1984.
48. **Fraser, P. and Chilvers, C.,** Health aspects of nitrate in drinking water, *Sci. Total Environ.*, 18, 103, 1981.
49. **Caygill, C.,** unpublished results.

Chapter 10

# GUT BACTERIA AND THE ENTEROHEPATIC CIRCULATION OF FOREIGN COMPOUNDS

**A. G. Renwick**

## TABLE OF CONTENTS

# I. INTRODUCTION

It has been recognized for many years that the bile represents a major route for the elimination of foreign compounds and/ or their metabolites from the body.[1] Similarly the metabolism of xenobiotics by the gut microflora has been known for many years and has been the subject of a number of reviews.[2-4] Although microbial metabolism of foreign compounds may occur as part of a first-pass phenomenon, as a result of incomplete absorption following oral administration,[5] of equal or greater importance is metabolism occurring as a result of delivery of the drug via the bile. Reabsorption of the products of bacterial metabolism into the portal circulation provides the potential for an enterohepatic circulation of the compound. In an example of an enterohepatic circulation the compound would undergo:

1. Hepatic metabolism
2. Biliary excretion of metabolites
3. Bacterial metabolism back to the parent compound
4. Reabsorption from the hind gut
5. Transport back to the liver, where a proportion would escape extraction and re-enter the general circulation

Clearly such a process would increase the time during which the chemical was present in the body and possibly the duration of its effects. The implications of this in the fields of pharmacology and therapeutics are obvious, and most of the examples considered in the chapter are drugs or potential therapeutic agents. Many studies on drug metabolism utilize radio-labeled compounds to allow identification and quantification of all metabolic products. Such studies frequently have identified an enterohepatic circulation of radioactivity that may be unrelated to the presence of the parent compound but due to metabolites. Although not a true enterohepatic circulation of the drug, these examples will be considered in this chapter, since in many cases metabolites are responsible for the pharmacologic and toxic properties and their incorporation into an enterohepatic circulation can affect their potential for activity.

A consideration of the metabolism of xenobiotics performed by the gut flora shows that the major reactions are hydrolyses and reductions,[2,3,5] and thus these figure prominently in enterohepatic circulations. Conversely, the liver performs largely oxidation and conjugation reactions. In view of the minimum molecular weight requirements for extensive biliary excretion,[1,6] hepatic conjugation plays an important role in the enterohepatic circulation of most chemicals, especially when the conjugation is via a bulky substituent such as glucuronic acid or glutathione.

# II. METABOLIC CONSIDERATIONS

## A. Glucuronic Acid and Sulfate Conjugates

Many foreign compounds are excreted in the bile as glucuronic acid conjugates (See Table 1). Glucuronides are highly polar molecules and undergo little absorption from the gut per se in adult animals, although absorption without hydrolysis may occur in very young animals due to difference in the permeability characteristics of the gut wall.[7] Thus, hydrolysis by a β-glucuronidase is essential for completion of the enterohepatic circulation of drugs excreted in the bile as glucuronides. In normal animals most β-glucuronidase activity in the gut lumen is found in the contents of the cecum and hind gut.[5,8] The low levels of β-glucuronidase present in the upper intestine are present also in germfree rats suggesting that this is of tissue origin, however the very high cecal activity was not found.[8] This demonstrates that the majority of β-glucuronidase in the gut lumen is bacterial in origin. β-Glucuronidase activity has been detected in various strains of intestinal bacteria in vitro[9] with the highest activity

**Table 1**
**THE BILIARY EXCRETION AND ENTEROHEPATIC CIRCULATION OF GLUCURONIC ACID CONJUGATES**

| Compound | Species | % in bile (donor)[a] | Duration (hr) | % in bile (recipient)[b] | Duration (hr) | Ref. |
|---|---|---|---|---|---|---|
| Morphine | Rat | 45 | 3 | 41 | 24 | 12 |
| Buprenorphine | Rat | 91 | 3—4 | 42—72 | 24 | 14 |
| Hexachlorophene | Rat | 31 | 72 | 95 | 72 | 15 |
| Phenolphthalein | Rat | 90 | 3 | 85 | 24 | 12 |
| Oxazepam glucuronide | Guinea pig | — | — | 32 | 3 | 16 |
| HWA 923[c] | Rat | 68 | 48 | 55 | 24 | 18 |
| Dichlorophen | Rat | 79 | 4[d] | 13 | 3 | 19, 20 |

[a] Percentage of dose given to donor animal by parenteral injection.
[b] Percentage of the drug related material infused into the duodenum of the recipient animal.
[c] HWA 923 or O-[3-(4-<2-methoxyphenyl>-1-piperazinyl)-2-hydroxypropyl]-3-methoxybenzaldoxim dihydrochloride was given orally to donor rats and about 20% of the drug related material in bile was present as a glucuronide conjugate of the parent compound.
[d] Bile was collected from 1 to 4 hr only and contained 36% of the dose; an additional 43% was present in the gut lumen and gut wall probably due to biliary excretion in the 0 to 1 hr time interval.

in the aerobic *Enterobacteria,* and lower amounts in anaerobes such as *Bacteroides* spp. (see also Chapter 3, Table 6). However, the latter are present in the intestines in much higher numbers and are probably of greater importance in vivo. This conclusion is supported by various findings:[10]

1. The development of bacterial β-glucuronidase, which is absent in newborn rats and human infants, reflects the increase in numbers of anaerobes.
2. Mixed cultures of rat cecal organisms contain β-glucuronidase if grown under anaerobic but not aerobic conditions.
3. Cecal β-glucuronidase can be eliminated in vivo by giving antibiotics, such as lincomycin, which are active against anaerobes.

The above conclusions regarding the essential role of the gut microflora in the enterohepatic circulation of compounds excreted in the bile as glucuronides are incorporated into a generalized scheme (Figure 1). The scientific literature contains numerous examples of drugs excreted in the bile as glucuronic acid conjugates. However, such information cannot be taken as adequate evidence for an enterohepatic circulation since the aglycone may be unabsorbed. Foreign compounds which undergo enterohepatic circulation as glucuronides may be subdivided into those involving the parent compound and those involving oxidative metabolites.

*1. Enterohepatic Circulation as Parent Compound and Its Glucuronide*

A requisite for compounds in this group is the presence of a suitable functional group for conjugation. Published examples include phenols, alcohols, and carboxylic acids (Figure 2) although it is likely that *N*- and *S*- glucuronides could also undergo enterohepatic circulation.

Morphine is excreted in the bile of rats primarily as the 3-glucuronide, which accounts for about 45% of the dose.[11,12] Little reabsorption occurred in the upper intestine[11] but extensive uptake occurred from the cecum following hydrolysis of the conjugate. The dependence of this hydrolysis on the gut flora was demonstrated by reduced hydrolysis and

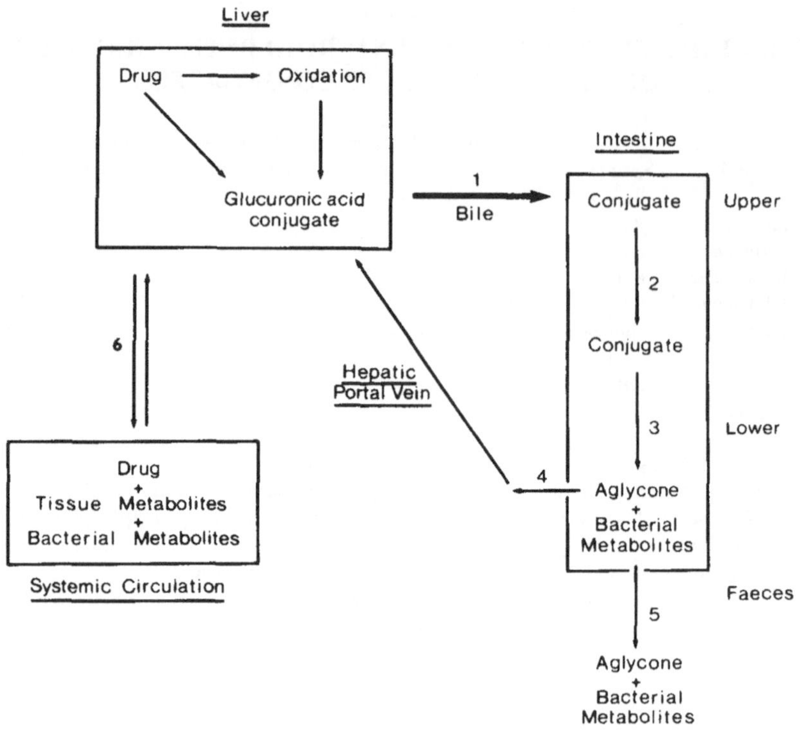


FIGURE 1. Scheme for the enterohepatic circulation of glucuronic acid conjugates. The complete enterohepatic circulation requires a finite time before the drug is transported back to the liver, via the hepatic portal vein, and is able to re-enter the systemic circulation. Delays include: (1) intermittent emptying of the gall bladder in appropriate species; (2) intestinal transit time; (3) bacterial β-glucuronidase; and (4) reabsorption from the cecum, colon, and rectum. The extent of enterohepatic circulation is determined also by, (5) loss of drug and bacterial metabolites in the feces. The systemic effects of recirculating drug and metabolites are determined also by, (6) the proportion of the absorbed material which escapes first pass uptake by the liver.

reabsorption in animals treated with lincomycin,[11] with the resultant elimination of the conjugate as such in the feces.

The excretion of $^{14}C$ in the bile collected from rats which received intraduodenal infusion of the bile from other animals dosed with [$^{14}C$] morphine, was decreased from 41% in 24 hr in normal to 9% in recipient rats treated with antibiotics.[12] This experiment provided clear evidence of the importance of the gut flora in the complete enterohepatic circulation of morphine in the rat. The reabsorption of morphine following the biliary excretion of a single dose given orally or intravenously, results in a secondary peak in the plasma concentration time curve about 6 hr after the dose.[13] Approximately one third of the morphine absorbed from the gut lumen leaves the enterohepatic recirculation[12] and contributes to this secondary peak. The structural analogue buprenorphine is excreted extensively in the feces of rat, dog, monkey, and man after intramuscular administration (84, 95, 64, and 68% respectively) and appears in the bile as a glucuronide conjugate which undergoes enterohepatic circulation.[14] However, the role of the gut flora in this process was not defined.

A number of other phenolic compounds have been shown to undergo enterohepatic circulation primarily as glucuronic acid conjugates of the parent compound (Table 1).

Phenolphthalein, like morphine undergoes an extensive enterohepatic circulation. Evidence for an important role for the gut flora includes a 4 hr delay between intraduodenal instillation of biliary metabolites and their re-elimination, and a suppression of this re-

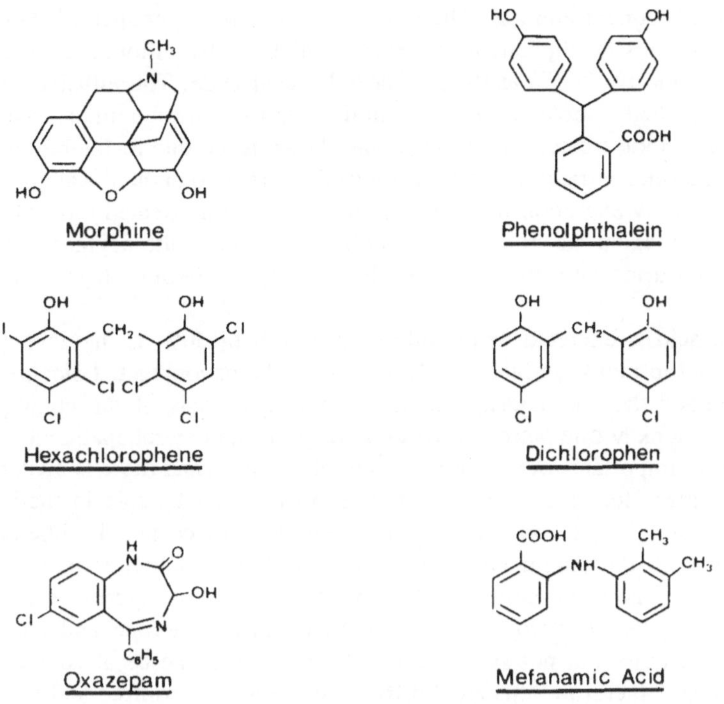


FIGURE 2. Drugs undergoing enterohepatic circulation largely as the parent compound and its conjugates.

elimination by antibiotic treatment, which was shown not to affect the absorption of the aglycone and thus, acted by inhibition of microbial hydrolysis.[12] The results with hexachlorophene [2,2′-methylene-*bis* (3,4,6-trichlorophenol)], which is eliminated slowly from the body, suggested that the proportion of the dose which is initially distributed to the liver and excreted in the bile is retained within the enterohepatic circulation. This implies essentially complete hydrolysis of the hexachlorophene-glucuronide at each passage down the gut. Evidence for the role of the gut flora in this process was not provided but may be inferred.[15] The *O*-glucuronide of oxazepam which is a major metabolite in bile of guinea pigs is readily re-excreted in the bile after intraduodenal instillation.[16] The importance of this in vivo in both guinea pigs[16] and dogs[17] was shown by the presence of lower plasma levels of free drug, a higher plasma clearance and a reduced renal elimination in animals undergoing bile collection compared with those with an intact enterohepatic circulation.

The results which dichlorophen are interesting in that a greater proportion of the dose underwent biliary excretion at the first passage across the gut wall[19] than at subsequent cycles.[20] The principal metabolite in the bile was the monoglucuronide and it is possible that this was formed in the wall of the duodenum following oral dosing. However, the aglycone when released in the hind gut following bacterial hydrolysis, would undergo reduced glucuronidation in the gut wall.[20] Thus, more free drug would reach the liver where it would form sulfate and diglucuronide conjugates which are excreted in the urine.

The hepatic formation and bacterial hydrolysis of acyl glucuronides can also result in an enterohepatic circulation. DDT is slowly excreted primarily in the bile (8% in 3 days) as a glucuronic acid conjugate of DDA (*bis*-(*p*-chlorophenyl)acetic acid).[21] After intravenous administration of the metabolite DDA, 97% was recovered in the bile within 24 hr as the glucuronide while after intraduodenal instillation of this bile into a recipient rat, 67% was re-excreted in the bile. The glucuronide was shown to be a substrate for bacterial β-glucuronidase in vitro, but the only in vivo evidence was a 2 to 3 hr delay in the biliary excretion of DDA by the recipient animals. Direct evidence of the importance of the gut flora was obtained with studies on diphenylacetic acid[12] which is also eliminated in the bile of rats (70% in 24 hr), mostly (87%) as diphenylacetyl-glucuronide. The infusion of this bile into recipient rats resulted in 66% re-excretion in the bile for control animals, but only 20% re-excretion in antibiotic treated rats.[12] This was shown to be due to inhibition of hydrolysis of the conjugate since intraduodenal infusion of the parent compound was followed by rapid and extensive biliary excretion in both control and antibiotic treated rats. The nonsteroidal anti-inflammatory agent carprofen is metabolized in man to the acylglucuronide which is excreted in both urine and bile and is probably subject to some degree of enterohepatic circulation.[22]

Mefanamic acid is excreted in the bile of dogs (about 30% as the acyl glucuronide of parent drug) and in monkeys (about 27% mostly as the unconjugated alcohol).[23] The feces of dogs contained the parent drug, demonstrating hydrolysis of the conjugate while the metabolites in monkey bile were shown to undergo a true enterohepatic circulation.

The bile is an important route of elimination of steroids, and there is evidence from both animal and human studies that the gut microflora play a vital role in their enterohepatic circulation. Hydrolysis of conjugates is of major importance but the liberated aglycones which can be both the parent compound and oxidative metabolites may be metabolized subsequently.[24,25] The cecal contents of conventional rats can hydrolyze both sulfate and glucuronic conjugates of steroids such as dehydroepiandrosterone and testosterone.[26] β-Glucuronidase activity but not sulfatase was found also in the cecal contents of germfree rats. The authors[26] therefore attributed differences between germfree and conventional rats to the fate of sulfate conjugates. For example, the larger amounts of conjugated [$^{14}$C] corticosterone in the feces of germ free rats given an intraperitoneal dose were attributed to the absence of sulfate hydrolysis.[26] However, glucuronic acid conjugates are of considerable importance in the enterohepatic circulation of steroids in the conventional rat. Glucuronides are important biliary metabolites following the intragastric or intracecal administration of pregnenolone, (3β-hydroxy-5-pregnen-20-one), pregnenolone-3-sulfate and deoxycorticosterone-21-sulfate.[27,28] For example following administration of [$^{14}$C] pregnenolone to bile duct cannulated female rats about 70% of the dose was recovered in the bile of which 12% was glucuronides, 24% as monosulfates and 15% as disulfates.[28] In contrast negligible amounts of the glucuronides were found in the feces or urine indicating their subsequent hydrolysis. Based on in vitro studies[26] it was concluded that the hydrolysis was mediated by β-glucuronidase of the intestinal wall, although the gut microflora could also be involved. (See various examples discussed later.) Studies in which five rats were connected in series with the bile from one rat draining into the duodenum of the next showed that 7% of the dose of [$^{14}$C] pregnenolone underwent five enterohepatic circulations. Each subsequent cycle was accompanied by an increased proportion of total activity eliminated in the feces. Thus, for the first rat in the series only 1% of the metabolites in urine and feces were found in the feces, whereas over 95% of the $^{14}$C excreted by the fifth rat was in the feces.[27] This suggests that the processes of hydrolysis further bacterial metabolism[29] reabsorption, and further hepatic metabolism yields a different pattern of biliary metabolites. This altered metabolism during enterohepatic recycling was supported by the more uniform labeling of steroid when [$^{14}$C] pregnenolone was given to intact rats compared with bile duct cannulated animals.[28]

In contrast to these data with pregnenolone in the rat, the analogue pregnanolone (Figure 3) was excreted rapidly in the bile of rats as glucuronides with little or no sulfate conjugates.[30] Following administration of [$^{3}$H] pregnanolone, the $^{3}$H content of bile was 400 times that in plasma. The presence of an enterohepatic circulation was confirmed by the biliary excretion of radioactivity following intragastric administration of the radiolabeled metabolites in bile (mainly pregnanolone-3-glucuronide).[30] Analysis of the radioactivity in the cecal contents

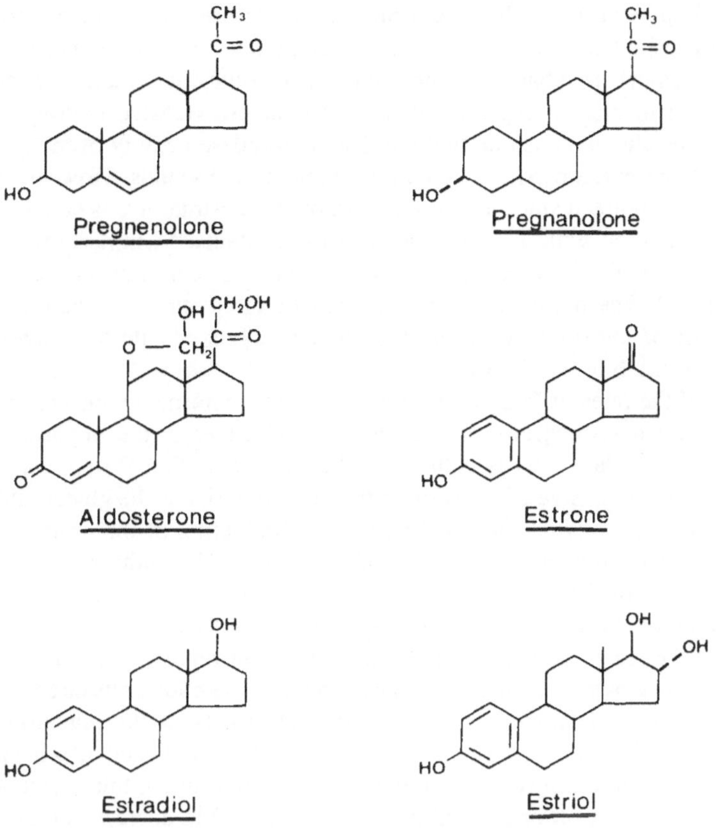


FIGURE 3. Steroids discussed in the text.

and feces 6 hr after the dose showed that less than 5% was present as the polar glucuronide compared with about 95% in the administered bile. The differences reported between pregnenolone and pregnanolone in the extent of glucuronidation compared with sulfation probably arise from the dose administered. The pregnenolone studies[28] employed very low doses (about 70 μg/kg) while pregnanolone was studied[30] at a dose of 40 mg/kg.

The bile is the principal route of elimination of aldosterone in the rat.[31] Following intraduodenal administration of the bile from rats given a tracer dose (0.2 μg/rat) of [$^3$H] aldosterone, twice as much was excreted in the bile of male rats (about 20% of dose) compared with females. This sex difference in the extent of enterohepatic circulation was due to the nature of the biliary metabolites, since infusion of bile from females given [$^3$H] aldosterone into the duodenum of male rats gave a lower re-excretion of radioactivity. The bile of male rats given [$^3$H] aldosterone intravenously consisted largely of glucuronides and monosulfates while that of females contained mostly disulfates, smaller amounts of monosulfates, and negligible glucuronides.[31] Thus, the extent of enterohepatic circulation may be affected by both the dose and sex of animal studied.

The hormone estrone is excreted in the bile of the bull as estrone-3-glucuronide plus the glucuronide of 17-α-estradiol,[32] however evidence of the role of the gut flora in an enterohepatic circulation was not given. The hydroxy analogue, 17-β-estradiol (Figure 3) was excreted in the bile (about 75% of the dose in 6 hr) of female rats after intraduodenal administration of [$^{14}$C] estradiol.[33] Intraduodenal administration of the bile of rats given [$^{14}$C] estradiol resulted in two phases of excretion of $^{14}$C in the bile of recipient rats. The slower initial phase, which proceeded almost linearly from the start of the study, was detected both in control rats and in animals given neomycin. In contrast the more rapid later phase, which started about 3 hr after administration, was suppressed by neomycin administration. A very similar profile was obtained with the synthetic hormone, mestranol (17α-ethynylestradiol-3-methyl ether). Incubation of the biliary metabolites of estradiol or mestranol with rat cecal microorganisms or β-glucuronidase resulted in extensive hydrolysis to the free steroid.[33] These results demonstrate that the β-glucuronidase activity of the gut flora are an essential part of the enterohepatic circulation of these compounds. However the aglycone may be further metabolized since about one third of [$^{14}$C] estradiol was converted to estrone on anaerobic incubation with a culture of rat cecal contents.[34] Similarly the hydroxylated hormone estriol (Figure 3) is excreted mostly as the 3-glucuronide in the bile of guinea pigs[35] and rabbits.[36] The presence of an enterohepatic circulation may be inferred from the recovery of most of the dose in the urine of intact rabbits despite the excretion of about 30% of a dose in the bile within 3 hr.

Comparison of the fates of four contraceptive steroids (chlormadinone acetate, mestranol, norethynodrel, and norethindrone) in rats demonstrated clearly the importance of the gut microflora β-glucuronidase activity in the enterohepatic circulation of steroids.[37] Each steroid underwent rapid and extensive elimination in the bile primarily as the glucuronide conjugate. Enterohepatic circulation was confirmed by the re-elimination of the drug metabolites following intraduodenal infusion of the metabolites in bile. The authors concluded that glucuronides were the most important metabolites for the enterohepatic circulation, since polar metabolites which could not be hydrolyzed by β-glucuronidase were not absorbed from the intestine or re-excreted in bile. Similar results have been reported for ethinylestradiol and norethisterone[38] since over 79% of a dose was excreted in the bile of rats largely as glucuronic acid conjugates. Ethynylestradiol is excreted in bile largely as the glucuronide of the 2-hydroxy metabolite.[39] For both compounds about 15% of radioactive biliary metabolites infused into the duodenum of recipient rats was re-excreted in the bile. Pretreatment of the recipient rats with neomycin or ampicillin caused a three fold decrease in the re-excretion, providing direct evidence for the importance of the gut microflora.

In these studies on the fate of steroids in animals the importance of the gut microflora frequently has to be assumed based on the excretion of glucuronides and sulfates in the bile. In contrast clinical studies on the fate of endogenous steroids have attempted to prove the presence of an enterohepatic circulation by use of antibiotics to interrupt the cycle and thereby have provided direct evidence of the involvement of the microflora. Thus, the administration of ampicillin during pregnancy inhibited the microbial hydrolysis of conjugates and resulted in the elimination of greatly increased amounts of conjugated estrogens in the feces.[40] This was accompanied by a decrease in the concentrations of conjugated estrone, estradiol, and estriol in the plasma[44] and in the urinary excretion of estriol-3-glucuronide[42] and other estrogens.[43]

*2. Enterohepatic Circulation as an Hydroxylated Metabolite and Glucuronide*

In the following examples the drug is eliminated in the bile as an hydroxylated metabolite and the enterohepatic circulations discussed refer to the recycling of metabolites, not of the parent compound. Glucuronic acid conjugation and subsequent hydrolysis by the intestinal flora are important steps in the enterohepatic circulation of the metabolites of a number of therapeutic drugs.

Metabolism of the anticoagulant warfarin in rats was altered considerably by pretreatment with neomycin, bacitracin, and tetracycline orally.[44] The antibiotics caused a decrease in fecal β-glucuronidase activity which was associated with an increased elimination of conjugates in the feces. There was also a decrease in the proportion of the radioactive dose excreted as certain hydroxylated metabolites, indicating that these continue to be formed during each enterohepatic circulation. A detailed comparison of the fate of warfarin in conventional, antibiotic treated, germfree, and ex-germfree rats[45] confirmed that the gut

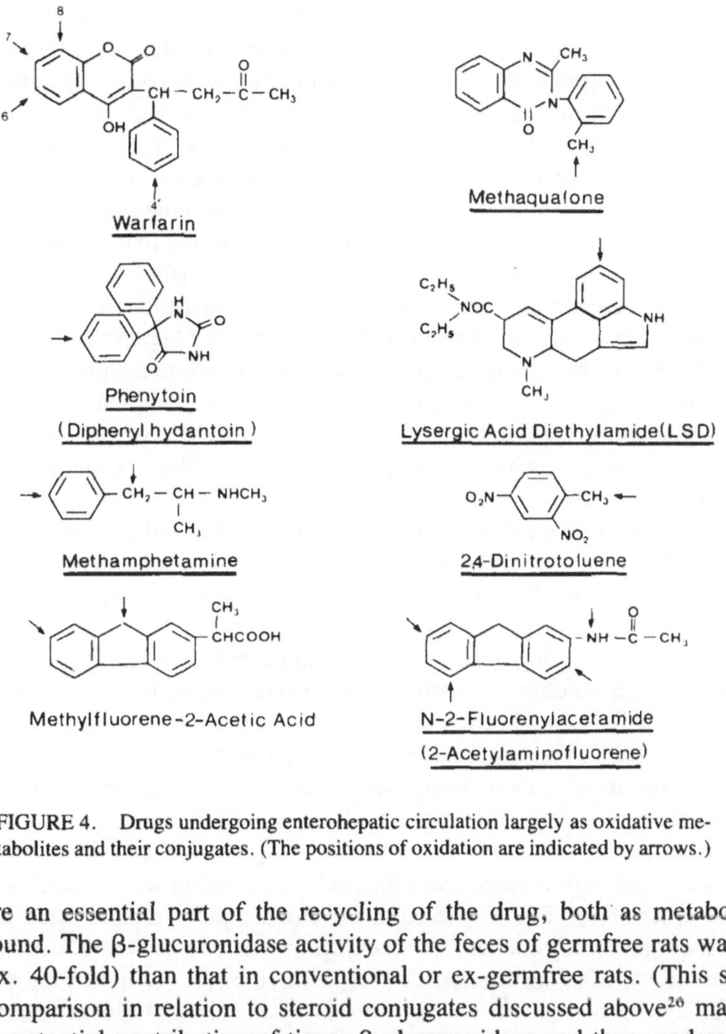


FIGURE 4. Drugs undergoing enterohepatic circulation largely as oxidative metabolites and their conjugates. (The positions of oxidation are indicated by arrows.)

microflora are an essential part of the recycling of the drug, both as metabolites and as parent compound. The β-glucuronidase activity of the feces of germfree rats was very much lower (approx. 40-fold) than that in conventional or ex-germfree rats. (This suggests that the similar comparison in relation to steroid conjugates discussed above[26] may have over estimated the potential contribution of tissue β-glucuronidase and thus, underestimated the importance of glucuronides in the enterohepatic circulation of steroids.) The feces of germfree rats contained both more of the radioactive dose of warfarin and more unhydrolyzed glucuronide conjugates than the ex-germfree animals. When bile from donor rats given [$^{14}$C] warfarin was instilled into the duodenum of recipients, the activity was re-eliminated in the bile after an interval of about 3 hr. The ratio of the concentration of $^{14}$C in the bile of the recipient to that in the donor was low and showed marked inter-individual variation from about 0.05 to 0.4. The composition of donor and recipient bile was similar and consisted of conjugates of 7-hydroxywarfarin (40 to 60%), warfarin (25 to 45%) and smaller amounts of minor metabolites including the pharmacologically active warfarin alcohols.[45] Differences in the extent of enterohepatic circulation of the parent drug may have contributed to the inter-individual variations in the elimination half-life reported in intact exgermfree rats (10 to 20 hr). The half-life in germfree rats (12 hr) was significantly less than that in ex-germfree animals (18 hr) suggesting that the enterohepatic circulation is an important determinant of the elimination half-life.

Methaqualone (Figure 4) is a sedative hypnotic which undergoes extensive biliary excretion in the rat, with 77% of the dose of [$^{14}$C] labeled drug recovered in the bile within 24 hr.[46] The glucuronide of the hydroxymethyl metabolite accounted for 95% of the biliary radioactivity. After infusion of the labeled metabolites in bile into the duodenum of recipient rats, 52% of the activity was re-excreted in the bile. However in intact animals given the biliary metabolites only 21% was excreted in the urine within 24 hr. indicting an extensive recirculation. Although the role of the gut flora was not investigated, hydrolysis of the biliary metabolites would be required for completion of their enterohepatic circulation. A similar pattern was reported with the anti-convulsant phenytoin.[47] Although over 50% of a dose of [$^{14}$C] phenytoin was recovered in the urine of rats within 72 hr, much of this must have originated from an enterohepatic circulation since 69% of a dose could be recovered in the bile of cannulated animals within 8 hr. The presence of an enterohepatic circulation was confirmed by measurement of the rate of excretion of $^{14}$C in bile in cannulated rats in which the bile was allowed to drain into the duodenum and in animals in which all the bile was collected. The rates were similar for the first 3 hr after which the rates of excretion of $^{14}$C in bile was considerably higher in animals with an intact enterohepatic circulation. Similarly the infusion of the bile from a donor rat given [$^{14}$C] phenytoin into the duodenum of a second rat resulted in an excretion of 52% of the infused $^{14}$C by the recipient. The importance of the gut flora β-glucuronidase activity was demonstrated by the suppression of biliary excretion of $^{14}$C by recipients which were pretreated with either saccharo-1,4-lactone (an inhibitor of β-glucuronidase) or neomycin. The majority of the radioactivity in bile was as the *p*-hydroxy metabolite (Figure 4) (about 70%) and unchanged phenytoin accounted for only 2%. Interruption of the enterohepatic circulation by collection of the bile decreased the total $^{14}$C in plasma and liver, but did not affect the distribution or elimination of phenytoin per se.

L-α-Acetylmethadol is a long acting narcotic analgesic which is excreted extensively in the bile of rats as water soluble metabolites.[48] However despite the demonstration of reabsorption and an extensive enterohepatic circulation of these metabolites as glucuronides they did not contribute either to the prolonged analgesic activity[48] or to the maintenance of tissue $^{3}$H concentrations following subcutaneous administration of the radiolabeled drug.[49]

Lysergic acid diethylamide (LSD) is excreted extensively in the bile of rats mainly as the 13-hydroxy-glucuronide.[12] Infusion of the bile of rats given [$^{14}$C] LSD into the duodenum of cannulated recipient rats demonstrated marked re-excretion as the label in bile (28%) compared with urine (7%). The re-excretion increased rapidly after a delay of about 3 hr after instillation of the biliary metabolites (conjugates). However, there was no time lag in the re-excretion of $^{14}$C after instillation of [$^{14}$C] 13-hydroxy-LSD indicating that the delay was due to passage to the cecum and hydrolysis by bacterial β-glucuronidase. Methamphetamine, another drug of abuse, similarly undergoes an extensive enterohepatic circulation involving the glucuronic acid conjugate of an hydroxy metabolite (Figure 4).[50] Oral administration of [$^{3}$H] methamphetamine to rats fitted with a T-tube in the bile duct (to allow bile sampling, while most bile drained into the duodenum) and in intact animals showed the presence of two peaks of radioactivity in the plasma. The early peak, at about 2 hr corresponded to absorption from the gut, while the later peak, at 8 hr was assigned to enterohepatic circulation. The second peak was detected also in the bile of animals fitted with T-tubes, but not those cannulated to allow collection of all the bile. The bile contained mostly the glucuronide conjugates of 4-hydroxymethamphetamine, 4-hydroxyamphetamine, and 4-hydroxynorephedrine, with only traces of methamphetamine and amphetamine. However, the radioactivity in the second peak in plasma consisted of relatively larger amounts of methamphetamine (0.04 μg equiv./mℓ) than total (free and conjugated) 4-hydroxymethamphetamine (0.15 μg equiv./mℓ) compared with the activity excreted in bile. This could have arisen from either a rapid clearance of the metabolites from plasma or the reformation of methamphetamine from the 4-hydroxyglucuronide by the gut microflora. The latter has not been studied but bacterial demethylation of methamphetamine to amphetamine has been reported.[51] In contrast to the above findings, Caldwell et al.[52] reported that the main metabolite in the bile of cannulated rats was 4-hydroxynorephedrine (13% of dose), with little 4-hydroxymethamphetamine or 4-hydroxyamphetamine. The urine of intact rats contained 11% of the dose as 4-hydroxynorephedrine, but only 2% was excreted in the urine of cannulated rats, demonstrating the enterohepatic circulation of this metabolite.

α-Methylfluorene-2-acetic acid is excreted extensively in the bile as the acylglucuronide of the parent compound and as conjugates of the 7-hydroxy metabolite.[53] Enterohepatic circulation was apparent since 83% of the (+) enantiomer was excreted in bile in 10 hr in cannulated rats and 32% was recovered in the urine in intact animals. However, a substantial amount of the dose (56%) was eliminated in the feces indicating poor reabsorption of the biliary metabolites.

The potent carcinogen *N*-2-fluorenylacetamide (acetylamino-fluorene; AAF) is excreted in the urine of rats as conjugates of the 7-, 5-, 3-, and *N*-hydroxy metabolites. In germfree animals less of the dose of radio-labeled AAF was eliminated in the urine compared with conventional rats. The activity in the cecal contents of germfree animals was present as glucuronide and sulfate conjugates of the metabolites, while in conventional rats lower levels of activity were found, largely as the free hydroxy metabolites.[54] These data suggest that enterohepatic recirculation of hydroxy metabolites is important in the overall fate of AAF, and that the gut microflora are essential for the hydrolysis of the glucuronide and sulfate conjugates. The *N*-hydroxy metabolite is of greatest interest since this is the active species. When [$^{14}$C] *N*-hydroxy AAF was administered to conventional and germfree rats,[55] similar amounts of the $^{14}$C dose were recovered in the urine and feces. However, the germfree animals showed a much higher proportion of conjugates to free hydroxy metabolites in the cecal contents. Although free and conjugated *N*-hydroxy AAF was a major labeled component in the cecal contents of germfree rats, almost none was detected in the cecal contents of conventional rats, which contained high levels of AAF instead. This suggests that the gut microflora were responsible for both hydrolysis of the conjugate and the subsequent *N*-dehydroxylation of *N*-hydroxy AAF. This enterohepatic recirculation resulted in a greater proportion of the urinary metabolites present as ring hydroxylation products of AAF in normal rats compared with germfree animals which excreted conjugates of *N*-hydroxy AAF. A strain of *Escherichia coli* was able to reduce *N*-hydroxy AAF.[56]

Dinitrotoluene, which is also an hepatocarcinogen in rats, undergoes similar glucuronic acid conjugation, biliary excretion, and microbial reduction. About 25% of a dose of labeled dinitrotoluene is excreted in the bile within 24 hr, largely as 2,4-dinitrobenzyl alcohol glucuronide.[57] Collection of bile was associated with a decreased urinary elimination of the radioactivity suggesting interference with an enterohepatic circulation. Comparison of the fate of dinitrotoluene in germfree and conventional rats[58] showed that the gut flora were essential to the enterohepatic circulation of this chemical. Germfree rats excreted similar amounts of 2,4-dinitrobenzoic acid and 2,4-dinitrobenzyl glucuronide, but about 10-fold less 4-*N*-acetylamino-2-nitrobenzoic acid and 2-amino-4nitrobenzoic acid. The toxicological importance of this was indicated by the lower covalent binding to hepatic macromolecules, detected in the germfree rats.

The analgesic paracetamol is excreted in the bile of rats (29% of dose) and enterohepatic circulation was demonstrated since 63% of the dose was excreted in the urine of bile duct cannulated rats and 84% in sham-operated animals.[53] Little drug was present unchanged in bile, and the glucuronide, sulfate, and glutathione conjugates (see below) were the major components. Enterohepatic recycling is unlikely to contribute greatly to the duration of action since only 6% of metabolites in bile infused into the duodenum, was re-excreted in the bile.

### B. Gluthathione Conjugation and Enterohepatic Circulation

The conjugation of drugs and foreign chemicals with glutathione increases the molecular weight by 306 and provides a suitable substrate for biliary excretion. Although microbial catabolism of a glutathione conjugate would be unlikely to regenerate the parent compound, the gut flora can produce lipid soluble, reabsorbable metabolites. These metabolites are not

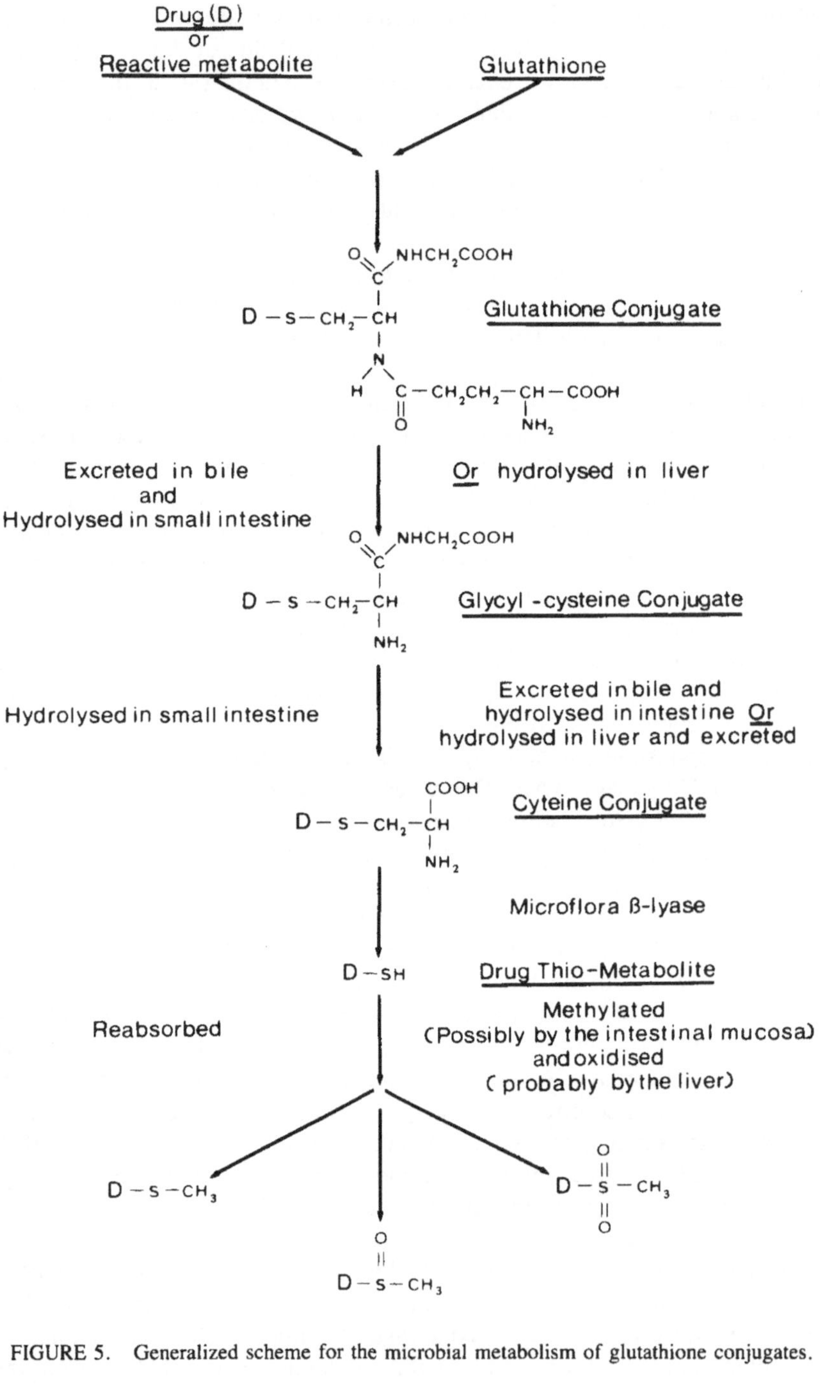


FIGURE 5. Generalized scheme for the microbial metabolism of glutathione conjugates.

recirculated as the glutathione conjugate, but may undergo enterohepatic circulation following further hepatic metabolism. The basic scheme for the enterohepatic circulation is given in Figure 5 (based on Reference 60). The liberated thiol can be reabsorbed and undergoes methylation in the liver. Methylthio metabolites have been reported for a number of drugs (see Stillwell[61] for review) although the role of the gut flora has not been clearly defined in most cases.

The thiomethyl ether of 2-acetamido-4-chloromethylthiazole (Figure 6) and the corresponding sulfoxide and sulfones were recognized as urinary metabolites in the rat[62] many

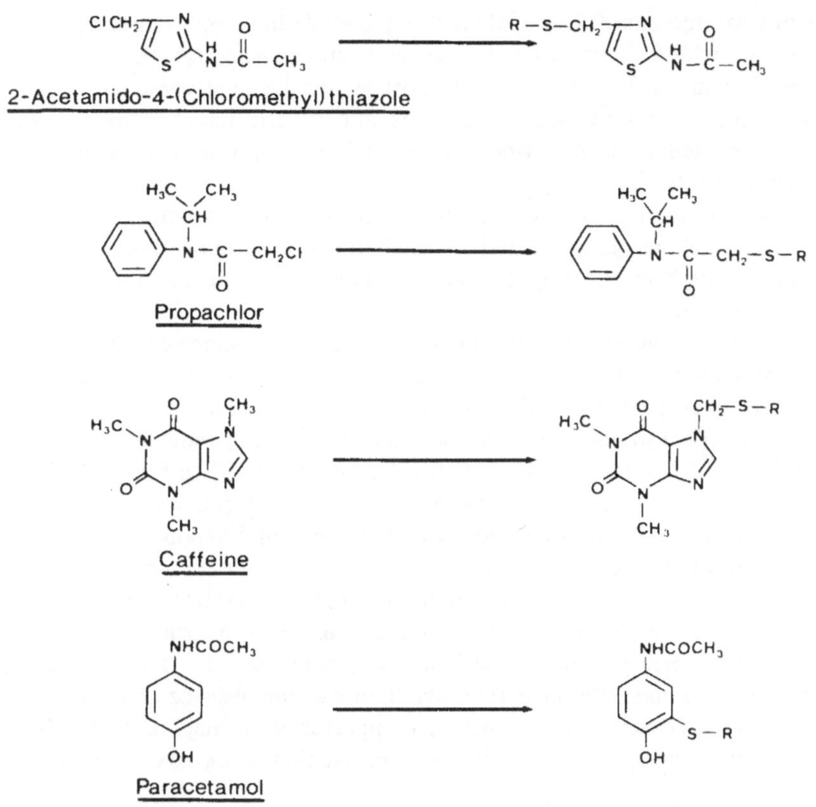


FIGURE 6. Drugs metabolized to thiomethyl analogues by the gut microflora during enterohepatic circulation.

years before the role of the gut flora in their formation was defined. These metabolites were not detected in the urine of germfree rats[63] indicating their bacterial origin. The presence of the *S*-glucuronide in the urine of germfree rats was assigned to the presence of an hepatic β-lyase[64] acting on the glutathione conjugate within the liver. The authors proposed that the enterohepatic circulation of the thio metabolites involved both the mercapturic acid (*N*-acetylcysteine conjugate) and the *S*-glucuronide conjugate. However, the mercapturate but not the *S*-glucuronide was detected in the feces of the germfree rats. This suggests that the *S*-glucuronide does not contribute significantly to the formation of thiomethylethers.

The most extensively documented example is the herbicide propachlor (Figure 6). The agent is excreted in the bile as metabolites of the glutathione pathway (i.e., glutathione, mercapturate, and cysteine conjugates, and the sulfoxide of the cysteine conjugate) which accounted for about 60% of the dose.[65,66] Intravenous infusion of the cysteine conjugate resulted in the biliary excretion of the mercapturate and its sulfoxide, while oral administration resulted in the biliary excretion of glucuronide conjugates of hydroxylated metabolites of the methysulfonyl compound.[65] This route of administration difference indicates that the gut and/or its flora are essential in the formation of the methylsulfonyl analogue. The role of the gut flora was established by the suppression of methylsulfonyl formation by feeding antibiotics.[65] Antibiotic administration also resulted in extractable residues of propachlor in the feces, in contrast to the unextractable residues in control rats. This indicates that the liberated thiol may be the source of the irreversible binding to normal fecal constituents. The methylsulfonyl glucuronide derived from hepatic metabolism of the methylsulfonyl analogue would be subject to its own enterohepatic circulation since 70% of a dose was recovered in the bile.[65] The definitive proof of the involvement of the gut flora was provided by studies in germfree rats.[66] The methylsulfonyl metabolites were not detected in the urine of germfree rats which contained only the mercapturate and its sulfoxide. The feces of germfree rats contained 30% of the dose most as the cysteine conjugate, while those of normal rats contained nonextractable residues only.[66] The metabolism of propachlor in germfree rats reverted to normal when the animals were given enemas of cecal contents from conventional rats.[67]

The gut flora are involved in the formation of methylthio metabolites of caffeine (Figure 6). The excretion of the methylsulfoxide metabolite was 42 times higher in conventional than in germfree rats,[68] implicating the gut flora; however, evidence of a re-excretion in the bile was not provided.

The toxicological potential of the glutathione/methylthio enterohepatic circulation was provided by studies on 2,4,5-trichlorobiphenyl[69] and paracetamol.[70] Administration of unlabeled 2,4,5-trichlorobiphenyl with [$^{35}$S] cysteine or [$^{35}$S] methionine to conventional and germfree mice resulted in about 30-fold more accumulation of radioactivity in the lungs of conventional mice. This activity was identified as methyl [$^{35}$S] sulfonyl-trichlorobiphenyl, and the gut flora were identified as the principal site of formation of this lung seeking metabolite. Comparison of the toxic potential of the parent compound in conventional and germfree animals would provide an analysis of the toxicity of the microbial metabolite. The glutathione conjugate of paracetamol (acetaminophen) is formed from the reactive metabolite produced in the liver. Its excretion in the bile and subsequent microbial metabolism formed 3-methylthio- and 3-methylsulfoxide derivatives of paracetamol.[70] Evidence of the importance of the gut flora and enterohepatic circulation was the absence of these metabolites in animals with bile duct cannulae and their late appearance in intact animals. The excretion of the 3-thio metabolites gave a good index of production of the toxic metabolite.

### C. Enterohepatic Circulation of Sulfoxides/Sulfides

The antiplatelet activity of sulfinpyrazone, which contains a sulfoxide moiety, is due largely to its sulfide metabolite. This is formed exclusively by the gut flora in the rat since no sulfide was formed in germfree animals.[71] Sulfinpyrazone is excreted in the bile of rats and rabbits (about 20% of dose; [72]) and in guinea pigs (about 58% of dose;[73]). The formation of the more lipid soluble sulfide by the gut flora (which is absorbed and re-oxidized to the sulfoxide in both rats and rabbits) should result in an enterohepatic circulation. In rat[71] and rabbit[72] considerably more sulfide is formed after oral compared with intravenous administration. This indicates that much of the substrate for microbial reduction is derived from the unabsorbed dose. However, in man equal amounts of sulfide were formed after oral and intravenous dosing,[74] suggesting either the involvement of biliary excretion, or that the sulfoxide is reduced in the tissues. Patients who had undergone surgical removal of the lower bowel absorbed sulfinpyrazone normally, but the drug was not reduced to the sulfide either in vivo or in vitro by the ileostomy effluent.[74] These data suggest that the active sulfide metabolite of sulfinpyrazone is probably formed in man from sulfinpyrazone excreted in the bile and reduced by the gut flora. Sulfoxide reducing activity has been demonstrated in a wide variety of microorganisms.[75]

Evidence of the role of the gut flora in the in vivo reduction of other sulfoxide containing drugs is less clear. Sulindac is reduced to an active sulfide metabolite in animals[76] and man.[77] The parent drug is eliminated extensively in the bile of animals[76] and man.[78,79] Since human feces possessed considerable ability to convert sulindac to its sulfide[77] it is likely that an enterohepatic circulation of the sulfoxide and sulfide (formed by the gut flora) exists. However, unlike sulfinpyrazone which undergoes negligible reduction by the tissues,[71,72] sulindac can also be reduced by the liver of rat[80] and guinea pig.[81] Thus, an enterohepatic circulation with bacterial reduction could contribute to the total sulfide production but would be unlikely to be the sole source. This could be clarified by suitable studies in germfree animals. The muscle relaxant tolmesoxide is excreted extensively in the bile of rats as sulfone derivatives.[82] Intraduodenal infusion of the metabolites in bile resulted in extensive re-excretion (50 to 55%) demonstrating the presence of a true enterohepatic circulation. However, it appears unlikely that microbial alteration to the redox state of the sulfur is involved in the recirculation of the drug, since the sulfoxide or sulfide were not detected in animals dosed with the sulfone. Sulfinalol is another sulfoxide which is converted to both sulfide and sulfone metabolites in vivo.[83]

## III. METHODOLOGICAL CONSIDERATIONS

### A. Evidence for an Enterohepatic Circulation

If the total amount of drug or radioactivity excreted in the bile and urine exceeds 100% of the dose then biliary material must be able to re-enter the systemic circulation, to be eliminated in the urine. A number of methods of cannulation of the bile duct to allow collection of bile in anesthetized or conscious animals have been described.[84-87]

Enterohepatic circulation can be confirmed by the infusion of bile containing radioactive metabolites collected from one animal into the duodenum of a second animal. The elimination of activity in the bile of the recipient animal provides direct and convincing evidence of an enterohepatic circulation.

Enterohepatic circulation can result in prolonged plasma levels of a drug and thus, interruption of the circulation should result in an increased elimination rate and a decreased pharmacologic response. Collection of bile provides one method of interruption and examples have been cited in this chapter. An alternative approach is to give an oral dose of a compound which interferes with the cycle. Cholestyramine is a nonabsorbed resin and its administration increased the clearance and decreased the anticoagulant effect of phenprocoumon,[88] thus providing evidence for both the presence and importance of the enterohepatic circulation. Cholestyramine also protects against lethal digitoxin intoxication by binding the drug within the gut.[89] Similarly, D-glucaro-1,4-lactone, an inhibitor of β-glucuronidase, decreased the pharmacological response of rats to phenobarbitone and progesterone.[90]

Circumstantial evidence of an enterohepatic circulation is provided by the presence of a secondary peak in the plasma concentration time curve 3 hr or more after drug administration. This is the time-lag associated with the elimination in the bile, passage down the gut, bacterial metabolism, and reabsorption. Which of these steps is rate limiting is unclear. Studies on the fate of diethylstilbestrol in young rats[7] suggested that the rate limiting step may be reabsorption across the wall of the distal gut and not the hydrolysis of its glucuronide conjugate which is extremely rapid. This contrasts with the conclusion of Parker et al.[12] that bacterial β-glucuronidase could be the rate limiting step in the hydrolysis of phenolphthalein conjugates.

The mathematical models used to describe the enterohepatic circulation of foreign compounds usually incorporate some time-lag component into the equations.[91,92]

### B. Evidence for the Importance of the Gut Flora

Comparison of the fate of the compound in normal and germfree animals provides the definitive proof of bacterial involvement in an enterohepatic circulation. However, due to possible differences in organ blood flow and tissue distribution, subtle differences cannot be assigned unequivocally to the gut flora. Frequently published studies have compared germfree and conventional rats of different strains and again slight differences should be viewed with caution. Recontamination of germfree rats with a conventional gut flora is a good way of avoiding this problem, but the time taken to re-establish normality can vary widely from a few days for nitro-reduction[94] to many weeks.[67]

The other commonly used method is to compare the fate of the drug in animals given high oral doses of antibiotics. However, the use of antibiotics may produce spurious results due to either decreased absorption from the gut that is not related to inhibition of bacterial metabolism,[93] or decreased apparent volume of distribution[45] which would result in higher plasma concentrations and suggest a higher area under curve (AUC) and greater amounts of compound in the body than were really present. An example of this was shown by studies in the fate of sulfinpyrazone in the rabbit.[72] Oral antibiotics caused a 30-fold decrease in the amounts of the *p*-hydroxysulfide metabolite excreted. This suggested that the gut flora were the main site of reduction to the sulfide as had been found in the rat.[71] However, the area under the plasma concentration time curve for the sulfide decreased only fourfold suggesting a more limited bacterial involvement. In comparison the AUC for the parent compound increased fivefold indicating that the plasma measurements were providing an overestimate of true body exposure of both sulfide and parent compound.

## IV. CONCLUSIONS

The gut flora are an essential part of the enterohepatic circulation of many pharmacologically and toxicologicaly active compounds. In most cases the gut flora convert highly polar biliary metabolites into more lipid soluble forms capable of reabsorption. However, some metabolic reactions performed by the gut flora, such as ester hydrolysis, may produce more polar metabolites and thereby increase elimination in the feces. An enterohepatic circulation can prolong the duration of drug action and thus may contribute to the effects produced, especially during chronic administration. Interspecies and interindividual differences in an enterohepatic circulation may arise from a differences in the metabolism of the compound in the liver, the molecular weight threshold for biliary excretion and the nature and metabolic capacity of the gut flora.

## REFERENCES


1. **Smith, R. L.,** *The Excretary Function of Bile: The Elimination of Drugs and Toxic Substances in Bile,* Chapman and Hall, London, 1973.
2. **Scheline, R. R.,** Metabolism of foreign compounds by gastrointestinal microorganisms, *Pharmacol. Rev.,* 25, 451, 1973.
3. **Renwick, A. G.,** Microbial metabolism of drugs, in *Drug Metabolism — From Microbe to Man,* Parke, D. V. and Smith, R. L., Eds. Taylor and Francis, London, 1977, 169.
4. **Boxenbaum, H. G., Bekersky, I., Jack, M. L., and Kaplan, S. A.,** Influence of gut microflora on bioavailability, *Drug. Metab. Rev.,* 9, 279, 1979.
5. **Renwick, A. G.,** First pass metabolism within the lumen of the gastro intestinal tract, in *Presystemic Drug Elimination,* George, C. F., Shand, D. G., and Renwick, A. G., Eds., Butterworths, London, 1982, 28.
6. **Hirom, P. C., Millburn, P., Smith, R. L., and Williams, R. T.,** Molecular weight and chemical structure as factors in the biliary excretion of sulphonamides in the rat, *Xenobiotica,* 2, 205, 1972.
7. **Fischer, L. J., Kent, T. H., and Weissinger, J. L.,** Absorption of diethylstilboestrol and its glucuronide conjugate from the intestine of five- and twenty-five-day-old rats, *J. Pharmacol. Exp. Ther.,* 185, 163, 1973.
8. **Rod, T. O. and Midtvedt, T.,** Origin of intestinal β-glucuronidase in germ free, monocontaminated and conventional rats, *Acta Pathol. Microbiol. Scand. Sect. B.,* 85, 271, 1977.
9. **Hawksworth, G., Drasar, B. S., and Hill, M. J.,** Intestinal bacteria and the hydrolysis of glycosidic bonds, *J. Med. Microbiol.,* 4, 451, 1971.
10. **Kent, T. H., Fischer, L. J., and Marr, R.,** Glucuronidase activity in intestinal contents of rat and man and relationship to bacterial flora, *Proc. Soc. Exp. Biol. Med.,* 140, 590, 1972.
11. **Walsh, C. T. and Levine, R. R.,** Studies on the enterohepatic circulation of morphine in the rat, *J. Pharmacol. Exp. Ther.,* 195, 303, 1975.
12. **Parker, R. J., Hirom, P. C., and Millburn, P.,** Enterohepatic recycling of phenolphthalein, morphine, lysergic acid diethylamide (LSD) and diphenylacetic acid in the rat. Hydrolysis of glucuronic acid conjugates in the gut lumen, *Xenobiotica,* 10, 689, 1980.

13. **Dahlstrom, B. E. and Paalzow, L. K.**, Pharmacokinetic interpretation of the enterohepatic recirculation and first-pass elimination of morphine in the rat, *J. Pharmacokinet. Biopharm.*, 6, 505, 1978.
14. **Brewster, D., Humphrey, M. J., and McLeavy, M. A.**, Biliary excretion, metabolism and enterohepatic circulation of buprenorphine, *Xenobiotica*, 11, 189, 1981.
15. **Gandolfi, A. J. and Buhler, D. R.**, Biliary metabolites and enterohepatic circulation of hexachlorophene in the rat, *Xenobiotica*, 4, 693, 1974.
16. **Bertagni, P., Bianchi, R., Marcucci, F., Mussini, E., and Garattini, S.**, The enterohepatic circulation of oxazepam-O-glucuronide in guinea-pigs, *J. Pharm. Pharmacol.*, 30, 185, 1978.
17. **Alvan, G., Jonsson, M., Sundwall, A., and Vessman, J.**, First pass conjugation and enterohepatic recycling of oxazepam in dogs: intravenous tolerance of oxazepam in propylene glycol, *Acta Pharmacol. Toxicol. Suppl.*, 40, 16, 1977.
18. **Illing, H. P. A. and House, E. S. A.**, Enterohepatic circulation in the rat and dog of $^{14}$C-O-[3-(4-<2-methoxyphenyl>-1-piperazinyl)-2-hydroxypropyl]-3 methoxybenzaldoxim dihydrochloride and it's demethylated metabolite, *Eur. J. Drug Metab. Pharmacokinet.*, 6, 303, 1981.
19. **Dixon, P. A. F. and Caldwell, J.**, The fate of dichlorophen in the rat, *Eur. J. Drug Metab. Pharmacokinet.*, 2, 95, 1978.
20. **Dixon, P. A. F.**, Sulphation and diglucuronidation as constraints to enterohepatic circulation of dichlorophen in rats, *J. Pharm. Pharmacol.*, 34, 596, 1982.
21. **Gingell, R.**, Enterohepatic circulation of bis(p-chlorophenyl) acetic acid in the rat, *Drug Metab. Dispos.* 3, 42, 1975.
22. **Rubio, F., Seawall, R., Pocelinko, R., DeBarbieri, B., Benz, W., Berger, L., Morgan, L., Pao, J., Williams, T. H., and Koechlin, B.**, Metabolism of carprofen, a nonsteroidal anti-inflammatory agent, in rats, dogs, and humans. *J. Pharm. Sci.*, 69, 1245, 1980.
23. **Glazko, A. J.**, Metabolic disposition, *Ann. Phys. Med.*, 8, 23, 1966.
24. **Eriksson, H. E. and Gustafsson, J.**, Excretion of steroids in adults. Steroids in faeces from adults, *Eur. J. Biochem.*, 18, 146, 1971.
25. **Winter J., O'Rourke, S., Bokkenheuser, V. D., Hylemon, P. B., and Glass, T. L.**, 16-α-Dehydroxylation of corticosteroids by bacteria isolated from rat feacal flora, *J. Steroid Biochem.*, 16, 231, 1982.
26. **Eriksson, H. and Gustafsson, J.**, Steroids in germ free and conventional rats. Sulpho- and glucuronohydrolase activities of caecal contents from conventional rats, *Eur. J. Biochem.*, 13, 198, 1970.
27. **Eriksson, H.**, Absorption and enterohepatic circulation of neutral steroids in the rat, *Eur. J. Biochem.*, 19, 416, 1971.
28. **Cronholm, T., Eriksson, H., and Gustafsson, J.**, Excretion of endogenous steroids and metabolites of [4-$^{14}$C] pregnenolone in bile of female rats, *Eur. J. Biochem.*, 19, 424, 1971.
29. **Eriksson, H., Gustafsson, J., and Sjovall, J.**, Studies in the structure, biosynthesis and bacterial metabolism of 15-hydroxylated steroids in the female rat, *Eur. J. Biochem.*, 19, 433, 1971.
30. **Long, R. J. and Soyka, L. F.**, Biliary excretion and enterohepatic circulation of pregnanolone in the rat, *Biochem. Pharmacol.*, 24, 1067, 1975.
31. **Morris, D. J., Hantoot, M. S., and DeConti, G. A.**, The enterohepatic circulation of aldosterone metabolites and its sex dependence in rats, *Endocrinology*, 101, 1776, 1977.
32. **Leung, B. S., Pearson, J. R., and Martin, R. P.**, Enterohepatic cycling of $^{3}$H-estrone in the bull. Identification of estrone-3-glucuronide, *J. Steroid Biochem.*, 6, 1477, 1975.
33. **Brewster, D., Jones, R. S., and Symons, A. M.**, Effects of neomycin on the biliary excretion and enterohepatic circulation of mestranol and 17β-oestradiol, *Biochem. Pharmacol.*, 26, 943, 1977.
34. **Amland, M. D. and Stoa, K. F.**, Metabolism of oestradiol-17β by intestinal bacteria in rats, *Horm. Res.*, 6, 366, 1975.
35. **Kirdani, R. Y. and Sandberg, A. A.**, The metabolism of estriol in the guinea pig, *J. Steroid Biochem.*, 7, 439, 1976.
36. **Kirdani, R. Y., Musey, P., Verkarakis, M. J., and Sandberg, A. A.**, The metabolism of estrol in rabbits, *Steroids*, 22, 1, 1973.
37. **Hanasono, G. K. and Fischer, L. J.**, The excretion of tritium — labelled chlormadinone acetate, mestranol, norethindrone and norethyndrel in rats and the enterohepatic circulation of metabolites, *Drug Metab. Dispos.*, 2, 159, 1974.
38. **Back, D. J., Breckinridge, A. M., Challiner, M., Crawford, F. F., Orme, M. L., Rowe, P. H., and Smith, E.**, The effect of antibiotics on the enterohepatic circulation of ethinylestradiol and norethisterone in the rat, *J. Steroid Biochem.*, 9, 527, 1978.
39. **Maggs, J. L., Grabowski, P. S., and Park, B. K.**, The enterohepatic circulation of the metabolites of 17-α-ethynyl [$^{3}$H]estradiol in the rat, *Xenobiotica*, 13, 619, 1983.
40. **Adlercreutz, H., Margin, F., Pulkkinen, M., Dencker, H., Rimer, U., Sjoberg, N., and Tikkanen, M. J.**, Intestinal metabolism of estrogens, *J. Clin. Endocrinol. Metab.*, 43, 479, 1976.

41. **Aldercreutz, H., Martin, F., Lehtinen, T., Tikkanen, M. J., and Pulkkinen, M. O.,** Effect of ampicillin administration on plasma conjugated and unconjugated oestrogen and progesterone levels in pregnancy, *Am. J. Obstet. Gynecol.*, 128, 266, 1977.
42. **Tikkanen, M. J., Pulkkinen, M. O., and Adlercreutz, H.,** Effect of ampicillin treatment on the urinary excretion of estriol conjugates in pregnancy, *J. Steroid Biochem.*, 4, 439, 1973.
43. **Adlercreutz, H. F., Martin, M. J. T., and Pulkkinen, M.,** Effect of ampicillin administration on the excretion of 12 oestrogens in pregnancy urine, *Acta Endocrinol.*, 80, 551, 1975.
44. **Remmel, R. P. and Elmer, G. W.,** The effect of broad-spectrum antibiotics on warfarin excretion and metabolism in the rat, *Res. Commun. Chem. Pathol. Pharmacol.*, 34, 503, 1981.
45. **Remmel, R. P., Pohl, L. R., and Elmer, G. W.,** Influence of the intestinal microflora on the elimination of warfarin in the rat. *Drug Metab. Disp.* 9, 410, 1981.
46. **Polk, A., Delong, A. F., Smyth, R. D., Herczeg, T., Burns, B. J., Klurk, L. J., and Reavey-Contwell, N. H.,** Biliary excretion and enterohepatic recirculation of methaqualone in the rat, *Drug Metab. Disp.*, 2, 365, 1974.
47. **El-Hawari, A. M. and Plaa, G. L.,** Role of the enterohepatic circulation in the elimination of phenytoin in the rat, *Drug Metab. Disp.*, 6, 59, 1978.
48. **Roerig, D. L., Hasegawa, A. T., and Wang, R. I. H.,** Role of enterohepatic circulation in the analgesic action of *l-α-acetylmethadol* in the rat, *Drug Metab. Disp.*, 7, 306, 1979.
49. **Roerig, D. L., Hasegawa, A. T., and Wang, R. I. H.,** Enterohepatic circulation of 1-α-acetylmethadol in the rat, *J. Pharmacol. Exp. Ther.*, 213, 284, 1980.
50. **Sakai, T., Niwaguchi, T., and Murata, T.,** Distribution and excretion of methamphetamine and its metabolites in rats. I. Time-course of concentrations in blood and bile after oral administration. *Xenobiotica*, 12, 233, 1982.
51. **Caldwell, J. and Hawksworth, G. M.,** The demethylation of methamphetamine by intestinal microflora, *J. Pharm. Pharmacol.*, 25, 422, 1973.
52. **Caldwell, J., Dring, L. G., and Williams, R. T.,** Biliary excretion of amphetamine and methamphetamine in the rat, *Biochem. J.*, 129, 25, 1972.
53. **Dean, A. V., Lan, S. J., Kripalani, K. J., Difazio, L. T., and Schreiber, E. C.,** Metabolism of the (±), (+), and (−)− enantiomers of α-methylfluorene-2-acetic acid (cicloprofen) in rats, *Xenobiotica*, 7, 459, 1977.
54. **Grantham, P. H., Horton, R. E., Weisburger, E. K., and Weisburger, J. H.,** Metabolism of the carcinogen N-2-fluorenylacetamide in germ-free and conventional rats, *Biochem. Pharmacol.*, 19, 163, 1970.
55. **Weisburger, J. H., Grantham, P. H., Horton, R. E., and Weisburger E. K.,** Metabolism of the carcinogen N-hydroxy-N-2-fluorenylacetamide in germ free rats, *Biochem. Pharmacol.*, 19, 151, 1970.
56. **Williams, J. R., Grantham, P. H., Marsh, H. H., Weisburger, J. H., and Weisburger, E. H.,** Participation of liver fractions and of intestinal bacteria in the metabolism of N-hydroxy-N-2-fluorenylacetamide in the rat, *Biochem. Pharmacol.*, 19, 173, 1970.
57. **Medinsky, M. A. and Dent, J. G.,** Biliary excretion and enterophepatic circulation of 2,4-dinitrotoluene metabolites in Fischer-344 rats, *Toxicol. Appl. Pharmacol.*, 68, 359, 1983.
58. **Rickert, D. E., Long, R. M., Krakowka, S., and Dent, J. G.,** Metabolism and excretion of 2,4-[$^{14}$C] dinitrotoluene in conventional and axenic Fischer-344 rats, *Toxicol. Appl. Pharmacol.*, 59, 574, 1981.
59. **Siegers, C. P., Rozman, K., and Klaassen, C. D.,** Biliary excretion and enterohepatic circulation of paracetamol in the rat, *Xenobiotica*, 13, 591, 1983.
60. **Larsen, G. L., Larson, J. P., and Gustafsson, J. A.,** Cysteine conjugate β-lyase in the gastrointestinal bacterium *Fusobacterium necrophorum*, *Xenobiotica*, 13, 689, 1983.
61. **Stillwell, W. G.,** Methylthiolation: a new pathway of drug-metabolism, *Top. Pharm. Sci.*, 2, 250, 1981.
62. **Chatfield, D. H. and Hunter, W. H.,** The metabolism of aceta-midothiazoles in the rat. 2-Acetamido-4-chloromethylthiazole, *Biochem. J.*, 134, 879, 1973.
63. **Bakke, J. E., Rafter, J. J., Lindeskog, P., Feil, V. J., Gustafsson, J. A., and Gustafsson, B. E.,** Metabolism of 2-acetamido-4-(chloromethyl)-thiazole in germ free and conventional rats, *Biochem. Pharmacol.*, 30, 1839, 1981.
64. **Tateishi, M., Suzuki, S., and Shimizu, H.,** Cysteine conjugate β-lyase in rat liver. A novel enzyme catalysing formation of thiol-containing metabolites of drugs, *J. Biol. Chem.*, 253, 8854, 1978.
65. **Larsen, G. L. and Bakke, J. E.,** Enterohepatic circulation in formation of propachlor (2-chloro-N-isopropylacetanilide) metabolites in the rat, *Xenobiotica*, 11, 473, 1981.
66. **Bakke, J. E., Gustafsson, J. A., and Gustafsson, B. E.,** Metabolism of propachlor by the germ free rat, *Science*, 210, 433, 1980.
67. **Rafter, J. J., Gustafsson, J. A., Bakke, J. E., Larsen, G. L., Norin, K. E., and Gustafsson, B. E.,** Studies on the re-establishment of the intestinal microflora in germ-free rats with special reference to the metabolism of N-isopropyl-α-chloroacetanilide (propachlor), *Xenobiotica*, 13, 171, 1983.

68. **Rafter, J. J. and Nilsson, L.**, Involvement of the intestinal microflora in the formation of sulphur-containing metabolites of caffeine, *Xenobiotica,* 11, 771, 1981.
69. **Bergman, A., Beissmann, A., Brandt, I., and Rafter, J.**, Metabolism of 2,4′,5-trichlorobiphenyl: role of the intestinal microflora in the formation of bronchial seeking methyl sulphone metabolites in mice, *Chem. Biol. Interact.*, 40, 123, 1982.
70. **Gemborys, M. W. and Mudge, G. H.**, Formation and disposition of the minor metabolites of acetaminophen in the hamster, *Drug Metab. Disp.*, 9, 340, 1981.
71. **Renwick, A. G., Evans, S. P., Sweatman, T. W., Cumberland, J., and George, C. F.**, The role of the gut flora in the reduction of sulphinpyrazone in the rat, *Biochem. Pharmacol.*, 31, 2649, 1982.
72. **Strong, H. A., Renwick, A. G., and George, C. F.**, The site of reduction of sulphinpyrazone in the rabbit, *Xenobiotica,* 14, 815, 1984.
73. **Dieterle, W. and Faigle, J. W.**, Species differences in the disposition and metabolism of sulphinpyrazone, *Xenobiotica,* 11, 559, 1981.
74. **Strong, H. A., Oates, J., Sembi, J., Renwick, A. G., and George, C. F.**, Role of the gut flora in the reduction of sulphinpyrazone in human, *J. Pharmacol. Exp. Ther.*, 230, 726, 1984.
75. **Zinder, S. H. and Brock, T. D.**, Dimethyl sulphoxide reduction by micro-organisms, *J. Gen. Microbiol.*, 105, 335, 1978.
76. **Duggan, D. E., Hooke, K. F., Noll, R. M., Hucker, H. B., and Van Arman, C. G.**, Comparative disposition of sulindac and metabolites in five species, *Biochem. Pharmacol.*, 27, 2311, 1978.
77. **Duggan, D. E., Hare, L. E., Ditzler, C. A., Lei, B. W., and Kwan, K. C.**, The disposition of sulindac, *Clin. Pharmacol. Ther.*, 21, 326, 1977.
78. **Dobrinska, M. R., Furst, D. E., Spiegel, T., Vincek, W. C., Tompkins, R., Duggan, D. E., Davies, R. O., and Paulus, H. E.**, Bilary secretion of sulindac and metabolites in man, *Biopharm. Drug Disp.*, 4, 347, 1983.
79. **Dujovne, C. A., Pitterman, A., Vincek, W. C., Dobrinska, M. R., Davies, R. O., and Duggan, D. E.**, Enterohepatic circulation of sulindac and metabolites, *Clin. Pharmacol. Ther.*, 33, 172, 1983.
80. **Ratnayake, J. H., Hanna, P. E., Anders, M. W., and Duggan, D. E.**, Sulfoxide reduction. In vitro reduction of sulindac by rat hepatic cytosolic enzymes, *Drug Metab. Disp.*, 9, 85, 1981.
81. **Kitamura, S., Tatsumi, K., and Yoshimura, H.**, Metabolism *in vitro* of sulindac. Sulfoxide-reducing enzyme systems in guinea pig liver, *J. Pharm. Dyn.*, 3, 290, 1980.
82. **Greenslade, D., Havler, M. E., Humphrey, M J., Jordan, B. J., Lewis, C. J., and Rance, M. J.**, Biotransformation of tolmesoxide in animals and man, *Xenobiotica,* 11, 89, 1981.
83. **Benziger, D. P., Fritz, A., and Edelson, J.**, Metabolism and disposition of sulfinalol in laboratory animals, *Drug Metab. Disp.*, 9, 493, 1981.
84. **Waynforth, H. B.**, *Experimental and Surgical Techniques in the Rat,* Academic Press, New York, 1980.
85. **Lipsky, M. H. and Berkley, S.**, Prolonged biliary fistulization in the rat without interruption of the enterohepatic cycle, *J. Surg. Res.*, 22, 65, 1977.
86. **Johnson, P. and Rising, P. A.**, Technique for assessment of biliary excretion and enterohepatic circulation in the rat, *Xenobiotica,* 8, 27, 1978.
87. **Caldwell, J. H., Thornford, N. R., and Kakumoto, Y. L.**, Assessment of enterohepatic circulation of $^{3}$H-digoxin with a minimal interruption technique, *J. Pharm. Sci.*, 67, 415, 1978.
88. **Meinertz, T., Gilfrich, H. J., Groth, U., Jonen, H. G., and Jahnchen, E.**, Interruption of the enterohepatic circulation of phenprocoumon by cholestyramine, *Clin. Pharmacol. Ther.*, 21, 731, 1977.
89. **Caldwell, J. H. and Greenberger, N. J.**, Interruption of the enterohepatic circulation of digitoxin by cholestyramine. I. Protection against lethal intoxication, *J. Clin. Invest.*, 50, 2626, 1971.
90. **Marselos, M., Dutton, G., and Hanninen, O.**, Evidence that D-glucaro-1,4-lactone shortens the pharmacological action of drugs being disposed via the bile as glucuronides, *Biochem. Pharmacol.*, 24, 1855, 1975.
91. **Colburn, W. A., Hirom, P. C., Parker, R. J., and Millburn, P.**, A pharmacokinetic model for enterohepatic recirculation in the rat: phenolphthalein, a model drug, *Drug Metab. Disp.*, 7, 100, 1979.
92. **Steimer, J. L., Plusquellec, Y., Guillaume, A., and Boisvieux, J. F.**, A time lag model for pharmacokinetics of drugs subject to enterohepatic circulation, *J. Pharm. Sci.*, 71, 297, 1982.
93. **Gardner, D. M. and Renwick, A. G.**, The reduction of nitrobenzoic acids in the rat, *Xenobiotica,* 8, 679, 1978.
94. **Renwick, A. G.**, unpublished data.

Chapter 11

# THE METABOLISM OF ORAL BACTERIA IN HEALTH AND DISEASE

**P. D. Marsh and C. W. Keevil**

## TABLE OF CONTENTS

## I. INTRODUCTION

Impetus for much of the current research on oral microorganisms has stemmed from the finding of a relationship between the metabolism of the bacteria associated with the teeth (dental plaque) and two of the most prevalent diseases affecting man: caries and periodontal disease. Dental caries results from the dissolution of enamel (demineralization) by acids produced from the bacterial degradation of dietary carbohydrates. Periodontal disease is a collective term describing several conditions that affect the supporting tissues of the teeth. Except in late stages of periodontal disease, invasion of host tissues is rare. Tissue destruction results from the production of enzymes and toxic end products of metabolism by bacteria, and is also attributable to a destructive host inflammatory response to these products and to antigenic components on the surface of these organisms.

In the following sections, aspects of the metabolism of oral bacteria considered to be important in both health and disease will be discussed. The literature on these subjects is often large and many issues remain contentious. Consequently, in a review of this length, some topics have been omitted while others have been simplified for clarity. As it is now well-documented that the metabolism of bacteria is influenced by and will vary with the environment, the types of bacteria and the environment in which they grow will be described first.

## II. THE MOUTH AS A MICROBIAL HABITAT

The mouth does not provide a single uniform environment but consists of several distinct habitats each of which, because of their different biological and physical properties, will be colonized by a characteristic microbial community. Mucosal surfaces include those with either keratinized (palate) or unkeratinized cells (cheek, gum) while the tongue has a highly papillated surface and so provides a large surface area for microbial colonization. Relatively few bacteria are found on epithelial surfaces due to desquamation and to the fact that only a limited number of species of bacteria can attach to receptors on these mucosal cells.[1] In contrast, large masses of bacteria accumulate on the protected surfaces of teeth. Teeth are unique surfaces for microbial colonization because they are nonshedding. The film of microorganisms together with the polymers of salivary and bacterial origin found on the tooth surface is termed dental plaque. Dental plaque is present in the healthy mouth but is also implicated in the etiology of caries and periodontal disease, and it has been calculated that plaque contains $10^{11}$ cells/g wet weight. Teeth provide several distinct habitats each of which is suitable for colonization by only certain types of bacteria. The areas between adjacent teeth (approximal) and in the gingival crevice afford protection from the removal forces (saliva flow, mastication, etc.) that operate in the mouth (Figure 1). Both sites are anaerobic and the gingival crevice is also bathed in the nutritionally-rich crevicular fluid. Crevicular fluid resembles serum in composition and is important in the defense of the oral cavity by providing components for both the humoral and cell mediated immune response including IgG, IgM, IgA, complement, leukocytes (particularly polymorphonuclear leukocytes), monocytes, and B and T lymphocytes.[2] Crevicular fluid also contains vitamins and polypeptides essential to the growth of several fastidious anaerobes such as hemin for *Bacteroides* spp. and $\alpha$-2-globulin for *Treponema denticola*. In disease, the gingival crevice enlarges to form a periodontal pocket, due to the migration of the junctional epithelium down the root of the tooth and the flow of crevicular fluid is increased. The combination of an anaerobic site and the provision of growth factors results in the microbial community from these sites being the most complex. Pits and fissures of the biting (occlusal) surfaces of teeth also offer protection from the environment but support a microbial community with a more limited composition. The environmentally protected regions of the teeth, therefore, are associated with the largest and most diverse microbial communities and generally, the most disease.[3,4]

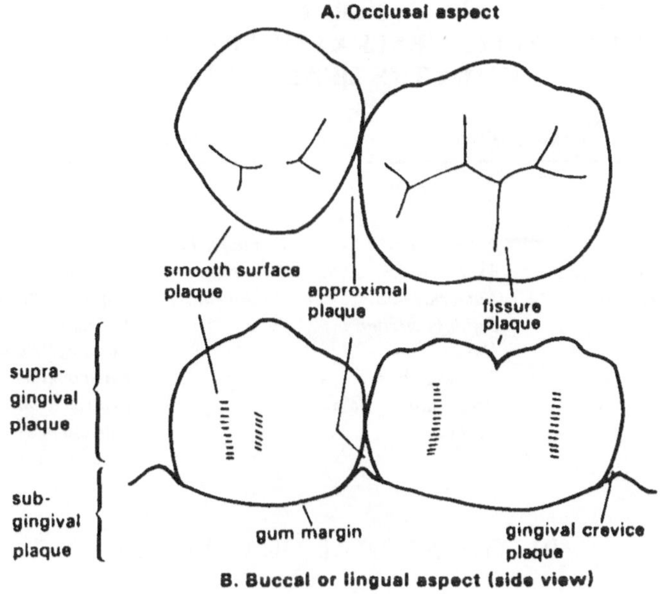


FIGURE 1. Diagram illustrating the different surfaces of a tooth and the terminology used to describe dental plaque sampling sites.[4]

The mouth is kept moist by saliva which flows over all internal surfaces of the oral cavity. Saliva contains several ions including sodium, potassium, calcium, phosphate, and bicarbonate, some of which contribute to the buffering capacity of saliva.[5] The major organic components of saliva are proteins and in particular mucin. Many of the proteins are associated with carbohydrate to form glycoproteins which can aggregate some species of oral bacteria. Certain glycoproteins can adsorb onto enamel to form the acquired pellicle. Dental plaque forms by the adhesion of bacteria from saliva in a characteristic sequence by means of specific molecular interactions between the cell surface and the acquired pellicle initially, and later to already adsorbed bacteria.[1] During plaque development bacteria can modify the pellicle enzymically and the strength of bacterial attachment can be increased by the synthesis of a variety of polysaccharides from carbohydrates and in particular, sucrose. The early colonizers of the tooth surface are facultatively anaerobic but during plaque development the metabolism of these pioneer species produces an environment with a lower Eh facilitating the establishment of obligate anaerobes.[6] Saliva also contains several antibacterial factors, for example, secretory IgA, lysozyme, lactoferrin, and the salivary peroxidase system which generates the toxic ion hypothiocyanate from hydrogen peroxide of bacterial origin,[7] and so plaque organisms can be under some degree of host control. The chemical composition, pH (range 5.6 to 7.6), and flow rate of saliva varies between individuals, and within a mouth the flow rate and, for example, the calcium and phosphate concentrations have circadian rhythms.[5] The importance of saliva can be gauged from the presence of dental plaque in patients fed solely by stomach-tube intubation,[8,9] and from the frequent occurrence of rampant caries in subjects with xerostomia.[10] Although saliva can support the growth of a variety of microorganisms, it has been argued that bacterial growth in the oral cavity is carbohydrate-limited because in vitro studies have shown that the yield of organisms in human saliva is proportional to the concentration of added glucose.[11] However, as shall be seen in the following section, the characteristic high species diversity of the normal oral flora implies that it is unlikely that microbial growth in the mouth is being limited by a single nutrient.[4]

**Table 1**
**THE BACTERIAL GENERA COMMONLY ISOLATED FROM THE ORAL CAVITY**

| Gram-positive | | Gram-negative | |
|---|---|---|---|
| **Cocci** | **Rods** | **Cocci** | **Rods** |
| *Micrococcus* | *Actinomyces* | *Branhamella* | *Actinobacillus* |
| *Peptostreptococcus* | *Arachnia* | *Neisseria* | *Bacteroides* |
| *Streptococcus* | *Bacterionema* | *Veillonella* | *Capnocytophaga* |
| | *Bifidobacterium* | | *Eikenella* |
| | *Eubacterium* | | *Fusobacterium* |
| | *Lactobacillus* | | *Haemophilus* |
| | *Propionibacterium* | | *Leptrotrichia* |
| | *Rothia* | | *Selenomonas* |
| | | | *Treponema* |
| | | | *Wolinella* |

## III. THE NORMAL ORAL FLORA

A wide range of bacterial species including facultatively and obligately anaerobic populations can be readily isolated from the oral cavity. The mouth has a characteristic microbial flora and several species (e.g. *Micrococcus mucilagenosus, Bacterionema matruchotii, Actinomyces* spp., *Rothia dentocariosa,* and *Bacteroides gingivalis)* apparently have the oral cavity as their sole habitat.[3] Also, groups of organisms such as *Staphylococci* and enteric bacteria that are commonly isolated from neighboring ecosystems (skin and gut, respectively) are not regarded as inhabitants of a normal mouth. A list of the principal bacterial genera found in the mouth is shown in Table 1. This list is constantly being extended as new bacterial populations are being recognized, particularly following taxonomic studies of organisms being recovered from periodontal pockets. The use of rigorous anaerobic sampling and processing procedures coupled with improved methods of plaque dispersion and cultivation have significantly increased the numbers of bacteria that can be recovered from periodontal pockets.

Most of the genera listed in Table 1 consist of several species many of which can act as opportunist pathogens, e.g., *Streptococcus sanguis* and *S. mitior* in infective endocarditis, and *Actinomyces israelii* in actinomycosis. The intra-oral distribution of bacteria is influenced by their affinity for particular surfaces, the Eh of a site, and the provision of growth factors by crevicular fluid.[1,3,4] Variations can occur in the microbial composition of plaque at different sites on the same tooth and at the same site both between and within mouths. The same site can also undergo fluctuations in its microbial composition with time.[4] Some bacterial populations including *Actinomyces* spp., *S. sanguis,* and *S. mutans* are found predominantly in dental plaque. An example of the flora of approximal plaque from school children is shown in Table 2.[12] Certain species, such as *Bacteroides gingivalis* and oral spirochetes are limited to specific sites on the tooth surface, namely, the gingival crevice and periodontal pocket.[3] In a film such as dental plaque, gradients will develop of pH and Eh, and in the concentrations of essential nutrients and toxic end products of metabolism. This will result in the production of a mosaic of microhabitats which is responsible for the high species diversity of plaque. Although many species are fermentative, some are asaccharolytic, for example, *Veillonella* spp. utilize lactic acid as an energy and carbon source and several *Bacteroides* spp. are incapable of metabolizing carbohydrates. Other microorganisms regularly isolated from the oral cavity include *Candida* spp., mycoplasmas, and protozoa; viruses can also appear in saliva during infection.[4]

## Table 2
## PREDOMINANT CULTIVABLE FLORA FROM APPROXIMAL PLAQUE

| | Percentage of total cultivable flora | Range | Percentage isolation frequency |
|---|---|---|---|
| | ***Streptococcus*** | | |
| | 22.9 | 0.4—70.0 | 100 |
| Gram-positive rods (predom. *Actinomyces*) | 42.1 | 4.0—81.0 | 100 |
| Gram-negative rods (predom. *Bacteroides*) | 7.8 | 0—66.0 | 93 |
| *Neisseria* | 1.5 | 0—44.0 | 76 |
| *Veillonella* | 13.1 | 0—59.0 | 93 |
| *Fusobacterium* | 0.4 | 0— 5.4 | 55 |
| | **Individual species** | | |
| *S. mutans* | 2.2 | 0—23.0 | 66 |
| *S. sanguis* | 5.9 | 0—64.0 | 86 |
| *S. milleri* | 0.5 | 0—33.0 | 45 |
| *S. salivarius* | 0.7 | 0— 7.0 | 54 |
| *A. israelii* | 16.5 | 0—78.0 | 72 |
| *A. viscosus/naeslundii* | 19.1 | 0—74.0 | 97 |

*Note:* Viable counts were derived from 58 samples of approximal plaque from 10 subjects (upper first premolar, left and right side, 12-year-old school children).[12]

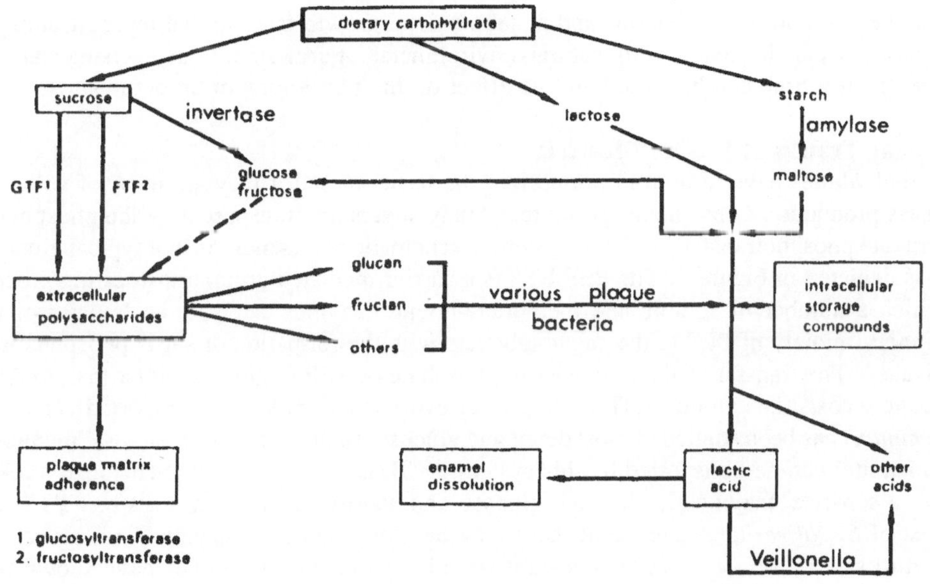


FIGURE 2. Simple diagram to show the possible fate of dietary carbohydrates.[4] Some metabolic interactions involving plaque bacteria are also indicated. Factors relating to the pathogenesis of dental caries are boxed.

## IV. METABOLISM OF CARBOHYDRATES

Figure 2 summarizes in a simplified form the fate of dietary carbohydrates in plaque. Starches will be digested into their constituent sugars by amylases of salivary origin. Milk is the major source of lactose in the diet while sucrose occupies a key position in bacterial metabolism in the oral cavity. It is the most widely used sweetening agent and in many industrialized societies, consumption is approximately 50 kg/person/year. Sucrose is the only well-documented example of a dietary component affecting the composition of the oral flora. Increased consumption of sucrose results in a rise in the numbers of acidogenic and aciduric (acid-tolerating) species in dental plaque, notably *S. mutans* and *Lactobacillus* spp.[13,14] Sucrose can be

1. Broken down by extracellular bacterial invertases (α-glucosidases) and the resultant glucose and fructose molecules taken up directly by plaque organisms
2. Transported intact as the disaccharide or disaccharide phosphate and cleaved inside the cell by an intracellular invertase or a sucrose phosphate hydrolase
3. Utilized extracellularly by glycosyltransferases. Glucosyltransferases (GTF) produce both soluble and insoluble glucans (with the release of fructose) which are important in plaque formation and in the consolidation of adherence by bacteria, while fructosyltransferases (FTF) produce fructans (and liberate glucose) which are frequently labile and can be utilized by other plaque organisms

The high consumption of sucrose by man in industrialized societies coupled with its ready utilization by plaque bacteria to extracellular polysaccharides and to acidic fermentation products has meant that it (and other fermentable carbohydrates) is strongly implicated in the pathogenesis of dental caries. Several aspects of the metabolism of carbohydrates by plaque bacteria will be considered in the following sections. In many studies the unique advantages of continuous culture techniques have been exploited to grow plaque bacteria under controllable and defined conditions. The chemostat allows organisms to be grown under specific nutrient limitations and at submaximal growth rates so that more meaningful comparisons can be made with natural environments. Furthermore, in a chemostat one parameter at a time can be varied and its effect on the physiology of an organism studied.

### A. Sugar Transport by Oral Bacteria

Carbohydrates have to be transported into the bacterial cell if they are to be of value for biomass production or as an energy source. Many sugars are transported by the phosphoenolpyruvate phosphotransferase (PTS) system; a schematic representation of a typical glucose-PTS is depicted in Figure 3. The PEP PTS is a carrier-mediated group translocating process in which a number of soluble and membrane-bound enzymes catalyze the transfer of the phosphoryl moiety of PEP to the sugar substrate with the formation of sugar-phosphate and pyruvate.[15] The transfer of the high energy phosphate of PEP to glucose rather than to ADP effectively costs the cell one ATP molecule for every sugar molecule transported. The PTS of *S. mutans* has been studied in most detail and glucose, sucrose, fructose, lactose, mannose, and mannitol can be transported by this system.[16-20] It is not clear yet if each sugar is taken up by a separate enzyme II, although glucose and sucrose operate on a distinct PTS and studies of *S. salivarius* suggest that there may be more than one enzyme II for glucose.[21] The glucose-PTS of *S. mutans* is constitutive although chemostat studies have shown that the activity of the PTS is modulated by growth conditions. The glucose and sucrose-PTS was repressed when substrate was in excess and highest activities were found in slow growing cells under sugar-limitation.[22,23] It is likely that intracellular sucrose and lactose phosphates are subsequently hydrolyzed to glucose-6-P and fructose, or galactose-6-P and glucose in reactions catalyzed by an invertase or phospho-β-galactosidase respectively. PTS systems have been found in other oral bacteria including *S. sanguis, S. mitior, S. mitis, S. milleri, S. salivarius, Actinomyces viscosus,* and *A. naeslundii* and some *Lactobacillus* spp.[24-29]

The uptake of sugars by a PTS mechanism has been proposed as a possible rate-limiting step of carbohydrate metabolism for many microorganisms. However, during growth in the

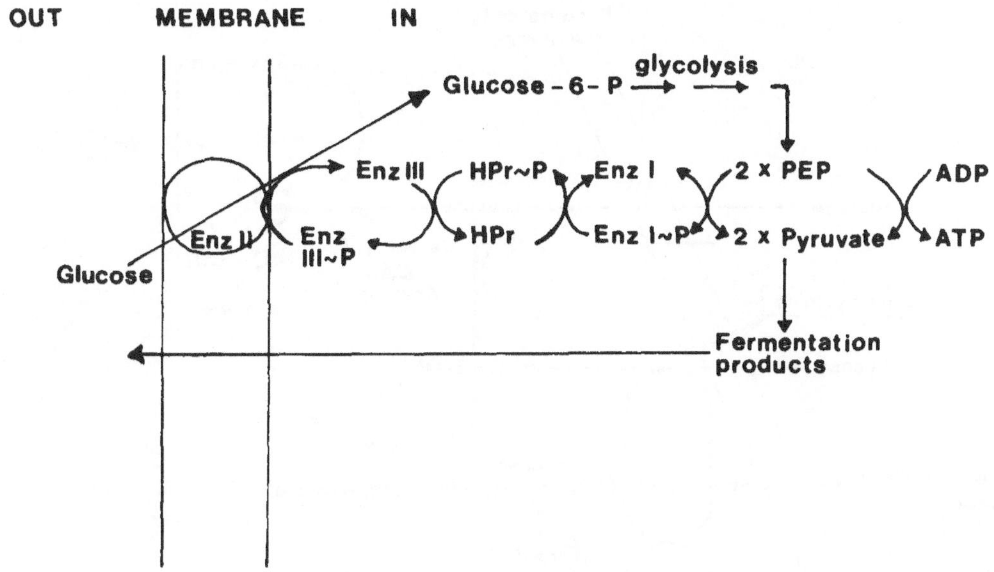


FIGURE 3. Diagrammatic representation of the phosphoenolpyruvate (PEP)-mediated glucose phosphotransferase system. Enz I, II, and III represent enzymes I, II, and III, respectively; HPr is a heat stable protein. $Mg^{2+}$ is required as a cofactor for this transport system. Other carbohydrates can be transported by this system.

chemostat under glucose-limitation or conditions of glucose-excess, the activity of the glucose-PTS of *S. mutans* or *S. sanguis* was not commensurate with the concentration of glucose utilized except at very slow growth rates.[22] Furthermore, when *S. mutans* was grown at a constant pH 6.5, 6.0, and 5.5 in a chemostat, the activity of the glucose-PTS was repressed at a low pH and yet washed cells grown at pH 5.5 had the highest glycolytic activity when pulsed with substrate.[28] Collectively, this evidence suggested that an alternative sugar transport system must operate in these cells. Recent studies of glycolysis by a PTS-defective mutant of *S. mutans*[29] and of a range of oral *Streptococci* in which the activity of the PTS had been abolished by a bacteriostatic concentration of chlorhexidine,[26] have provided strong evidence that the second uptake system is driven by the energy associated with the bacterial membrane (proton motive force, pmf) (Figure 4). In both studies, inhibitors of membrane activity (uncouplers, ionophores) abolished the residual glycolysis that could not be attributed to sugar transport by the PTS.[26,29,30] The alternative system transports sugars as the free sugar which then have to be phosphorylated intracellularly in order to be made available for use by the cell. Free glucose can be phosphorylated directly to glucose-6-P by glucokinase in which ATP acts as the phosphoryl donor[31] (Figure 4) and other phosphorylation reactions will be described in Section IV.C. Kinetic studies have shown that the alternative transport system driven by the energized membrane is the low affinity system of *S. mutans*[29] while the PTS acts as the high affinity scavenger system operating principally in substrate-limited environments.[22] The alternative system will be favored by environments with a low pH as this will generate a larger transmembrane proton gradient.[32] As the activity of the PTS is repressed below pH 6.0 in *S. mutans*, and if the likely conditions in plaque following carbohydrate consumption are considered, the alternative system might be of great significance in the pathogenicity of cariogenic organisms.

## B. Acid Production from Carbohydrate Metabolism

### *1. Fermentation Products*

Dental caries results from the demineralization of enamel by acids produced from the metabolism of plaque bacteria. Consequently, there is much interest in the types, concen-

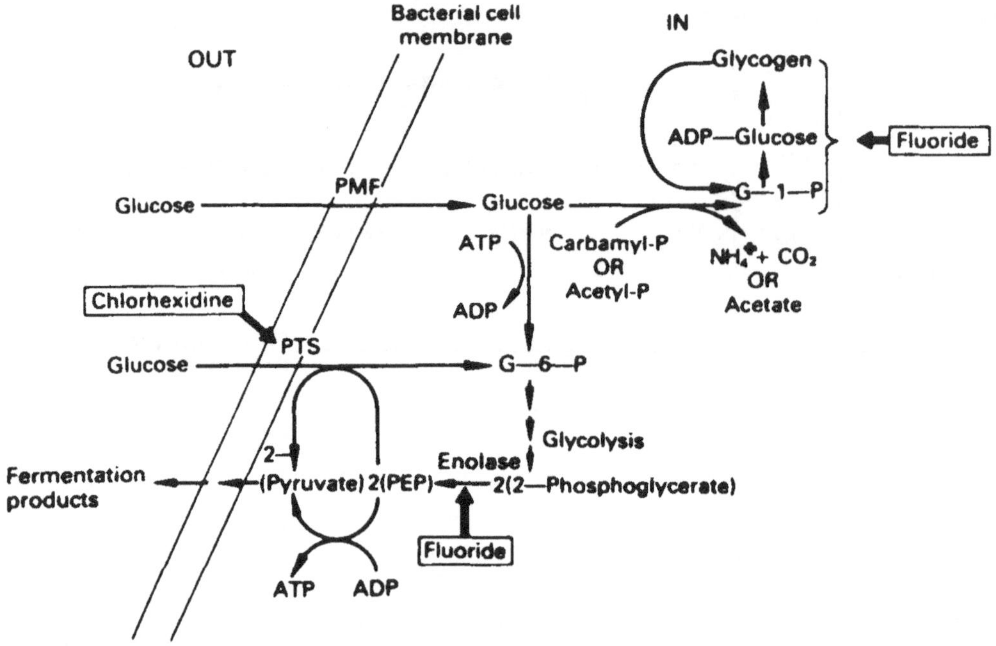


FIGURE 4. Routes of glucose uptake and metabolism in oral *Streptococci*. Possible sites of fluoride and chlorhexidine action are also indicated. Abbreviations: G-1-P = glucose-1-phosphate; G-6-P = glucose-6-phosphate; PEP = phosphoenolpyruvate; PMF = protonmotive force; PTS = phosphotransferase sugar transport system; and CHX = chlorhexidine.

trations, and rates of acid produced by plaque organisms. Oral *Streptococci* were considered to be homolactic fermenters until chemostat studies showed that for several species the pattern of their metabolic products was a function of their conditions of growth. In general, under glucose-limitation, acetate, formate, and ethanol were the major fermentation products while under an excess of glucose (and at fast growth rates under glucose-limitation) the predominant product was lactate.[33,34] These acids vary in strength ($pK_a$) and there is some evidence to suggest that the types of acids might influence the development of a carious lesion.[35] The mechanism controlling the switch from a homolactic to a mixed pattern of fermentation products has been determined for *S. mutans* and shown to be due to the regulation of two enzymes, lactate dehydrogenase and pyruvate formate-lyase (Figure 5).[36] Other oral species show a similar variation in their fermentation products.[37] For example, *Actinomyces viscosus* produced lactate and an extracellular polysaccharide when glucose was in excess but little of either product with a limiting supply of glucose. Under these latter conditions the predominant fermentation product was acetate.

## 2. *Acids in Plaque*

Variations in acid profiles from plaque fluid (a free fluid phase that can be obtained from plaque by centrifugation) are found at different times of the day. Acetic, propionic and *n*-butyric acids were found in studies with human samples of overnight (starved) plaque.[38] Similar results were found with monkey plaque except that iso-butyric, iso-valeric, caproic, and succinic acids were also detected.[39] These patterns reflect heterolactic fermentation and amino acid catabolism. Following exposure to sucrose, the concentration of volatile acids fell, while that of lactate increased markedly in both animal and human plaque.[38,40,42] This switch to a homolactic fermentation pattern suggests that the regulatory mechanism determined from in vitro studies of oral *Streptococci* might well operate in plaque. Many of the volatile acids found in plaque will be produced from the metabolism of amino acids and

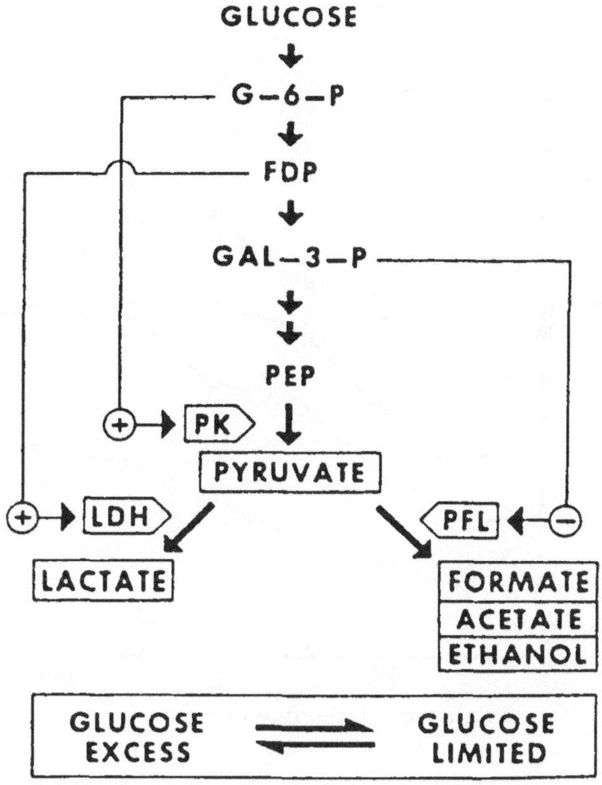


FIGURE 5. Control of fermentation products by *Streptococcus mutans* growing under limiting or excess concentrations of glucose.[36] Abbreviations: G-6-P = glucose-6-phosphate; FDP = fructose-1,6-diphosphate; Gal-3-P = glyceraldehyde-3-phosphate; PEP = phosphoenolpyruvate; PK = pyruvate kinase; LDH = lactate dehydrogenase; PFL = pyruvate formate-lyase; + = positive control; and − = negative control.

polypeptides by some of the asaccharolytic obligately anaerobic species, particularly *Bacteroides* spp. and *Fusobacterium nucleatum*.

*3. Acid Production in Response to a Carbohydrate Pulse*

In addition to the generation of fermentation products during growth, acids can also be produced by resting cells both in vitro and in vivo when exposed to a pulse of fermentable carbohydrate. Changes in pH in plaque can be measured by touching the plaque with a glass or antimony micro-electrode, or by in-dwelling telemetric methods in which glass or ion-sensitive field effect transistors are implanted at caries-prone retentive sites in a restoration such as a crown. The advantages and disadvantages of each method have recently been reviewed[43] and these techniques are proving of value in studies aimed at developing food products with a decreased acidogenic (and presumably a reduced cariogenic) potential. After a sucrose challenge, the pH of plaque may fall from around neutrality to below 5.0 in only a few minutes. This is followed by a slower return to the resting pH of plaque when all of the carbohydrate is utilized and saliva begins to buffer the acids formed and base production occurs. When plotted, this type of pH-response by plaque is called a Stephan curve (Figure 6) as it was Stephan who noticed first that the drop in plaque pH following a glucose rinse was more pronounced in individuals with active caries.[44] It is now well-established that a strong correlation exists between a high frequency consumption of sucrose and a high caries

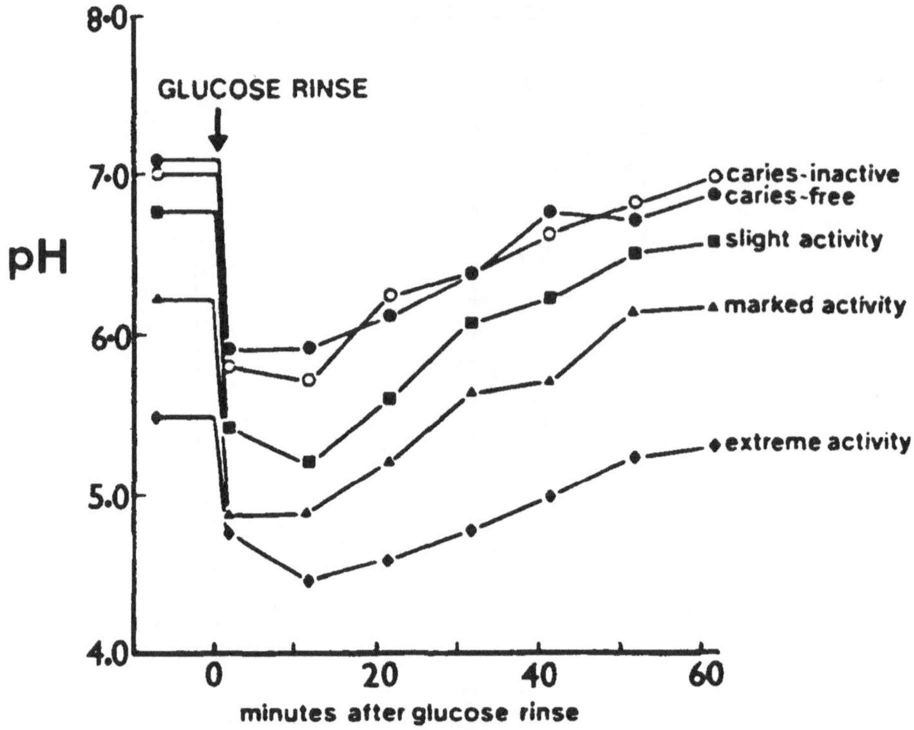


FIGURE 6. The fall in pH of plaque after a rinse with 10% glucose solution in individuals scored for caries activity.[112]

incidence. The pH of plaque of individuals taking frequent snacks falls to below the critical pH (enamel demineralization accelerates below pH 5.5) significantly more often than plaque from individuals who limit food consumption to main meals only. Recent studies have shown that little or no pH drop occurs with high molecular weight substances such as starch due to problems of diffusion of these molecules into plaque.[45] Similarly, sweet-tasting compounds such as sorbitol or xylitol that cannot be metabolized rapidly by plaque bacteria do not produce a significant pH change in vitro.[46] Thus, the measurement of acid production in this way can serve as a useful indicator of the potential cariogenicity of particular foodstuffs and as a means of assessing the possibility of using alternative, nonmetabolizable sweetening agents, and of the value of inhibitors of bacterial metabolism in preventive dentistry.

Acid production by plaque bacteria can also be measured and compared in vitro as either a rate (by continuous titration with alkali to maintain a constant pH) or as a pH-fall (customarily from pH 7.0) where the terminal pH is recorded after a time interval of 15 or 20 min. Such studies have enabled the influence of environmental parameters relevant to plaque ecology to be investigated with respect to acid production. The use of the chemostat has meant that comparisons can be made of the acidogenicity of plaque bacteria grown under different but standardized conditions. Such studies have shown that washed-cells from glucose- or sucrose-limited cultures produce more acid than do cells grown in an excess of substrate. The growth rate of bacteria and the pH and ionic composition of the environment also affect rates of acid production, while sugars such as glucose and sucrose are more rapidly converted into acid than are carbohydrates such as fructose, lactose, and maltose.[22,23,28,47-49] These variations have been attributed to alterations in the activity of the appropriate sugar transport system induced by changes in the environment. When acid production by washed cells of a range of batch-culture grown oral bacteria were compared, *S. mutans* (considered to be a major pathogen in the etiology of dental caries) did not

**Table 3**
**ACID PRODUCTION BY WASHED CELLS OF ORAL BACTERIA**

| Bacterial population | Terminal pH | | Rate (nmoℓ/min/mg) | |
|---|---|---|---|---|
| | Glucose | Sucrose | Glucose | Sucrose |
| *S. mutans* | 4.01 | 4.07 | 278 | 185 |
| *S. sanguis* | 4.36 | 4.26 | 59 | 59 |
| *S. mitis (mitior)* | 4.34 | 4.51 | 167 | 157 |
| *S. milleri* | 4.47 | 4.70 | ND[a] | ND[a] |
| *S. salivarius* | 3.89 | 4.06 | 92 | 66 |
| *A. naeslundii* | 4.57 | 4.78 | 57 | 38 |
| *A. viscosus* | 4.44 | 4.33 | 76 | 56 |
| *L. casei* | 3.86 | 6.29 | ND[a] | 0 |

*Note:* Acid production following a pulse of glucose or sucrose was measured as either the terminal pH after 15 min (starting at pH 7.0) or as a rate at pH 7.0. Cells were harvested in late log phase and washed and resuspended in 200 m*M* KCl.[26,50]

[a] ND = not determined.

necessarily produce the lowest terminal pH values although it invariably made acid at the fastest rate (Table 3), particularly at low pH values.[28,50]

An important property in the pathogenesis of dental caries that has been neglected in terms of research is the aciduricity (acid-tolerating ability) of plaque bacterial populations.[51] Many bacteria are inhibited or even killed by the acidic environments they can generate. Both *S. mutans* and *Lactobacillus* spp. are capable of growing in low pH environments and this ability may grant these bacterial populations an advantage in plaque associated with a carious lesion because, as discussed above, the pH of plaque is lower for longer time periods in caries-susceptible individuals. It may be that it is the combination of acidogenic with aciduric properties that characterize cariogenic bacteria.

### C. Intracellular Polysaccharide Production

Many populations of oral *Streptococci* are able to store and degrade intracellular polysaccharides (IPS). These polysaccharides are usually glycogen (1,4-α-glucan) although other carbohydrate polymers might also be formed. In a rat-caries test system, the virulence of IPS-defective mutants of a serotype c strain of *S. mutans* were compared with those of the parent strain. Although the mutants were able to colonize and persist on tooth surfaces, they differed from the parent strains in causing fewer carious lesions both in fissures and on smooth surfaces. This difference in pathogenicity was attributed to the inability of the mutants to produce acid from IPS reserves in the absence of exogenous supplies of carbohydrate.[52] Early epidemiological studies of man indicated an association between the presence of IPS-producing populations in plaque and the incidence of dental caries.[53] However, care must be taken in any attempt to correlate IPS synthesis and pathogenicity. Strains of *S. mutans* serotype d characteristically have low IPS levels and although this serotype is not as widely distributed as serotype c, a correlation was recently found between the presence of serotype d strains and the development of progressive carious lesions in man,[54] and serotype d strains can produce carious lesions in experimental animals.

It has been generally assumed that glycogen synthesis and degradation in oral *Streptococci* followed the conventional pathway:

$$\text{glucose-6-phosphate} \xrightleftharpoons{\text{phosphoglucomutase}} \text{glucose-1-phosphate}$$

$$\text{glucose-1-phosphate} + \text{ATP} \xrightleftharpoons{\text{ADP-glucose synthase}} \text{ADP-glucose} + \text{PPi}$$

$$\text{ADP-glucose} + (\text{glucose})_n \xrightleftharpoons{\text{ADP-glucose transferase}} (\text{glucose})_{n+1}$$

$$(\text{glucose})_{n+1} + \text{Pi} \xrightleftharpoons{\text{glycogen phosphorylase}} \text{glucose-1-phosphate} + (\text{glucose})_n$$

However, in a recent survey of a range of oral *Streptococci*, little phosphoglucomutase activity could be detected and a novel first step in glycogen synthesis was discovered.[55] Glucose, transported across the energized membrane by the pmf-driven uptake system as free glucose, could be phosphorylated by PEP or ATP to glucose-6-phosphate for glycolysis. However, other compounds could also act as donors and carbamyl phosphate or acetyl phosphate produced glucose-1-phosphate directly for glycogen synthesis (Figure 4). Carbamyl phosphate is generated from the metabolism of arginine and acetyl-P is an intermediate in acetic acid formation during glucose-limited growth. In a range of oral *Streptococci*, high glucose-1-phosphate-forming activity correlated well with species known to contain large concentrations of glycogen.[55] In these bacteria, acid production from endogenous metabolism could be significant; for example, the terminal pH after 15 min was pH 4.98 with *S. mitis* compared with only pH 6.29 by *S. mutans*. Thus, the regulation of glucose uptake and phosphorylation during glucose-limited and glucose-excess growth leads to the stimulation of either glycolysis or glycogen formation.

**D. Extracellular Polysaccharide Production and Degradation**

Many oral bacterial populations produce extracellular polysaccharides from carbohydrates and particularly from sucrose (Table 4). The insoluble polysaccharides make a major contribution to the structural integrity of plaque and are important in the consolidation of attachment of weakly adherent bacteria to the tooth surface or to bacteria already adsorbed. *S. mutans* produces several extracellular polysaccharides from the enzymic breakdown of sucrose. Sucrose has a unique property in that the bond between the glucose and fructose moieties has sufficient energy on cleavage to support the synthesis of a polysaccharide. There is some doubt at present about the role of primers in polymer production; the rate of synthesis of polysaccharide is increased in the presence of primer with *S. mutans* but not with *S. sanguis* or *S. salivarius*. The enzymes involved are termed glycosyltransferases; a glucosyltransferase (GTF-S) of *S. mutans* synthesizes a soluble glucan while another (GTF-I) is responsible for insoluble glucan formation.[56] The former molecule is a highly branched dextran with 32% 1,3-α-branch linkages while GTF-I produces a linear 1,3-α-glucan that has been termed mutan. Low concentrations of GTF-S stimulate the activity of GTF-I and it is believed that the synthesis of glucans by *S. mutans* involves the concerted action of both GTFs (Figure 7). The molecule so produced is highly branched and capable of extending long distances into the environment.[56] There is some debate as to whether these enzymes are glycoproteins, but their properties are certainly complex. The activity of these enzymes vary between strains, between serotypes of *S. mutans*, with the conditions of growth and also with the purification procedures. Moreover, strains of *S. mutans* serotype c secrete several different GTFs and also at least three FTFs of different molecular weights and isoelectric points. Much of the perplexing and sometimes contradictory evidence on glucosyltransferases has been reviewed recently.[58]

**Table 4**
**EXTRACELLULAR POLYSACCHARIDE-PRODUCING BACTERIAL POPULATIONS FOUND IN THE ORAL CAVITY[4]**

| Population | Carbohydrate substrate | Type of polymer (with predominant linkages[a]) |
|---|---|---|
| *Streptococcus mutans* | Sucrose | Water-insoluble glucan (mutan) 1,3-α; 1,3-α- + 1,6-α- |
| | Sucrose | Water soluble glucan (dextran) 1,6-α- |
| | Sucrose | Fructan 2,1-β- |
| *Streptococcus sanguis*[c] | Sucrose | Water-insoluble glucan 1,3-α- + 1,6-α- |
| | Sucrose | Water soluble glucan (dextran) 1,6-α- |
| *Streptococcus mitior* | Sucrose | as for *S. sanguis* |
| *Streptococcus salivarius*[d] | Sucrose | Fructan (levan) 2,6-β- |
| *Actinomyces viscosus* | Sucrose | Fructan (levan) 2,6-β- |
| | —[b] | Heteropolysaccharide (60% *N*-acetyl glucosamine) |
| *Lactobacillus* sp. | —[b] | Glucan Heteropolysaccharide |
| *Eubacterium ingens* | —[b] | Heteropolysaccharide (predominantly acetate) |
| *Rothia dentocariosa* | Glucose | Heteropolysaccharide |
| | Sucrose | Levan |
| *Micrococcus mucilagenosus* | —[b] | Heteropolysaccharide (hexoses, hexosamines, amino acids) |
| *Neisseria* sp. | Sucrose | Glycogen-like |

[a] Where known
[b] No specific substrate required.
[c] Some strains produce a fructan.
[d] Some strains produce a glucan.

Receptors (dextran-binding proteins) on the surface of oral *Streptococci* can bind these glucans and this reaction is important in the aggregation of bacteria.[59] *S. mutans* also produces an unusual water-soluble fructan with predominantly 2,1-β-linkages.[60] This is in contrast to the fructan produced by other bacteria, including *S. salivarius,* which are levans with a characteristic 2,6-β-linkage. Other oral *Streptococci* produce insoluble glucans from sucrose (Table 4). Although a mixed-linkage polymer is produced by *S. sanguis,* the predominant bond is 1,6-α-. The reason for this may be a difference in activity between the GTF-S and GTF-I, or it may be due to the fact that *S. sanguis* does not produce a dextranase which in *S. mutans* probably trims the (1-6)-glucan chains.[56]

Several other bacterial populations produce extracellular polysaccharides (Table 4). While some also depend on sucrose, many have no specific substrate requirement for polymer production. The hetero-polysaccharides are generally complex in composition; for example, the polymer produced by a strain of *A. viscosus* contains *N*-acetyl glucosamine (62%), galactose (7%), glucose (4%), uronic acids (3%) and small amounts of glycerol, rhamnose, arabinose, and xylose.[61]

As mentioned above, some of these bacterial extracellular polysaccharides can be metabolized by other plaque organisms in the absence of exogenous supplies of carbohydrates. The detailed literature on this was recently reviewed.[3] The levans of *S. salivarius* and the

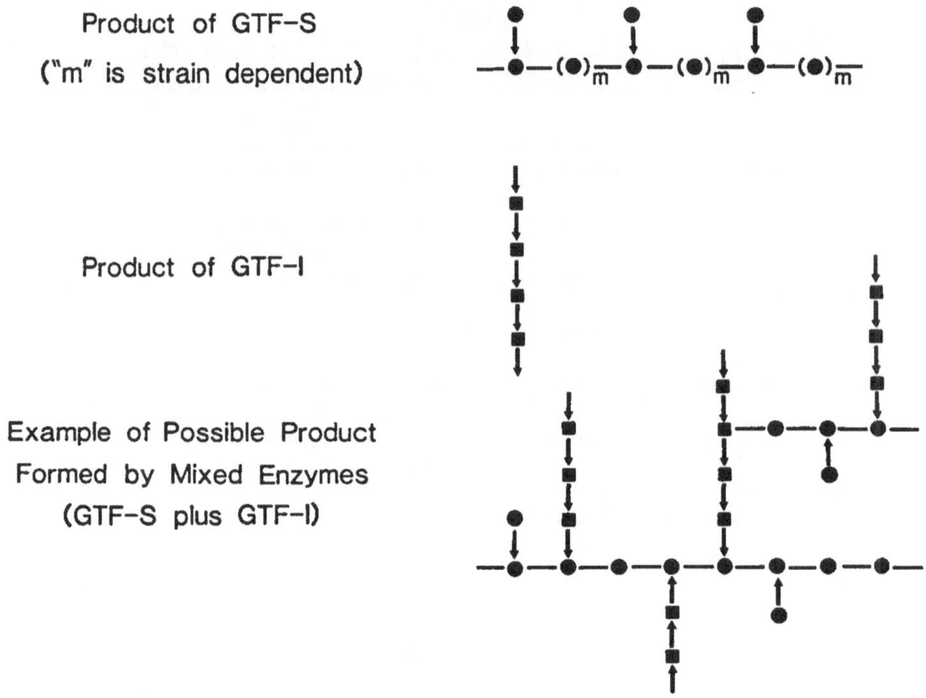


FIGURE 7. Production of branched extracellular polysaccharides from sucrose by the concerted action of GTF-S and GTF-I from *Streptococcus mutans*. GTF-S produces an α-1,6-glucan (●) with α-1,3-branch linkages while GTF-I produces a linear α-1,3-insoluble glucan (■) that has been termed mutan. The concerted action of these enzymes produces a highly branched mixed-linkage polysaccharide.

glycogen-like polymer of a strain of *Neisseria* are particularly labile and can be utilized in vitro by a variety of oral *Streptococci* and *Lactobacilli* that produce fructanases. In addition, *S. mutans, S. mitior, S. salivarius, A. israelii, Capnocytophaga* spp., and *Fusobacterium* spp., posses exo- and/or endo-hydrolytic activity and can metabolize streptococcal glucans, although mutan is not readily attacked. The apparent paradox of strains possessing both polysaccharide synthesizing and degrading enzymes might be resolved by the recent finding in culture filtrates of *S. mutans* of a protein capable of inhibiting dextranase activity.[62] Also, the cell-associated fructosyltransferase of *S. salivarius* can be inactivated by proteolytic degradation.[63] The mechanism might involve protein modification and it appears that the enzyme involved is also cell-associated. It is apparent that there is still much to be learned about the regulation of these synthesizing and degradative enzymes.

## V. METABOLISM OF NITROGENOUS COMPOUNDS BY ORAL BACTERIA

In contrast to the vast literature on carbohydrate catabolism, comparatively few studies have been made of the metabolism of nitrogenous compounds by oral bacteria. This is particularly surprising in view of the fact that saliva will be a major determinant of the composition of the oral flora, while base production from the metabolism of nitrogenous components will play an important role in plaque ecology by neutralizing acids produced by bacteria from carbohydrates. Furthermore, many of the putative pathogens in periodontal disease are asaccharolytic, and protease production by these and other organisms is involved in tissue destruction. Apart from casein, dietary proteins are not readily utilized and the major sources of nitrogenous material for oral bacteria are from saliva and crevicular fluid.

### A. Metabolism of Salivary Components

Saliva contains little free-carbohydrate but has a wide range of amino acids, peptides, proteins, and glycoproteins[5] (carbohydrates are generated in saliva from the metabolism of glycoproteins). Urea is also present at a high concentration (20 mg%) and many oral bacteria can rapidly degrade urea to carbon dioxide and ammonia which can raise the plaque pH.[64] It has been postulated that urea nitrogen might be stored in the form of an as yet unknown compound which can be subsequently broken down to give ammonia. In this way a high pH could be maintained in plaque for long periods in an analogous (but opposite) way to the metabolism of intracellular polysaccharides to acid.[65] Ammonia generated in this way could also be incorporated into glutamate and by transamination, into other amino acids or amino sugars (see Chapter 7).

Approximately 18 free-amino acids can be detected in whole saliva although the concentration of many is low.[5] Studies of a range of oral *Streptococci* have shown that the low molecular weight constituents of saliva (amino acids and peptides) are unable to support bacterial growth[66] and it is now well-established that many bacteria are dependent on the protein fraction of saliva for growth. Thus, the hydrolysis of salivary and other proteins and peptides by extracellular bacterial proteases will generate free-amino acids and small peptides essential for growth. Oral *Streptococcal* species have different amino acid requirements[67,68] although little is known of the energy yield or metabolic pathways of amino acid breakdown by these bacteria. It has been suggested that amino acid degradation is an important method by which oral organisms might counter the extremes of pH caused by the metabolism of carbohydrates or urea.[65] At acidic pH values, decarboxylation reactions would yield carbon dioxide and amines[69] while at high pH values deamination would produce ammonia and keto acids which can be converted to acetic, propionic, and possibly iso- and *n*-butyric acids.[65] Arginine, which is required by several oral *Streptococci* for growth including *S. sanguis* and *S. milleri*, is metabolized by arginine dihydrolase (which is probably more than one enzyme) to citrulline and then to ornithine. The reaction also generates ammonia, carbon dioxide, and a high energy intermediate, carbamyl phosphate, which can act as a phosphoryl donor in some bacteria, converting glucose to glucose-1-phosphate (a precursor of glycogen formation, see Section IV.C). Ornithine can be decarboxylated to putrescine and further metabolized to succinate. Other basic amino acids are metabolized by different pathways, for example, lysine is decarboxylated to cadaverine. Arginine is the most efficient amino acid at producing a pH-rise in plaque through base production. Free-amino acids are generally taken up more slowly than peptides, and arginine in peptide form can be transported almost three times more rapidly than free arginine. Peptide uptake in oral bacteria is poorly documented but studies of other organisms suggest that small peptides (up to four or five amino acids) can be transported intact directly into the cell where they are hydrolyzed to their constituent free-amino acids by intracellular peptidases. Bacteria often possess separate systems for the transport of free-amino acids and for both di- and oligo-peptides.

The metabolism of salivary proteins by oral bacteria has aroused considerable interest. *S. sanguis* is more proteolytic than *S. mutans* and analysis of proteins before and after incubation with cells suggests that the pattern of attack of each species is different.[68,70] Furthermore, it has been found that saliva from subjects with a high incidence of dental caries is better at supporting the growth of *S. mutans* than saliva from caries-free individuals.[71] This has been attributed to the presence of specific growth supportive proteins in the saliva of the caries active individuals that were absent from the saliva of caries-free subjects. Subsequent work, however, has identified a protein inhibitory to the growth of *S. mutans* in the saliva from two of the caries-free individuals.[72] These proteins are anionic and do not appear to be related to other known antimicrobial factors in saliva such as sIgA, IgG, lysozyme, lactoferrin, or lactoperoxidase.

Another group of peptides and low molecular weight proteins that has been studied extensively are those that have been found to enhance the glycolytic activity of plaque. The properties of these peptides have recently been reviewed.[73] One peptide that can increase lactic acid production by plaque bacteria can also give rise to a significant pH-rise effect in plaque and therefore could be important in reducing the risk of caries development.[74] This peptide is sialin and has the composition glycine-glycine-lysine-arginine.[75] Following the metabolism of sialin more putrescine than cadaverine is formed suggesting that arginine is degraded more readily than lysine.[65] The pH-rise effect of sialin in plaque is believed to be due to the enhancement of glycolysis which will result in a rapid clearance of dietary carbohydrate from the mouth, coupled with base production which will both neutralize the acids formed from the increase in glycolytic activity and raise the pH quickly to values around neutrality.[65,75] Several bacterial populations including serotype b *S. mutans* (a serotype not regularly isolated from the human mouth), *A. naeslundii*, and heterofermentative *Lactobacilli* have pH-rise activity when incubated with a combination of glucose and either arginine or lysylarginine.[76] Addition of such bacteria to samples of pooled plaque metabolizing carbohydrates in vitro can result in an alteration to the shape of the Stephan curve causing a rise in the minimum pH reached and a faster return of the pH to resting values. Indeed it has been argued that caries results not so much from an over-production of acid but more from a deficiency in base production by plaque bacteria.[75]

### B. Metabolism of Crevicular Fluid

In comparison to the studies of the bacterial utilization of salivary components, much less is known about the metabolism of crevicular fluid and yet this exudate provides factors (for example, hemin) that are essential for the growth of anaerobes implicated in the etiology of periodontal disease. Many of the organisms isolated from the gingival crevice are proteolytic and asaccharolytic, and are dependent for their growth on their ability to utilize the nutrients present in crevicular fluid although an alternative source of amino acids and peptides for the subgingival microflora might be neutrophils and desquamated epithelial cells. The gingival crevice and periodontal pocket harbor the most complex flora of the mouth[3,4] and the existence of many species is dependent on the development of food chains between groups of organisms.[77,78] Information on such interactions is sparse but some are shown in Figure 8. Other interactions known to exist have been reviewed recently[3,4] and include the provision of hydrogen and formate by *Fusobacterium* and *Bacteroides* spp. for the growth of vibrio-like organisms. Lactate can be utilized by *Veillonella* spp.[78] and *Treponema oralis*, while other oral spirochetes are dependent on the production of isobutyrate and putrescine or spermine by *Fusobacterium* spp., and on succinate by various Gram-positive rods. Black-pigmented *Bacteroides* utilize vitamin K which is produced by a number of subgingival organisms. One member of this group of organisms, the putative periodontal pathogen *B. gingivalis*, has been shown to deaminate asparagine and glutamine to aspartate and glutamate which may be further deaminated to yield additional ammonia.[79] The pH of the periodontal pocket is high and the possession of alkaline-stable enzymes by *B. gingivalis* may be an important factor in its ability to compete and dominate during disease.[80]

## VI. METABOLISM OF ORAL BACTERIA ON A SURFACE

The rate of flow of saliva dictates that bacteria have to adhere to oral surfaces in order to remain in the mouth. However, there have been few studies of the influence of attachment on bacterial metabolism although those that have been reported suggest that the physiology of surface-associated organisms may be different to that predicted from studies of cells growing in liquid culture.[81] Early studies of *Bacillus* spp., enteric and marine microorganisms in nutrient-depleted environments showed that their metabolic activity was raised in the presence of a surface. Growth rates of an estuarine bacterium were also higher initially on surfaces immersed in a chemostat than in the liquid culture.[81] More recently, lactic acid production by oral *Streptococci* was found to be influenced by the addition of hydroxyapatite

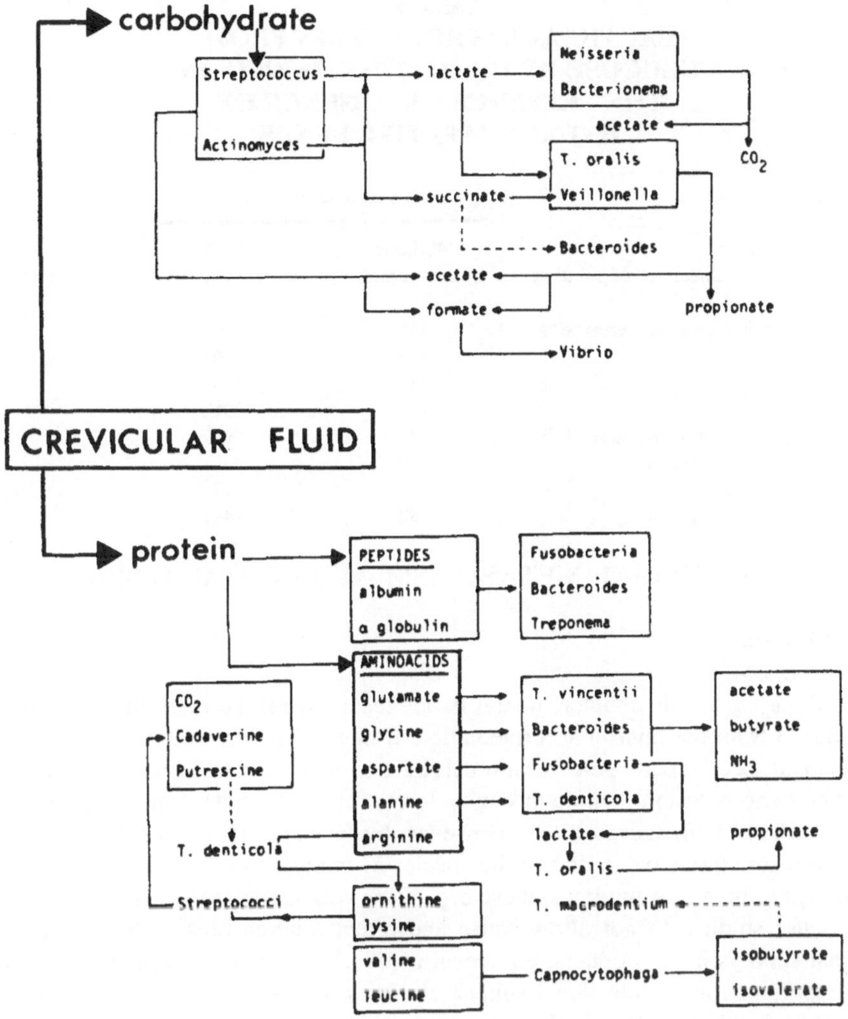


FIGURE 8. Some examples of bacterial food chains that may exist in subgingival plaque following the metabolism of crevicular fluid.[77] Proteins are present at high concentrations in crevicular fluid. The subgingival concentration of carbohydrates is much lower than that found in supragingival plaque.

beads to bacterial suspensions. All nine strains tested of *S. mutans* but only two out of five strains of *S. sanguis* showed an enhanced glycolytic rate in the presence of beads while other strains exhibited a reduced metabolic activity.[82] Some results are shown in Table 5. Several hypotheses have been proposed to explain the enhanced metabolism of bacteria on a surface. One widely accepted theory is that there is an increased concentration of nutrients at an interface between a surface and a liquid. Another more recent explanation is based on the hypothesis that energy-conserving and energy-sharing interactions between bacterial cells would be more likely to occur on a surface,[81] particularly during the development of a microbial film such as dental plaque. The clinical relevance of these findings have yet to be determined.

**Table 5**
**LACTIC ACID PRODUCTION FROM GLUCOSE BY ORAL *STREPTOCOCCI* IN THE PRESENCE OR ABSENCE OF HYDROXYAPATITE BEADS**

| | Lactic acid (μg) | |
|---|---|---|
| Bacterial population | Without hydroxyapatite | With hydroxyapatite |
| *S. mutans* serotype a | 355 | 400 |
| b | 353 | 426 |
| c | 345 | 524 |
| d | 249 | 396 |
| *S. sanguis* strain 167 | 222 | 295 |
| Challis | 469 | 446 |
| *S. mitis* | 17 | 17 |
| *S. salivarius* | 92 | 44 |

# VII. BACTERIAL METABOLISM AND DENTAL DISEASE

## A. Dental Caries

### *1. Bacterial Etiology*

As stated earlier in this chapter, dental caries results from the dissolution of enamel by acids produced from the microbial fermentation of dietary carbohydrates. Experiments with gnotobiotic and specific pathogen-free animals have shown that a wide variety of oral bacteria can produce carious lesions although cavity formation is dependent on the provision of a cariogenic (usually high sucrose) diet. However, by far the most virulent organisms in this model system are strains belonging to the species *S. mutans* and cross-sectional epidemiological surveys of man also implicate these organisms with caries (for a review see Reference 45). The latter studies do not allow cause and effect relationships to be established and longitudinal surveys have produced equivocal results.[54,83,84] Thus, although *S. mutans* will play an important role in the development of caries at a site it is probably not the sole pathogen in lesion formation. Furthermore, bacterial interactions might modulate the cariogenicity of bacteria in plaque. Fewer carious lesions were obtained in rats inoculated with either *S. mutans* or *S. sanguis* and *V. alkalescens* than in animals inoculated with either *Streptococcus* alone. This was attributed to the conversion of the streptococcal-produced lactic acid to weaker acids (acetic, propionic) by *V. alkalescens*.[78] Caries is a multifactorial disease, and in addition to the bacterial components of disease both the diet and the provision of a susceptible host are important in the development of a lesion. Caries is favored by diets rich in fermentable carbohydrates, particularly sucrose, and the frequency of consumption is important. Most caries are associated with the retentive sites (pits and fissures, and approximal surfaces) in the oral cavity but little is known of host factors which favor disease. In an attempt to predict caries-susceptible patients several caries activity tests have been devised. These include salivary counts of *S. mutans*,[85] *Lactobacillus* spp.,[85] or simply of acidogenic bacteria,[86] and assessments of the buffering capacity of saliva.[85] At present, there is no test that is satisfactory for use in general practice and the theoretical basis, advantages, and disadvantages of each method have been critically reviewed by Newbrun.[85] The failure of these simple tests probably reflects the multi-factorial nature of dental caries. Saliva is only one of the variables involved in dental caries and salivary levels of bacteria do not reflect their numbers in plaque at individual sites.

*2. Preventive Measures Involving Inhibition of Bacterial Metabolism*

The incidence of dental caries has reached pandemic proportions in industrialized societies. Consequently, a great deal of effort has been put into developing preventive measures to reduce this costly disease. These include the mechanical removal of plaque by effective oral hygiene procedures, improved education and dietary counselling, fissure sealants, and vaccines directed against *S. mutans*.[3,4,45,85] Other measures are designed to interfere with the metabolism of oral bacteria and they will be described now in more detail.

Fluoride has had a dramatic effect on reducing the incidence of dental caries both by systemic administration via the water supply or in the form of drops, tablets, or lozenges and by topical administration by toothpaste, gel or mouthrinse. Indeed, the marked drop in the incidence of dental caries in young children over recent years has been attributed to fluoride in toothpastes. The primary mode of action of fluoride is to increase the resistance of the enamel to dissolution by acid, the fluoride being incorporated into the enamel structures as fluorapatite. Fluoride is also an inhibitor of acid production and of several enzymes important in bacterial metabolism.[87] In particular fluoride can inhibit enolase which converts 2-phosphoglycerate to phosphoenolpyruvate. This not only inhibits glycolysis but also the transport of sugars by the PTS by reducing the availability of PEP. Fluoride also inhibits an enzyme in the synthesis, but not the degradation of intracellular storage material (Figure 4). Plaque has a fluoride-concentrating property and the sensitivity of cells is increased at low pH values. This is believed to be due to an increased uptake of fluoride as $H^+F^-$ under these conditions. Once inside the cell, it has been postulated that HF would dissociate; $F^-$ would specifically inhibit various enzymes as described above while protons would acidify the cytoplasm and so reduce

1. The activity of glycolytic enzymes having a pH optima around neutrality
2. The transmembrane pH gradient (and hence pmf-driven processes)
3. The aciduricity (the ability to metabolize at low pH values) of cells[88]

However, there is also evidence available suggesting that fluoride is not taken up in this way and therefore its mode of action might be different. For example, cells preincubated with fluoride at pH 5.5 or pH 5.0 and subsequently readjusted to pH 7.0 had a similar metabolic activity to cells not exposed to fluoride. Much of the literature on fluoride and bacterial metabolism has been reviewed recently.[87,89]

Chlorhexidine is an antimicrobial agent belonging to a group of $N^5$ derivatives of 1:6 *bis*-biguanidohexane that has been used extensively in the form of a mouthrinse to prevent the accumulation of dental plaque on teeth and to reduce the incidence of gingivitis and dental caries.[90] At high concentrations, chlorhexidine is bactericidal against a wide variety of Gram-positive and Gram-negative bacteria although its effectiveness is not believed to be related to its killing action. The benefit of chlorhexidine as a prophylactic measure is believed to be due to its long-term retention in the mouth with a subsequent slow release into saliva at a bacteriostatic concentration.[91] In addition to its antiplaque properties, low concentrations of chlorhexidine can inhibit acid production by plaque bacteria both in vitro and in vivo and its mode of action might include the inhibition of the PTS sugar transport system.[26]

The majority of members of modern industrialized societies have a preference for a sweet diet. In view of the strong association between sugar consumption (including amount and frequency of intake) and dental caries, alternative sweeteners are being sought which cannot be readily fermented by plaque bacteria.[92] These artificial sweeteners are of two types: the intense type many times sweeter than sucrose (e.g., cyclamates and saccharin) and the bulk agents such as the sugar alcohols (e.g., sorbitol and xylitol) which although not as sweet as sucrose, are of value to the confectionery industry because of their physical properties. Sorbitol and xylitol have been found to be particularly effective in reducing caries when they are incorporated in foods such as candies and chewing gum that are likely to be consumed between main meals. Indeed, not only are some of the bulk agents inert in terms of being nonmetabolizable by oral bacteria but also they appear to have the additional property of being able to reverse the process of enamel demineralization. This remineralization appears to be a consequence of the increased salivary flow that is associated with the consumption of these agents. However, much remains to be elucidated about the detailed effects of many sugar substitutes on the metabolism of individual plaque bacteria.

A variety of other compounds have been tried as anticaries agents. The fatty acid nonanoate inhibits only bacteria producing acid and its effect can last for up to 8 hr.[93] Surface active agents and detergents can inhibit GTF activity and there is a great deal of interest in molecules that might prevent the adherence of plaque bacteria. Crude glucan hydrolases interfere with the sucrose-mediated adherence of some oral *Streptococci* including *S. mutans* and when given to animals, reductions in caries have been claimed.[94] Attempts are being made to synthesize molecules that would condition the tooth surface in such a way as to prevent plaque formation. Dental disease would be eliminated if oral surfaces could be made uncolonizable by microorganisms.

**B. Periodontal Disease**

*1. Bacterial Etiology*

Periodontal disease is a collective term for several conditions which attack the supporting tissues of the teeth. More teeth are lost through periodontal disease than as a result of dental caries. Evidence for the role of bacteria in periodontal disease has come from studies of gnotobiotic animals. Germfree animals do not develop periodontal disease unless they are inoculated with specific bacterial populations isolated from infected sites in humans.[95] Cross-sectional and a limited number of longitudinal epidemiological studies of patients have provided strong evidence that each form of periodontal disease is associated with specific groups of bacteria. The complexity of the microflora from a periodontal pocket makes it unlikely that a single organism will be responsible. The disease will result from the concerted action of a consortia of bacteria, the components of which are still proving difficult to isolate due to the fastidiousness of their nutritional requirements and to the fact that many are obligately anaerobic.[96] As discussed earlier, the growth of many of the pocket bacteria is dependent on the nutrients provided by crevicular fluid and by the products of metabolism of the other organisms present.

Gingivitis is a common, mild, reversible inflammation of the gingivae. It is characterized by reddened and swollen gums that bleed easily during tooth brushing. Removal of plaque results in the rapid regression of the disease. The results of cultural studies suggest that gingivitis results from an increase in the mass of plaque through poor oral hygiene although more specific groups of organisms might be involved. During this build-up of plaque there is a shift in the composition and complexity of the bacterial flora. In health, plaque samples are dominated by *Streptococci* whereas gingivitis is associated with an increase in the numbers of actinomyces and Gram-negative anaerobic species including black-pigmented *Bacteroides* spp. *(B. gingivalis* and *B. intermedius),* fusobacteria, Veillonellae, and some *Spirochetes.* Longitudinal studies of experimental gingivitis also suggest that, with time, there is an increase in both the degree of gingival inflammation and in the complexity of the flora. This increase in complexity is a result of a reasonably predictable succession of bacterial populations so that some species while not associated with initiation are observed at later stages for example when gingivitis is associated with bleeding.[97,98] By comparison, much less is known about the microbial and host factors responsible for inflammation. The acidic end products of metabolism of many of these bacteria are known to act as gingival irritants (e.g., butyric and propionic acids) and undoubtedly, surface antigens together with other products will also be involved in the pathogenesis of this disease.

There is also an exaggerated form of gingivitis that is associated with pregnancy, puberty, menstruation, stress, and the use of oral contraceptives. The microbial factors responsible for gingivitis in pregnancy have been studied in detail. The composition of the subgingival flora fluctuates during pregnancy and during the second trimester the proportions of *Bacteroides intermedius* (formerly *B. melaninogenicus* subspecies *intermedius)* were found to increase significantly. Steroid hormones can be detected in crevicular fluid and following in vitro studies, the rise in numbers of *B. intermedius* was attributed to the preferential ability of this species to metabolize progesterone and estradiol. Both hormones were able to replace the requirement for growth by this species for vitamin K.[99] Other host factors that might regulate the proportions of bacteria in the subgingival flora remain to be determined.

Gingivitis, if left untreated, can develop into a more severe condition in which not only are the gingival tissues involved but also the periodontal fibers (important for the attachment of teeth) and alveolar bone. As a result of bacterial metabolism and a host inflammatory response to these bacteria and their products, the epithelium at the base of the crevice gradually migrates down the roots of the tooth and there is a concomitant bone loss and a decrease in the degree of attachment. An intriguing question in periodontal research is why gingivitis sometimes, but not always, progresses into more severe forms of periodontal disease. This question remains unanswered at present but may relate to either differences in the composition of the flora at individual sites or to an increased activation of the host defenses, or to both. The most common form of these severe forms of periodontal disease is chronic periodontitis which is the major cause of tooth loss after the age of 25. The microbial composition of the flora from chronic periodontitis is complex and may vary within and between pockets according to their depth. Pockets can be in excess of 10 mm deep and, as only those bacteria at the base may be involved in advancing the lesion, considerable technical problems are associated with sampling these discrete sites while avoiding contamination of the sample by organisms from other parts of the pocket. Not surprisingly therefore, there has been a wide range in the proportions and types of bacteria isolated from chronic periodontitis. A further difficulty has been in the lack of criteria to define the disease precisely in clinical terms and it is now believed that pocket formation is not a continuous process but has both active and inactive phases. The reasons for this are not yet clear but may relate to changes in bacterial metabolism or to alterations in the host response. The variations in the flora from various studies might be related to pocket depth, differences in the type of periodontal disease studied, the isolation and sampling procedures used, and whether the pocket was in an active or inactive phase. Nevertheless, several studies have indicated that one of the principal pathogens in advanced chronic periodontitis is *Bacteroides gingivalis* (formerly *B. melaninogenicus* subspecies *asaccharolyticus);* other organisms that are implicated include *B. intermedius* (but more so with pockets of intermediate depth, e.g., 4 to 6 mm), *Fusobacterium nucleatum, Wolinella recta,* a poorly defined fusiform *Bacteroides* spp., and some specific types of *Treponema* spp.[100-102] The properties of *B. gingivalis* and some other putative periodontal pathogens which may contribute to their pathogenicity have been studied in detail[103] although it should be remembered that pocket formation is also a consequence of the host inflammatory response. Apart from their obvious role in protection, the enzymes released by phagocytic cells (polymorphs, macrophages) can also damage host tissues.[104] Many organisms isolated from periodontal pockets have properties that enable them to evade the host defenses.[103] Black-pigmented *Bacteroides* spp. can survive in periodontal pockets and in experimental abscesses despite a massive migration of polymorphs. This ability to resist phagocytosis may be due to the presence of capsular material on the cell surface or to an impaired function of the neutrophils induced by these bacteria. It has been proposed that a capsule could mask lipopolysaccharide (endotoxin) and other cell surface antigens and thereby reduce polymorph chemotaxis. Several *Bacteroides* spp. and *Capnocytophaga* spp. produce IgG and IgA proteases while *B. gingivalis* can degrade $C_3$ and $C_5$ components of complement and is also a heavy producer of fibrinolysin.

**Table 6**
**BACTERIAL FACTORS WITH POTENTIAL TO DESTROY GINGIVAL TISSUE AND PERIODONTAL ATTACHMENT[103]**

| Item | *B. gingivalis* | *B. intermedius* | *Capnocytophaga* | *A. actinomycetemcomitans* |
|---|---|---|---|---|
| Abscess-former | Strong | Moderate | − | − |
| Enzymes | | | | |
| Collagenase | + | + | − | +/− |
| Trypsin-like | + | − | −/+ | − |
| Gelatinase | Strong | Moderate | − | − |
| Aminopeptidases | +/− | − | + | − |
| Phospholipase A | + | + | ? | ? |
| Alkaline phosphatase | + | + | + | + |
| Acid phosphatase | + | + | + | + |
| Toxic factors | | | | |
| Epitheliotoxin | + | + | + | + |
| Fibroblast growth inhibitors | + | + | + | + |
| Endotoxicity | Weak | Weak | Weak-to-moderate | Strong |
| Lipopolysaccharide-induced bone resorption | + | + | + | + |
| Other bone-resorbing factors | ? | ? | ? | + |
| Volatile sulfur compounds | + | + | ? | ? |
| Butyric and propionic acids | + | − | − | − |
| Indole | + | + | − | − |
| Ammonia | + | + | − | − |
| Polyclonal B-cell activators | + | + | + | + |

A striking feature of *B. gingivalis* is the wide range of proteolytic enzymes produced in culture with tissue-damaging activity (Table 6). These include collagenase, hyaluronidase, chondroitin sulfatase, and a range of phosphatases which may be involved in alveolar bone breakdown while phospholipase-A may produce prostaglandin precursors and so initiate prostaglandin-mediated bone resorption. These enzymes would destroy the integrity of the surrounding tissues facilitating the diffusion of cytotoxic substances such as indole, ammonia, $H_2S$, butyric acid, and methylmercaptan. The volatile sulfur compounds can increase the permeability of the oral mucosa and reduce collagen and noncollagenous protein synthesis.[103] Much research remains to be done, however, to determine the precise role of these enzymes and cytotoxins in periodontitis.

A rare form of periodontal disease that affects adolescents is juvenile periodontitis. This aggressive disease is associated with a scant plaque which contains only a few cells (approximately $10^6$ organisms per pocket) of a limited number of saccharolytic bacteria. The principal organisms have now been identified as *Actinobacillus actinomycetemcomitans* and *Capnocytophaga* spp.[100,105] Both organisms produce a range of enzymes that are capable of destroying periodontal tissues (Table 6) but the outstanding feature in the etiology of juvenile periodontitis is the production of a powerful leukotoxin by *A. actinomycetemcomitans.*[103] This enables it to avoid one of the most important host defense systems in the periodontal pocket and so far, *A. actinomycetemcomitans* is the only bacterial species in the mouth found to produce one. Leukotoxin production by *A. actinomycetemcomitans* is closely correlated with virulence and the majority of juvenile periodontis patients have neutralizing serum antibodies against the leukotoxin.

Acute necrotizing ulcerative gingivitis also referred to as Vincent's gingivitis, is one of the more common acute infections of the mouth. Characteristically, the gingival papillae between the teeth become ulcerated either in a localized or generalized manner. Smears of these areas when examined microscopically show a preponderance of spirochetes and fusiforms.[106] Recent cultural studies suggest that these latter organisms might not be fusobacteria but may be elongated forms of *B. intermedius*.[107] The properties of these organisms that are responsible for this form of periodontal disease have yet to be determined.

*2. Preventive Measures*

It is clear from the preceding sections that periodontal disease is a consequence of the metabolism of subgingival plaque bacteria and that a great deal of evidence is accumulating on the properties of these bacteria that contribute to their pathogenicity. However, in contrast to dental caries, it has not proved possible so far to convert this information into any practical applications for clinical purposes. At present, all forms of periodontal disease are based on the effective removal of the inflammatory agent (dental plaque) by a combination of oral hygiene, surgery, and the use of antimicrobial agents. While it is theoretically possible to eliminate all microorganisms from a pocket using antimicrobial agents it is not always practical to do so.[108,109] There can be problems of adequate penetration of the agents into the pocket at high enough concentrations while long-term use of, for example, antibiotics can lead to the development or selection of resistant organisms and the possible overgrowth of the resident flora by nonoral bacteria or yeasts. Antibiotics such as tetracycline have proved useful in temporarily reducing the pocket flora as an adjunct to nonsurgical therapy or in patients that do not respond to conventional therapy. This can encourage healing and repair although the clinical progress and antibiotic sensitivity of the pocket flora need careful and continual monitoring. The greatest success with an antimicrobial agent in periodontology has been with chlorhexidine in reducing gingivitis.[90] Chlorhexidine has been less successful in reducing chronic periodontal disease mainly because of limited penetration into the pocket. Studies using direct irrigation of the pocket with 0.2% chlorhexidine via a blunt hypodermic needle have shown encouraging results[110] as have the use of slow release devices (dialysis tubing) containing 20% chlorhexidine in patients suffering acute exacerbations of chronic periodontitis.[111] The mode of action of chlorhexidine in these studies is not known. Its beneficial action may be related to the bacteriocidal or antiplaque properties of chlorhexidine, although the inhibition of the metabolism of the pocket flora might also occur. In conclusion, knowledge of the pathogenesis of periodontal diseases is still being acquired and a rational approach to treatment involving either a reduction in the numbers of specific pathogens (for example, *A. actinomycetemcomitans* in juvenile periodontitis) or an enhancement in the host defenses to combat the plaque flora will be dependent on further research. The possibility that specific inhibitors of metabolism could be used to reduce the activity of periodontal pathogens also remains a goal for the future.

## REFERENCES


1. **Gibbons, R. J.,** Adhesion of bacteria to the surfaces of the mouth, in *Microbial Adhesion to Surfaces,* Berkeley, R. C. W., Lynch, J. M., Melling, J., Rutter, P. R., and Vincent, B., Eds., Ellis Horwood Ltd., Chichester, England, 1980, chap. 20.
2. **Cimasoni, G.,** *The Crevicular Fluid Updated,* Vol. 3, Myers, H. M., Ed., S. Karger, Basel, 1983.
3. **Bowden, G. H., Ellwood, D. C., and Hamilton, I. R.,** Microbial ecology of the oral cavity, in *Advances in Microbial Ecology,* Vol. 3, Alexander, M., Ed., Plenum Press, New York, 1979, 135.
4. **Marsh, P. D. and Martin, M. V.,** *Oral Microbiology,* 2nd ed., Van Nostrand Reinhold, New York, 1984, chap. 2.
5. **Jenkins, G. N.,** *Physiology of the Mouth,* 4th ed., Blackwell Scientific, Oxford, 1978, chap. 9.
6. **Kenney, E. B. and Ash, M.,** Oxidation-reduction potential of developing plaque, periodontal pockets and gingival sulci, *J. Periodontol.,* 40, 630, 1969.
7. **Mandel, I. D.,** In defense of the oral cavity, in *Saliva and Dental Caries,* Kleinberg, I., Ellison, S. A., and Mandel, I. D., Eds., Information Retrieval, New York, 1979, 473.

8. **Bowen, W. H. and Cornick, D. E.,** The microbiology of gingival-dental plaque. Recent findings from primate research, *Int. Dent. J.*, 20, 382, 1970.
9. **Littleton, N. W., McCabe, R. M., and Carter, C. H.,** Studies of oral health in persons nourished by stomach tube. II. Acidogenic properties and selected bacterial components of plaque material, *Arch. Oral Biol.*, 12, 601, 1967.
10. **Dreizen, S. and Brown, L. R.,** Xerostomia and dental caries, in *Microbial Aspects of Dental Caries*, Vol. 1, Stiles, H. M., Loesche, W. J., and O'Brien, T. C., Eds., Information Retrieval, New York, 1976, 263.
11. **van der Hoeven, J. S., de Jong, M. H., Rogers, A. H., and Camp, P. J. M.,** A conceptual model for the co-existence of *Streptococcus* spp. and *Actinomyces*, spp. in dental plaque, *J. Dent. Res.*, 63, 389, 1984.
12. **Bowden, G. H., Hardie, J. M., and Slack, G. L.,** Microbial variations in approximal dental plaque, *Caries Res.*, 9, 253, 1975.
13. **Dennis, D. A., Gawronski, T. H., Sudo, S. Z., Harris, R. S., and Folke, L. E. A.,** Variations in microbial and biochemical components of four-day plaque during a four-week controlled diet period, *J. Dent. Res.*, 54, 716, 1975.
14. **Staat, R. H., Gawronski, T. H., Cressey, D. E., Harris, R. S., and Folke, L. E. A.,** Effects of dietary sucrose levels on the quantity and microbial composition of human dental plaque, *J. Dent. Res.*, 54, 872, 1975.
15. **Hengstenberg, W.,** Enzymology of carbohydrate transport in bacteria, *Curr. Top. Microbiol. Immunol.*, 77, 97, 1977.
16. **Calmes, R.,** Involvement of phosphoenolpyruvate in the catabolism of cariesconducive disaccharides by *Streptococcus mutans:* lactose transport, *Infect. Immun.*, 19, 934, 1978.
17. **Maryanski, J. H. and Wittenberger, C. L.,** Mannitol transport in *Streptococcus mutans*, *J. Bacteriol.*, 124, 1475, 1975.
18. **Schachtele, C. F. and Mayo, J. A.,** Phosphoenolpyruvate-dependent glucose transport in oral streptococci, *J. Dent. Res.*, 52, 1209, 1973.
19. **Slee, A. M. and Tanzer, J. M.,** Phosphoenolpyruvate-dependent sucrose phosphotransferase system in *Streptococcus mutans* NCTC 10449, *Infect. Immun.*, 24, 821, 1979.
20. **St. Martin, E. J. and Wittenberger, C. L.,** Characterization of a phosphoenolpyruvate-dependent sucrose phosphotransferase system in *Streptococcus mutans*, *Infect. Immun.*, 24, 865, 1979.
21. **Vadeboncoeur, C. and Trahan, L.,** Glucose transport in *Streptococcus salivarius*. Evidence for the presence of a distinct phosphoenolpyruvate:glucose phosphotransferase system which catalyses the phosphorylation of α-methyl-glucoside, *Can. J. Microbiol.*, 28, 190, 1982.
22. **Ellwood, D. C., Phipps, P. J., and Hamilton, I. R.,** Effect of growth rate and glucose concentration on the activity of the phosphoenolpyruvate-phosphotransferase system in *Streptococcus mutans* grown in continuous culture, *Infect. Immun.*, 23, 224, 1979.
23. **Ellwood, D. C. and Hamilton, I. R.,** Properties of *Streptococcus mutans* Ingbritt growing on limiting sucrose in a chemostat: repression of the phosphoenolpyruvate phosphotransferase transport system, *Infect. Immun.*, 36, 576, 1982.
24. **Kanapka, J. A. and Hamilton, I. R.,** Fluoride inhibition of enolase activity *in vivo* and its relationship to the inhibition of glucose-6-P formation in *Streptococcus salivarius*, *Arch. Biochem. Biophys.*, 146, 167, 1971.
25. **Roberts, K. R. and Linder, L.,** Phosphoenolpyruvate-dependent glucose phosphotransferase activity in *Streptococcus mitis* ATCC 903, *Scand. J. Dent. Res.*, 88, 316, 1980.
26. **Marsh, P. D., Keevil, C. W., McDermid, A. S., Williamson, M. I., and Ellwood, D. C.,** Inhibition by the antimicrobial agent chlorhexidine of acid production and sugar transport in oral streptococcal bacteria, *Arch. Oral Biol.*, 28, 233, 1983.
27. **Vadeboncoeur, C., Bourgeau, G., Mayrand, D., and Trahan, L.,** Control of sugar utilization in the oral bacteria *Streptococcus salivarius* and *Streptococcus sanguis* by the phosphoenolpyruvate phosphotransferase system, *Arch. Oral Biol.*, 28, 123, 1983.
28. **Hamilton, I. R. and Ellwood, D. C.,** Effects of fluoride on carbohydrate metabolism by washed cells of *Streptococcus mutans* grown at various pH values in a chemostat, *Infect. Immun.*, 19, 434, 1978.
29. **Hamilton, I. R. and St. Martin, E. J.,** Evidence for the involvement of proton motive force in the transport of glucose by a mutant of *Streptococcus mutans* strain DR 0001 defective in glucose-phosphoenolpyruvate phosphotransferase activity, *Infect. Immun.*, 36, 567, 1982.
30. **Keevil, C. W., Williamson, M. I., Marsh, P. D., and Ellwood, D. C.,** Evidence that glucose and sucrose uptake in oral streptococci involves independent phosphotransferase and proton motive force-mediated mechanisms, *Arch. Oral Biol.*, in press.
31. **Porter, E. V., Chassy, B. M., and Holmbund, C. E.,** Partial purification and properties of a specific glucokinase from *Streptococcus mutans* SL-1., *Biochim. Biophys. Acta*, 611, 289, 1980.
32. **Padan, E., Zilberstein, D., and Schuldiner, S.,** pH Homeostasis in bacteria, *Biochim. Biophys. Acta*, 650, 151, 1981.

33. **Carlsson, J. and Griffiths, C. J.,** Fermentation products and bacterial yields in glucose-limited and nitrogen-limited cultures of streptococci, *Arch. Oral Biol.*, 19, 1105, 1974.
34. **Ellwood, D. C., Hunter, J. R., and Longyear, V. M. C.,** Growth of *Streptococcus mutans* in a chemostat, *Arch. Oral Biol.*, 19, 659, 1974.
35. **Featherstone, J. D. B. and Rodgers, B. E.,** Effect of acetic, lactic and other organic acids on the formation of artificial carious lesions, *Caries Res.*, 15, 377, 1981.
36. **Yamada, T. and Carlsson, J.,** Regulation of lactate dehydrogenase and change of fermentation products in streptococci, *J. Bacteriol.*, 124, 55, 1975.
37. **Ellwood, D. C.,** Chemostat studies of oral bacteria, in *Microbial Aspects of Dental Caries*, Vol. 3, Stiles, H. M., Loesche, W. J., and O'Brien, T. C., Eds., Information Retrieval, New York, 1976, 785.
38. **Geddes, D. A. M.,** Acids produced by human dental plaque metabolism *in situ*, *Caries Res.*, 9, 98, 1975.
39. **Cole, M. F., Bowden, G. H., Korts, D. C., and Bowen, W. H.,** The effect of pyridoxine, phytate and invert sugar on production of plaque acids *in situ* in the monkey *(M. fascicularis)*, *Caries Res.*, 12, 190, 1978.
40. **Geddes, D. A. M.,** The production of L(+) and D(−) lactic acid and volatile acids by human dental plaque and the effect of plaque buffering and acidic strength on pH., *Arch. Oral Biol.*, 17, 537, 1972.
41. **Gilmour, M. N., Green, G. C., Zahn, L. M., Sparmann, C. D., and Pearlman, J.,** The $C_1$-$C_4$ monocarboxylic and lactic acids in dental plaques before and after exposure to sucrose *in vivo*, in *Microbial Aspects of Dental Caries*, Vol. 2, Stiles, H. M., Loesche, W. J., and O'Brien, T. C., Eds., New York, 1976, 539.
42. **van der Hoeven, J. S., Franken, H. C. M., Camp, P. J. M., and Dellebarre, C. W.,** Analysis of bacterial fermentation products by isotachophoresis, *Appl. Environ. Microbiol.*, 35, 17, 1978.
43. **Schachtele, C. F. and Jensen, M. E.,** Comparison of methods for monitoring changes in the pH of human dental plaque, *J. Dent. Res.*, 61, 1117, 1982.
44. **Stephan, R. M.,** Intra-oral hydrogen-ion concentrations associated with dental caries activity, *J. Dent. Res.*, 23, 257, 1944.
45. **Silverstone, L. M., Johnson, M. W., Hardie, J. M., and Williams, R. A. D.,** *Dental Caries, Aetiology, Pathology and Prevention*, Macmillan, New York, 1981, chap. 5.
46. **Mäkinen, K. K.,** Microbial growth and metabolism in plaque in the presence of sugar alcohols, in *Microbial Aspects of Dental Caries*, Vol. 2, Stiles, H. M., Loesche, W. J., and O'Brien, T. C., Eds., Information Retrieval, New York, 1976, 521.
47. **Hamilton, I. R., Phipps, P. J., and Ellwood, D. C.,** Effect of growth rate and glucose concentration on the biochemical properties of *Streptococcus mutans* Ingbritt in continuous culture, *Infect. Immun.*, 19, 434, 1978.
48. **Marsh, P. D., Williamson, M. I., Keevil, C. W., McDermid, A. S., and Ellwood, D. C.,** Influence of sodium and potassium ions on acid production by washed cells of *Streptococcus mutans* Ingbritt and *Streptococcus sanguis* NCTC 7865 grown in a chemostat, *Infect. Immun.*, 36, 476, 1982.
49. **Beighton, D.,** The influence of manganese on carbohydrate metabolism and caries induction by *Streptococcus mutans* Ingbritt, *Caries Res.*, 16, 189, 1982.
50. **Marsh, P. D., Keevil, C. W., and Ellwood, D. C.,** Relationship of bioenergetic processes to the pathogenic properties of oral bacteria, *J. Dent. Res.*, 63, 401, 1984.
51. **Harper, D. S. and Loesche, W. J.,** Effect of pH upon sucrose and glucose catabolism by the various genogroups of *Streptococcus mutans*, *J. Dent. Res.*, 62, 526, 1983.
52. **Tanzer, J. M., Freedman, M. L., Woodiel, F. N., Eifert, R. L., and Rinehimer, L. A.,** Association of *Streptococcus mutans* virulence with synthesis of intracellular polysaccharide, in *Microbial Aspects of Dental Caries*, Vol. 3, Stiles, H. M., Loesche, W. J., and O'Brien, T. C., Eds., Information Retrieval, New York, 1976, 597.
53. **van Houte, J., Backer-Dirks, O., de Stoppelaar, J. D., and Jansen, H. M.,** Iodophilic polysaccharide-producing bacteria and dental caries in children consuming fluoridated and non-fluoridated drinking water, *Caries Res.*, 3, 178, 1969.
54. **Huis in't Veld, J. H. J., van Palenstein Helderman, W. H., and Backer Dirks, O.,** *Streptococcus mutans* and dental caries in humans: a bacteriological and immunological study, *Antonie van Leeuwenhoek; J. Microbiol. Serol.*, 45, 25, 1979.
55. **Keevil, C. W., Marsh, P. D., and Ellwood, D. C.,** Regulation of glucose metabolism in oral streptococci through independent pathways of glucose 6-phosphate and glucose 1-phosphate formation, *J. Bacteriol.*, 147, 560, 1984.
56. **Hare, M. D., Svensson, S., and Walker, G. J.,** Characterization of the extracellular, water-insoluble α-D-glucans of oral streptococci by methylation analysis, and by enzymic synthesis and degradation, *Carbohyd. Res.*, 66, 245, 1978.
57. **Hamada, S. and Slade, H. D.,** Mechanisms of adherence of *Streptococcus mutans* to smooth surfaces *in vitro*, in *Bacterial Adherence*, Beachey, E. H., Ed., Chapman & Hall, London, 1980.

58. **Doyle, R. J. and Ciardi, J. E., Eds.**, *Glucosyltransferases, Glucans, Sucrose and Dental Caries*, Information Retrieval, New York, 1983.
59. **McCabe, M. M., Hanelik, R. M., and Smith, E. E.**, Purification of dextran-binding protein from cariogenic *Streptococcus mutans*, *Biochem. Biophys. Res. Commun.*, 78, 273, 1977.
60. **Baird, J., Longyear, V. M. C., and Ellwood, D. C.**, Water-insoluble and soluble glucans produced by extracellular glucosyltransferases from *Streptococcus mutans*, *Microbios*, 8, 143, 1973.
61. **van der Hoeven, J. S.**, A slime producing microorganism in dental plaque of rats selected by glucose feeding. Chemical composition of extracellular slime elaborated by *Actinomyces viscosus* Ny 1, *Caries Res.*, 8, 193, 1974.
62. **Hamelik, R. M. and McCabe, M. M.**, An endodextranase inhibiter from batch cultures of *Streptococcus mutans*, *Biochem. Biophys. Res. Comm.*, 106, 875, 1982.
63. **Jacques, N. A. and Wittenberger, C. L.**, Inactivation of cell-associated fructosyltransferase in *Streptococcus salivarius*, *J. Bacteriol.*, 148, 912, 1981.
64. **Biswas, S. D. and Kleinberg, I.**, Effect of urea concentration on its utilization, on the pH and the formation of ammonia and carbon dioxide in a human salivary sediment system, *Arch. Oral Biol.*, 16, 759, 1971.
65. **Kleinberg, I., Kanapka, J. A., Chatterjee, R., Craw, D., D'Angelo, N., and Sandham, H. G.**, Metabolism of nitrogen by the oral mixed bacteria, in *Saliva and Dental Caries*, Kleinberg, I., Ellison, S. A., and Mandel, I. D., Eds., Information Retrieval, New York, 1979, 357.
66. **Cowman, R. A., Fitzgerald, R. J., Perella, M. M., and Cornell, A. H.**, Human saliva as a nitrogen source for oral streptococci, *Caries Res.*, 11, 1, 1977.
67. **Cowman, R. A., Perella, M. M., and Fitzgerald, R. J.**, Influence of incubation atmosphere on growth and amino acid requirements of *Streptococcus mutans*, *Appl. Microbiol.*, 27, 86, 1974.
68. **Cowman, R. A., Perella, M. M., Adams, B. O., and Fitzgerald, R. J.**, Amino acid requirements and proteolytic activity of *Streptococcus sanguis*, *Appl. Microbiol.*, 30, 374, 1975.
69. **Hayes, M. L. and Hyatt, A. T.**, The decarboxylation of amino acids by bacteria derived from human dental plaque, *Arch. Oral Biol.*, 19, 361, 1974.
70. **Cowman, R. A. and Fitzgerald, R. J.**, Effects of oral streptococci on electrophoretic properties of human salivary anionic proteins, *J. Dent. Res.*, 54, 298, 1975.
71. **Cowman, R. A., Schaefer, S. J., Fitzgerald, R. J., Rosner, D., Shklair, I. L., and Walker, R. G.**, Differential utilization of proteins in saliva from caries-active and caries-free subjects as growth substrates by plaque-forming streptococci, *J. Dent. Res.*, 58, 2019, 1979.
72. **Cowman, R. A., Baron, S. S., Fitzgerald, R. J., Danziger, J. L., and Quintana, J. A.**, Growth inhibition of oral streptococci in saliva by anionic proteins from two caries-free individuals, *Infect. Immun.*, 37, 513, 1982.
73. **Cowman, R. A., Fitzgerald, R. J., and Schaefer, S. J.**, Role of salivary nitrogenous constituents in enhancement of bacterial glycolysis and growth, in *Saliva and Dental Caries*, Kleinberg, I., Ellison, S. A., and Mandel, I. D., Eds., Information Retrieval, New York, 1979, 343.
74. **Kleinberg, I., Craw, D., and Komiyama, K.**, Effect of salivary supernatant on the glycolytic activity of the bacteria in salivary sediment, *Arch. Oral Biol.*, 18, 787, 1983.
75. **Kleinberg, I., Kanapka, J. A., and Craw, D.**, Effect of saliva and salivary factors on the metabolism of the mixed oral flora, in *Microbial Aspects of Dental Caries*, Vol. 2, Stiles, H. M., Loesche, W. J., and O'Brien, T. C., Eds., Information Retrieval, New York, 1976, 433.
76. **Kleinberg, I., Jenkins, G. N., Chatterjee, R., and Wijegeweera, L.**, The antimony pH electrode and its role in the assessment and interpretation of dental plaque pH, *J. Dent. Res.*, 61, 1139, 1982.
77. **Mikx, F. H. M. and van Campen, G. J.**, The microbial community on the teeth, with special reference to necrotizing ulcerative gingivitis in dogs, *Eur. J. Chemother. Antibiot.*, 2, 19, 1982.
78. **Mikx, F. H. M. and van der Hoeven, J. S.**, Symbiosis of *Streptococcus mutans* and *Veillonella alkalescens* in mixed continuous cultures, *Arch. Oral Biol.*, 20, 407, 1975.
79. **Shah, H. K. and Williams, R. A. D.**, Metabolic studies on *B. asaccharolyticus* and *B. melaninogenicus* subsp. *intermedius*, *J. Dent. Res.*, 60 (Abstr.), 1161, 1981.
80. **Williams, R. A. D. and Shah, H. N.**, Enzyme patterns in bacterial classification and identification, in *Microbiological Classification and Identification*, Goodfellow, M. and Board, R. G., Eds., Academic Press, New York, 1980.
81. **Ellwood, D. C., Keevil, C. W., Marsh, P. D., Brown, C. M., and Wardell, J. N.**, Surface-associated growth, *Philos. Trans. R. Soc. Lond. Ser. B*, 297, 517, 1982.
82. **Berry, C. W. and Henry, C. A.**, The effect of adsorption on the acid production of caries and non-caries-producing streptococci, *J. Dent. Res.*, 56, 1193, 1977.
83. **Hardie, J. M., Thomson, P. L., South, R. J., Marsh, P. D., Bowden, G. H., McKee, A. S., Fillery, E. D., and Slack, G. L.**, A longitudinal epidemiological study on dental plaque and the development of dental caries — interim results after two years, *J. Dent. Res.*, 56, C90, 1977.
84. **Loesche, W. J. and Straffon, L. H.**, Longitudinal investigation of the role of *Streptococcus mutans* in human fissure decay, *Infect. Immun.*, 26, 498, 1979.

85. **Newbrun, E.,** *Cariology,* Williams & Wilkins, Baltimore, 1978.
86. **Hayes, M. L., Carter, E. L., and Griffiths, S. J.,** The acidogenic microbial composition of dental plaque from caries-free and caries-prone people, *Arch. Oral Biol.,* 28, 381, 1983.
87. **Hamilton, I. R.,** Effects of fluoride on enzymatic regulation of bacterial carbohydrate metabolism, *Caries Res.,* 11, 262, 1977.
88. **Eisenberg, A. D., Bender, G. R., and Marquis, R. E.,** Reduction in the aciduric properties of the oral bacterium *Streptococcus mutans* GS-5 by fluoride, *Arch. Oral Biol.,* 25, 133, 1980.
89. **Edgar, W. M.,** Fluoride metabolism in dental plaque bacteria and man, in *The Environment of the Teeth,* Vol. 3, Ferguson, D. B., Ed., S. Karger, Basel, 1981.
90. **Leach, S. A.,** Mode of action of chlorhexidine in the mouth, in *The Borderland between Caries and Periodontal Disease,* Lehner, T., Ed., Academic Press, New York, 1977, 105.
91. **Gjermo, P., Bonesvoll, P., and Rölla, G.,** Relationship between plaque-inhibiting effect and retention of chlorhexidine in the human oral cavity, *Arch. Oral Biol.,* 19, 1031, 1974.
92. **Shaw, J. H. and Roussos, G. G., Eds.,** *Sweeteners and Dental Caries,* Information Retrieval, New York, 1978.
93. **Hayes, M. L.,** The inhibition of bacterial glycolysis in human dental plaque by medium-chain fatty acid-sugar mouth-washes, *Arch. Oral Biol.,* 26, 223, 1981.
94. **Guggenheim, B., Regolati, B., and Mühlemann, H. R.,** Caries and plaque inhibition by mutanase in rats, *Caries Res.,* 6, 289, 1972.
95. **Jordan, H. V., Keyes, P. H., and Bellack, S.,** Periodontal lesions in hamsters and gnotobiotic rats infected with *Actinomyces* of human origin, *J. Periodontal Res.,* 7, 21, 1972.
96. **Moore, W. E. C., Ranney, R. R., and Holdeman, L. V.,** Subgingival microflora in periodontal disease: cultural studies, in *Host-Parasite Interactions in Periodontal Diseases,* Genco, R. J. and Mergenhagen, S. E., Eds., American Society for Microbiology, Washington, D. C., 1982, 13.
97. **Loesche, W. J. and Syed, S. A.,** Bacteriology of human experimental gingivitis: effect of plaque and gingivitis score, *Infect. Immun.,* 41, 320, 1978.
98. **Moore, W. E. C., Holdeman, L. V., Smibert, R. M., Good, I. J., Burmeister, J. A., Palcanis, K. G., and Ranney, R. R.,** Bacteriology of experimental gingivitis in young adult humans, *Infect. Immun.,* 26, 498, 1982.
99. **Kornman, K. S.,** Age, supragingival plaque, and steroid hormones as ecological determinants of the subgingival flora, in *Host-Parasite Interactions in Periodontal Diseases,* Genco, R. J. and Mergenhagen, S. E., Eds., American Society for Microbiology, Washington, D.C., 1982, 132.
100. **Socransky, S. S., Tanner, S. C. R., Haffajee, A. D., Hillman, J. D., and Goodsen, J. M.,** Present status of studies on the microbial aetiology of periodontal disease, in *Host-Parasite Interactions in Periodontal Diseases,* Genco, R. J. and Mergenhagen, S. E., Eds., American Society for Microbiology, Washington, D.C., 1982, 1.
101. **Slots, J.,** The predominant cultivable micro-flora of advanced periodontitis, *Scand. J. Dent. Res.,* 85, 114, 1977.
102. **Moore, W. E. C., Holdeman, L. V., Cato, E. P., Smibert, R. M., Burmeister, J. A., and Ranney, R. R.,** Bacteriology of moderate (chronic) periodontitis in mature adult humans, *Infect. Immun.,* 42, 510, 1983.
103. **Slots, J. and Genco, R. J.,** Black-pigmented *Bacteroides* species, *Capnocytophaga* species, and *Actinobacillus actinomycetemcomitans* in human periodontal disease: virulence factors in colonization, survival and tissue destruction, *J. Dent. Res.,* 63, 412, 1984.
104. **Genco, R. J. and Slots, J.,** Host responses in periodontal diseases, *J. Dent. Res.,* 63, 441, 1984.
105. **Hammond, B. F. and Stevens, R. H.,** *Capnocytophaga* and *Actinobacillus actinomycetemcomitans:* occurrence and pathogenic potential in juvenile periodontitis, in *Host-Parasite Interactions in Periodontal Diseases,* Genco, R. J. and Mergenhagen, S. E., Eds., American Society for Microbiology, Washington, D.C., 1982, 46.
106. **Listgarten, M. A.,** Electron microscopic observations on the bacterial flora of acute necrotizing ulcerative gingivitis, *J. Periodontol.,* 36, 328, 1965.
107. **Loesche, W. J., Syed, S. A., Laughon, B. G., and Stoll, J.,** The bacteriology of acute necrotizing ulcerative gingivitis, *J. Periodontol.,* 53, 223, 1982.
108. **Lö, H. and Kornman, K.,** Strategies in the use of antibacterial agents in periodontal disease, in *Host-Parasite Interactions in Periodontal Diseases,* Genco, R. J. and Mergenhagen, S. E., Eds., American Society for Microbiology, Washington, D.C., 1982, 376.
109. **Genco, R. J.,** Antibiotics in the treatment of human periodontal diseases, *J. Periodontol.,* 52, 545, 1981.
110. **Soh, L. L., Newman, H. N., and Strahan, J. D.,** Effects of subgingival chlorhexidine irrigation on periodontal inflammation, *J. Clin. Periodontol.,* 9, 66, 1982.
111. **Coventry, J. and Newman, H. N.,** Experimental use of a slow release device employing chlorhexidine gluconate in areas of acute periodontal inflammation, *J. Clin. Periodontol.,* 9, 129, 1982.
112. **Stephan, R. M.,** Intra-oral hydrogen-ion concentrations associated with dental caries activity, *J. Dent. Res.,* 23, 257, 1944.

Chapter 12

# BACTERIAL METABOLISM AND ACUTE ENTERIC INFECTION

**B. S. Drasar**

## TABLE OF CONTENTS

# I. INTRODUCTION

The role of the normal intestinal flora in the development of diarrheal illness is much discussed but little understood. Normal intestinal bacteria are important in preventing colonization by intestinal pathogens and probably in eliminating those pathogens that do become established. These activities of the flora will be discussed under the general heading of colonization resistance; special aspects of the subject include the problem of infectious dose and question of breast feeding and weanling diarrhea. Some workers discuss colonization resistance as if it were a simple property of the flora but is undoubtedly the result of many complex events. Before proceeding to examine these topics I will first briefly outline causes of diarrhea.

# II. THE CAUSES OF DIARRHEA

Diarrhea and dysentery are among the major symptoms of acute enteric infection. The most important metabolic consequence is the osmotic imbalance that causes the characteristic water loss. There is a sense in which these effects are all attributable to microbial action either directly as a result of specific virulence factors or indirectly as a result of the fermentation of unabsorbed materials and products of mucosal damage. The normal bacterial flora may be involved both in the prevention and the amplification of the effects of infection.

Many noninfectious agencies are capable of producing intestinal dysfunction in man and animals which is manifest as diarrhea. The noninfectious forms may involve nutritional disturbances. The injudicious use of oral antibiotics can also create problems. Removal of the normal flora by antibiotics can allow overgrowth of microorganisms normally present in low numbers. With *Candida albicans* this produces intestinal irritation. A more serious problem follows overgrowth by *Clostridium difficile* following the use of clindamycin. This species elaborates a heat labile enterotoxin which contributes to the production of pseudomembranous enterocolitis in infants.

Diarrhea occurs in both developed and developing countries in all seasons and among people of all ages. The cause of most diarrhea remains uninvestigated and undiagnosed but it is assumed to be infective in origin. The major causes of diarrheal disease are listed in Table 1. Until recently even if every effort was made to determine the causes of diarrhea an infective agent would only be demonstrated in less than half of the cases. Recent advances have transformed our outlook with the demonstration of a number of previously unsuspected agents.[1-4]

Viral etiologies of diarrhea, although clinically and numerically very important have only been recently studied. Rotaviruses produce diarrhea in man and many domestic animals and fowl. They produce severe diarrhea in experimental animals. It is thought by producing stunting and loss of villi and microvilli, a transient malabsorption syndrome results with consequent impacts on bacterial metabolism in the gut. The normal bacterial flora may also contribute to the severity of the disease in other ways. Cross infection of different host species with viral strains from other hosts occurs and many identifiable strains exist indicating that the epidemiology is complex and at this stage the prospects for immunological control are poor. Many other viral agents such as astroviruses, caliciviruses and adenoviruses have been detected in the feces of children and pigs with diarrhea. A causal relationship has not been established in all cases. Experimental infection with some of these viruses has reproduced the diarrhea.[5]

Bacteriological studies of diarrhea have traditionally concentrated on the identification of bacterial pathogens. This approach has resulted in identification of many bacterial species associated with diarrheal illness. *Campylobacter jejuni* is the most recent success. This organism is now established as a major cause of diarrhea probably more important than Salmonellae.[6,7]

**Table 1**
**INFECTIOUS AGENTS OF DIARRHEA**

| | |
|---|---|
| Bacteria: part I | |
| *Bacillus cereus* | *Staphylococcus* aureus |
| *Campylobacter jejuni* | *Vibrio cholerae* |
| | *Vibrio cholerae* 01 |
| *Clostridia* | *Vibrio cholerae* non-01 |
| *Clostridium difficile* | *Vibrio parahaemolyticus* |
| *Clostridium perfringens* | *Yersinia enterocolitica* |
| *Escherichia coli* | |
| Enteroinvasive *E. coli* (EIEC) | |
| Enterotoxogenic *E. coli* (ETEC) | |
| *Salmonella* | |
| *Salmonella* paratyphi | |
| *Salmonella* typhi | |
| Serotypes other than *S. typhi* and *S. paratyphi* | |
| Shigellae | |
| Viruses: part II | |
| Rotavirus | Norwalk and Norwalk-like agents |
| Protozoa: part III | |
| *Balantidium coli* | *Entamoeba histolytica* |
| *Giardia lamblia* | |
| Helminths: part IV | |
| *Strongyloides stercoralis* | *Trichuris trichiura* |

However, since the improved understanding of the mechanisms by which diarrhea is produced by *Vibrio cholerae* attention has focused on the mechanisms of microbial pathogenicity (Table 2).

When discussing bacteria as a cause of diarrhea it is important to distinguish two groups of organisms in terms of their ecological association with the gut. Some organisms can produce diarrhea in man without colonizing the intestine as a result of the production of toxins in food or the release of toxin by ingested organisms (Table 2). These intestinal intoxications are characterized by their sudden onset, within a few hours of the ingestion of infected material. The symptoms are usually of short duration, person to person transmission cannot occur and outbreaks are associated with communal meals.

It is unlikely that the normal flora has any important influence on the initiation of these illnesses though there may be some effect on malabsorbed nutrients.

The initiation of infection by most intestinal pathogens is dependent on their ability to reach the intestine and to attack the intestinal mucosa. The interactions between the pathogen and the enterocyte are specific and are dependent upon the molecular architecture of the surface of the enterocyte and bacterial cell. Adhesive and colonization factors are the most obvious expression of these interactions. Other factors concerning bacteria including extracellular enzymes and motility may be equally important, while glycocalyx composition and mucus production are significant host determinants.[1,4]

Once established in the gut, enteric pathogens produce their effect either by the elaboration of a toxin or by invasion of the intestinal mucosa (Table 2). Symptoms of diarrhea result when the effect produced by the bacterial activity is able to overcome intestinal homeostasis. Thus, for example, the loss of water from the small intestine induced by enterotoxin producing bacteria such as *V. cholerae* and *Escherichia coli* may be compensated for by the adsorption of water in the colon and only when the reabsorptive mechanisms can no longer cope will symptoms result (Table 3).

**Table 2**
**BACTERIAL CAUSES OF DIARRHEA**

| Organism | Site of infection | Pathogenic mechanism |
|---|---|---|
| *Bacillus cereus* | Not applicable | Preformed toxin ingested |
| *Clostridium perfringens A* | Not applicable | Toxin released by ingested bacteria $10^9$-$10^{10}$ in small intestine |
| *Clostridium perfringens A*[a] | Large intestine | Toxin |
| *Staphylococcus aureus* | Not applicable | Preformed toxin ingested |
| *Escherichia coli* | Small intestine | Enterotoxins (LT and ST)[b] |
| | Large intestine | Invasion |
| *Salmonella* spp. | Small and large intestine | Invasion |
| *Shigella* spp. | Large intestine | Invasion enterotoxin[c] |
| *Yersinia enterocolitica* | Large intestine | Invasion enterotoxin? |
| *Vibrio cholerae* | Small intestine | Enterotoxin (LT)[b] |
| *Vibrio parahaemolyticus* | Small intestine | Enterotoxin |
| *Campylobacter jejuni* | ? | Invasion enterotoxin (ST)[b] |
| *Clostridium difficile*[c] | Large intestine | Toxins |

[a] Antibiotic treatment especially important as predisposing factor.
[b] Other enterotoxins described.
[c] *Shigella dysentieriae* type I.

**Table 3**
**FACTORS DETERMINING THE PRODUCTION OF DIARRHEAL DISEASE**

| Stage of infection | Host factors | Pathogen |
|---|---|---|
| Transmission and initiation of infection | Socio-economic factors<br>Food hygiene<br>Excreta disposal<br>Public health | Animal host range<br>Environmental survival<br>Infective dose |
| Nature of infection | Antimicrobial defenses<br>Gastric acidity<br>Intestinal receptors<br>Concurrent infections | Growth phase, adhesins and colonization factors |
| Pathological process | Immune system | Enterotoxin production |
| Water loss | Antitoxic immunity<br>Normal flora | Mucosal invasion |
| Diarrhea | Homeostatic mechanism of water absorption in the gut | Intestinal growth potential |

Fermentation of unabsorbed metabolites can potentiate the osmotic effects of intestinal infection. This is particularly important in those infections such as virus infections that cause structural damage to the enterocyte. The fermentative diarrhea associated with bacterial malabsorption may serve as a model for these effects. Genetically determined malabsorption of disaccharides associated with deficiencies of lactase, sucrase-isomaltase, and other disaccharidases[8] are well known: deficiency of trehalase has also been described. Secondary deficiencies also occur after an attack of acute gastroenteritis. Lactase deficiency is acquired on weaning by a large proportion of the world's population; this deficiency is related to ethnic/cultural groupings and may be related to the historical development of milk drinking.[8] Disaccharide malabsorption leads to fermentative diarrhea, the pH of the stool is lowered and the amount of lactic acid vastly increased.[8,9]

# III. COLONIZATION RESISTANCE

The gut bacteria function as part of the intestines' defenses against pathogenic microorganisms. To initiate disease, the pathogen must be able to compete successfully with the normal flora. Protection by the flora is most dramatically demonstrated in studies on germfree and antibiotic treated animals. Thus, the pretreatment of mice with streptomycin increases their susceptibility to oral challenge with Salmonellae.[10-14] In man the prolonged carriage of Salmonellae results.[15] Treatment of mice and guinea pigs with antibiotics increases their susceptibility to infection by Shigellae and in the case of guinea pigs also to *Vibrio cholerae*.[16] Germfree guinea pigs are susceptible to *Vibrio cholerae* and to Shigellae[17] while conventional animals are not.

Germfree mice are far more susceptible to oral challenge with Salmonellae and Shigellae than are conventional animals.[18,19]

The mechanisms of microbial interference are not understood. However, the production of volatile fatty acids is important in the inhibition of *Salmonella typhimurium* in the mouse gut. Savage[20] suggested that the spatial arrangement of microorganisms was important especially at the mucosal surface. Thus, organisms such as Shigellae might be excluded from the mucosa by the indigenous microbes adherent to the wall of the colon.

Except when successful invasion by a pathogen occurs the gut flora is probably a stable, self-regulating system. There is a natural resistance to the alteration of the flora by introduction of bacteria even if they are of species commonly encountered in the intestine. This resistance to change is understandable, the bacteria that comprise the flora are adapted to life in the intestine whereas freshly introduced strains require time to adapt.[21]

Although the intestinal bacteria have been the subject of intense study during the past 10 years our knowledge of the mechanisms controlling its composition and distribution has advanced little. The problem is largely technical. It has been estimated that some 400 distinct species of bacteria can normally be isolated from the intestine.[22] The labor of examining sufficient strains isolated from each sample obtained in order to detect this number of species renders ecological studies very difficult and expensive. The large intestine can be considered as a continuous culture fermenter. Indeed attempts have been made to explain the whole of host flora interactions in these terms. Dietary residues may be thought of as culture media.[23,24]

In practical terms this approach has led to the development of model systems that have allowed some of the factors controlling the flora to be identified.[25-28] The most interesting finding has been the demonstration that bacteria-bacteria interactions play a dominant role. Adhesion to specific receptors on the intestinal mucosa by the normal flora has been suggested to be an important mechanism by which pathogens are excluded and by implication an important mechanism controlling the flora.[20] However, studies in continuous culture suggest that most adhesion is between different types of bacteria, thus bacterial surfaces rather than host surfaces seem to be ecologically dominant.[28] These same studies suggest that metabolic poisons produced by bacteria, particularly $H_2S$, may be more significant in their impact on bacterial populations than the volatile fatty acids considered by most authorities.[16,29,30]

If the intestine is a continuous culture, a change in diet may be thought of as an alteration in the composition of the medium and would thus be expected to affect the bacterial growth.

Studies on people living on different diets in various countries under very different environmental circumstances have shown some differences in the relative numbers of some of the groups of bacteria present in their feces.[31,32] These differences may reflect diet but this is by no means certain. Attempts to change the composition of the flora by controlled changes in diet have, in general, been unsuccessful.[33-36]

Interactions between the bacteria must play an important role in determining both the total numbers and relative frequency of the various species. In any confined environment such as the large intestine space and nutrients are limited and bacteria that are able to transform

**Table 4**
**CHOLERA AND BLOOD GROUPS[39,40]**

| | Calcutta | | Philippines | |
|---|---|---|---|---|
| **Determinant** | **Population (%)** | **Patients (%)** | **Population (%)** | **Patients (%)** |
| A *N*-acetylgalactosamine | 24 | 20 | 26 | 10 |
| B D-galactose | 35 | 14 | 22 | 22 |
| O L fucose (H substance) | 31 | 61 | 45 | 64 |

nutrients into bacterial cells fastest under the prevailing conditions will occur in the greatest numbers. Bacteria able to utilize nutrients that are not used by other bacteria, will also possess an ecological advantage. The utilization of hydrogen by the *Methanobacteria* well illustrates this principle. Hydrogen is produced by the fermentative metabolism of many intestinal microbes in the absence of *Methanobacteria* the hydrogen is lost but these organisms convert it to methane which may be detected on the breath.[37]

The complexity of pathogen — normal flora — host interactions is well illustrated by consideration of the relationship of blood groups to clinical cholera (Table 4). Several studies have shown that cholera is more likely to occur in those with an O blood group.[38-40] This might be explained in terms of a congruence between the H substance and the fucose sensitive adhesion for *V. cholerae*.[41,42] The effectiveness of blood group substances as adhesins may in turn be dependent on secreter status of the host. The picture is further complicated by the demonstration that blood group substances are digested by the intestinal flora and that this activity is in part dependent on secreter status.[43,44] It may be objected that the activities of the flora are irrelevant in that *V. cholerae* is a small gut pathogen. However, as mentioned above, antibiotic treatment increases the susceptibility of animals to this organism.

### A. Infectious Dose

Bacteria able to cause intestinal infections are able to grow in the intestine. Thus, under the correct circumstances a single organism must be able to initiate a clinically significant infection. The infective dose to ensure that infection always occurs ($ID_{100}$), or the infective dose ensuring that an infection occurs half the time ($ID_{50}$), may be considered to represent, for a particular bacterial species, the probability that the single organism will reach the correct portion of intestine under the right circumstances to initiate a clinically apparent infection.

Among the factors that determine the ability of a Bacterium to initiate infection are the metabolic activities of the flora some of which have already been mentioned above. For several intestinal pathogens, attempts have been made to determine the number of bacteria needed to produce a clinically apparent infection or to produce intestinal colonization in human subjects (Table 5). The significance of the results of these studies is obscure. While they undoubtedly provide a league table of relative infectivity, it is doubtful if the actual infective doses are relevant to natural infections. The number of subjects exposed to infection in experimental studies is necessarily small and the experiments are designed to ensure that many of the subjects become infected. During an outbreak the number of subjects exposed may be very large, of whom only a small proportion become infected. Thus, the effective infective dose in an outbreak is probably much lower than in an experimental study. Indeed, studies in Nigeria on the ingestion of contaminated well water by infants have shown that less than 100 enterotoxogenic *E. coli* can cause infection.[45]

The influence of the intestinal flora on the effective infectious dose remains a matter of speculation. Significant factors might be

**Table 5**
**INFECTIVE DOSE**

| Organism | Infective dose |
|---|---|
| *Escherichia coli* | $10^6$—$10^8$ [a] |
| *Salmonella spp.* | $10^4$—$10^8$ |
| *Shigella spp.* | 10—200 |
| *Yersinia enterocolitica* | Unknown |
| *Vibrio cholerae* | $10^8$ |
| *Vibrio parahaemolyticus* | $10^6$—$10^8$ |
| *Campylobacter jejuni* | $10^2$—$10^5$ |
| *Clostridium difficile* | Unknown |

[a] Studies on children drinking contaminated well water in Africa indicate that fewer than 100 organisms will initiate an infection

1. Resistance of pathogens to metabolic inhibitors produced by the flora
2. Occupation of enterocyte receptors by the flora
3. Destruction of toxins and virulence factors by the flora

### B. *Bifidobacteria* and Breast Feeding

The interrelationship of breast feeding, infantile diarrhea and *Lactobacillus bifidus* (now more correctly *Bifidobacteria)* has been a subject of interest since the last century.

Infantile mortality from diarrheal disease has been recognized for a long time and has been reported under several names from different countries. Examination of historical records indicates that the epidemiology of the disease has changed as social and economic advance has occurred. In England and Wales the total mortality from diarrhea has decreased by some hundredfold during the present century. As total mortality fell the proportion of deaths contributed to this total, by the 0 to 5 year age group has fallen from about 90% to about 50%. Together with this reduction in mortality has occurred the disappearance of epidemics of summer diarrhea.

The reasons for this disappearance of summer diarrhea are not known but can be explained in terms of the improvements of excreta disposal and infant nutrition over this period. The relationship between environmental contamination and seasonal fluctuations in temperature and rainfall that precipitated an epidemic was not exactly determined. Newsholme[46] stated "The amount of epidemic or summer diarrhea varies directly with the summer temperature and inversely with the amount of rainfall in the summer months". In an earlier publication he laid stress also upon the excreta disposal systems and the structure of the subsoil, contending that these factors determined the amount and persistence of environmental contamination.[47] While Peters[48] included the date of the harvest among the important variables.

In Britain, summer diarrhea of infants was a major cause of infant mortality within living memory and had been recognized as a major epidemic disease for centuries.[49]

Similar epidemics have occurred in other developed countries and occur in the developing world at the present time. Endemic diarrhea of infants is common in most parts of the world and is often associated with weaning.[50] Breastfed infants are said to be resistant to infantile diarrhea whereas bottle-fed infants are susceptible.[51-55]

That *Bifidobacteria* are a major part of the fecal flora of breastfed infants has been known since the end of the 19th century. It was long assumed that these bacteria were absent from the gut of bottle-fed infants, older children and adults, and that the absence of *Bifidobacteria* in the intestine of the bottle-fed infant resulted in susceptibility to the agents of diarrhea. It has long been postulated that various bifidus factors in human breast milk stimulate the dominance of particular strains of *Bifidobacteria,* indeed some factors have been isolated.

*Bifidobacteria* are a prominent component of the fecal flora of breastfed infants in Nigeria, Guatemala and the U.K.[56-60] Examination of the results of these studies supports the suggestion that a breastfed flora is important. However, *Bifidobacteria* are also present in the feces of bottle-fed infants in the U.K. Mitsouka et al.[61-63] stated that they were unable to detect differences between the *Bifidobacteria* present in the feces of breast- and bottle-fed infants in Japan. A further complication was highlighted by the study of Simhon et al.[64] who showed that in some cases development of a Bifidobacterial flora may be delayed by a prolonged hospital stay even in breastfed infants.

In an attempt to resolve these problems Beerens et al.[65] concentrated their attention on the identity of the *Bifidobacteria* present in the feces of breastfed and bottle-fed infants. Their results suggest that *B. bifidum* and *B. longum* may be more common in the feces of breast than bottle-fed infants.

The whole question is far from clear. Systematic study of the fecal flora has shown that *Bifidobacteria* are present at all ages though the species may vary. Further, infantile diarrhea has declined in importance during the present century in the U.K. at the same time as breast-feeding has also declined. There is however, evidence from the developing world as to the importance of breast-feeding.

## REFERENCES


1. **Walker-Smith, J. A. and Manuel, P. D.,** *Infections of the Gastrointestinal Tract,* Churchill Livingstone, Edinburgh, 1984.
2. **Feachem, R.,** *Sanitation and Disease,* John Wiley & Sons, New York, 1983.
3. Ciba Foundation, Acute diarrhoea in childhood, *Ciba Foundation Symp.* (New Series), Elsevier, Amsterdam, 1976, 42.
4. **Dupont, H. and Pickering, L.,** *Infections of the Gastrointestinal Tract Microbiology, Pathophysiology, and Clinical Features,* Plenum Press, New York, 1980.
5. **Tyrrell, A. J.,** *Viral Infections of the Gastrointestinal Tract,* Dekker, New York, 1982.
6. **Pearson, A. D., Skirrow, M. B., Rowe, B., Davies, J. R., and Jones, D. M.,** *Campylobacter II,* Public Health Laboratory Service, London, 1983.
7. **Newell, D. G.,** *Campylobacter — Epidemiology, Pathogenesis and Biochemistry,* M.T.P. Press, England, 1982.
8. **Neale, G.,** Disaccharidases deficiencies, *J. Clin. Pathol.,* 24, 5, 1971.
9. **Fraser, A. C.,** *Malabsorption Syndromes,* Heinemann, London, 1968.
10. **Miller, C. P. and Bohnhoff, M.,** A study of experimental *Salmonella* infection in the mouse, *J. Infect. Dis.,* 111, 107, 1962.
11. **Bohnhoff, M., Miller, C. P., and Martin, W. R.,** Resistance of the mouse's intestinal tract to experimental Salmonella infection, *J. Exp. Med.,* 120, 817, 1964.
12. **Bohnhoff, M., Miller, C. P., and Martin, W. R.,** Resistance of a mousey's intestinal tract to experimental Salmonella infection, *J. Exp. Med.,* 120, 805, 1964.
13. **Meynell, G. G.,** Antibacterial mechanisms of the mouse gut. II. The role of Eh and volatile fatty acids in the normal gut, *Br. J. Exp. Pathol.,* 44, 209, 1963.
14. **Abrams, G. D. and Bishop, J. E.,** Effect of the normal microbial flora on the resistance of the small intestine to infection, *J. Bacteriol.,* 92, 1604, 1966.
15. Association for the Study of Infectious Diseases Report, Effect of neomycin in non-invasive salmonella infections of the gastrointestinal tract, *Lancet,* 2, 1159, 1970.
16. **Freter, R.,** Experimental enteric shigella and vibrio infections in mice and guinea pigs, *J. Exp. Med.,* 104, 411, 1956.
17. **Formal, S. B., Dammin, G., Sprinz, H., Kundel, D., Schneider, H., Horowitz, R., and Forbes, M.,** Shigella in germ-free guinea pigs, *J. Bacteriol.,* 82, 284, 1961.
18. **Bohnhoff, M., Drake, L., and Miller, C. P.,** Effect of streptomycin on susceptibility of intestinal tract to experimental Salmonella infection, *Proc. Soc. Exp. Biol.,* 86, 132, 1954.

19. **Maier, B. R. and Hentges, D. J.,** Experimental shigella infections in laboratory animals. I. Antagonism by human normal flora components in gnotobiotic mice, *Infect. Immun.*, 6, 168, 1972.
20. **Savage, D. C.,** Survival on mucosal epithelia, epithelial penetration and growth in tissues of pathogenic bacteria, *Microbial Pathogenicity in Man and Animals*, Smith, H. and Pearce, J. H., Eds., Cambridge University Press, New York, 1972.
21. **Ozawa, A. and Freter, R.,** Ecological mechanism controlling growth of *Escherichia coli* in continuous flow cultures and in the mouse intestine, *J. Infect. Dis.*, 114, 235, 1964.
22. **Holdeman, L. V., Good, I. J., and Moore, W. E. C.,** Human faecal flora: variation in bacterial composition within individuals and a possible effect of emotional stress, *Appl. Environ. Microbiol.*, 31, 359, 1976.
23. **Boni, P.,** Growth dynamics of the intestinal flora considered as a continuous flow culture, a thermodynamic and hydrodynamic qualitative approach, *Rev. Ist. Sieroter. Ital.*, 42, 261, 1967.
24. **Drasar, B. S.,** Some factors associated with geographical variations in the intestinal microflora, in *The Normal Microbial Flora of Man*, Skinner, F. A. and Carr, J. G., Eds., Academic Press, New York, 1974.
25. **Miller, T. L. and Wolin, M. J.,** Fermentation by the human large intestine microbial community in an in vitro semicontinuous culture system, *Appl. Environ. Microbiol.*, 42, 400, 1981.
26. **Freter, R., Stauffer, E., Cleren, D., Holdeman, L. V., and Moore, W. E. C.,** Continuous-flow cultures as in vitro models of the ecology of large intestinal flora, *Infect. Immun.*, 39, 666, 1983.
27. **Freter, R., Brickner, H., Fekete, J., Vickerman, M. M., and Carey, K. E.,** Survival and implantation of *Escherichia coli* in the intestinal tract, *Infect. Immun.*, 39, 686, 1983.
28. **Freter, R., Brickner, H., Botney, M., Cleven, D., and Aranki, A.,** Mechanisms that control bacterial populations in continuous-flow culture models of mouse large intestinal flora, *Infect. Immun.*, 39, 676, 1983.
29. **Hentges, D. J. and Freter, R.,** In vivo and in vitro antagonism of intestinal bacteria against *Shigella flexneri*. I. Correlation between various tests, *J. Infect. Dis.*, 110, 30, 1962.
30. **Maier, B. R., Onderdonk, A. B., Russell, M. S., Baskett, C., and Hentges, D. J.,** *Shigella*, indigenous flora interactions in mice, *Am. J. Clin. Nutr.*, 25, 1433, 1972.
30a. **Meynell, G. G. and Subbaiah, T. V.,** Antibacterial mechanisms of the mouse gut, *Br. J. Exp. Pathol.*, 44, 197, 1963.
31. **Aries, V., Crowther, J. S., Drasar, B. S., Hill, M. J., and Williams, R. E. O.,** Bacteria and the aetiology of cancer of the large bowel, *Gut*, 10, 334, 1969.
32. **Koornhof, H. J., Richardson, N. J., Wall, D. M., and Moore, W. E. C.,** Faecal bacteria in South African rural blacks and other population groups, *Isr. J. Med. Sci.*, 15, 335, 1979.
33. **Drasar, B. S., Jenkins, D. J. A., and Cummings, J. H.,** The influence of a diet rich in wheat fibre on the human faecal flora, *J. Med. Microbiol.*, 9, 423, 1976.
34. **Maier, B. R., Flynn, M. A., Burton, G. C., Tsutakawa, R. K., and Hentges, D. J.,** Effects of a high-beef diet on bowel flora: a preliminary report, *J. Clin. Nutr.*, 27, 1470, 1974.
35. **Cummings, J. H., Wiggins, H. S., Jenkins, D. J. A., Houston, H., Jivraj, T., Drasar, B. S., and Hill, M. J.,** Influence of diets high and low in animal fat on bowel habit, gastrointestinal transit time, faecal microflora, bile acid and fat excretion, *J. Clin. Invest.*, 61, 953, 1978.
36. **Drasar, B. S.,** The anaerobic flora of the human intestine — concepts and methods, *Symp. Inter.: Les Anaerobies*, Masson, New York, 1980, 109.
37. **Drasar, B. S., Tomkins, A. M., Wiggins, H., and Hudson, M.,** Breath methane levels and intestinal methanogenesis among rural Nigerians on a local diet, *Proc. Nutr. Soc.*, 43, 86a, 1986.
38. **Levine, M. M., Nalin, D. R., Rennels, M. B., Hornick, R. B., Sotman, S., Van Blerk, G., Hughes, T. P., O'Donnell, S., and Barua, D.,** Genetic susceptibility to cholera, *Ann. Hum. Biol.*, 6, 369, 1979.
39. **Barua, D. and Paguio, A. S.,** ABO blood groups and cholera, *Ann. Hum. Biol.*, 4, 489, 1977.
40. **Chaudhuri, A. and De, S.,** Cholera and blood groups, *Lancet*, 11, 404, 1977.
41. **Jones, G. W. and Freter, R.,** Adhesive properties of *Vibrio cholerae*. Nature of the interaction with isolated rabbit brush border membranes and human erythrocytes, *Infect. Immun.*, 14, 240, 1976.
42. **Freter, R. and Jones, G. W.,** Adhesive properties of *Vibrio cholerae:* nature of the interaction with intact mucusal surfaces, *Infect. Immunol.*, 14, 246, 1976.
43. **Hoskins, L. C., and Boulding, E. T.,** Degradation of blood group antigens in human colon ecosystems I, *J. Clin. Invest.*, 57, 63, 1976.
44. **Hoskings, L. C., and Boulding, E. T.,** Degradation of blood group antigens in human colon ecosystems II, *J. Clin. Invest.*, 57, 74, 1976.
45. **Essien, E. S.,** Quantitative Significance of Bacterial Index: Indicative of Faecal Pollution of Water in Malumfashi Area, Ph.D. thesis, Ahmadu Bello University, Nigeria, 1980.
46. **Newsholme, A.,** *The Elements of Vital Statistics*, Allen & Unwin, London, 1923.
47. **Newsholme, A.,** A contribution to the study of epidemic diarrhoea, *Public Health*, 12, 139, 1899.
48. **Peters, O. H.,** Season and disease, *Proc. R. Soc. Med.*, iii, 1, 1903.
49. **Creighton, C.,** *History of Epidemics in Britain*, 2nd ed., Frank Cass and Co., London, 1965.

50. **Schrumshaw, N. S., Taylor, C. E., and Gordon, J. E.,** *Interactions of Nutrition and Infection,* World Health Organization, Geneva, 1968.
51. **Ironside, A. G., Tuxford, A. F., and Heyworth, B.,** A survey of infantile gastroenteritis, *Br. Med. J.,* 3, 20, 1970.
52. **France, G. L., Marmer, D. J., and Steele, R. W.,** Breast-feeding and salmonella infection, *Am. J. Dis. Child.,* 134, 147, 1980.
53. **Bullen, C. L., Tearle, P. V., and Stewart, M. G.,** The effect of "humanised" milks and supplemented breast feeding on the faecal flora of infants, *J. Microsc.,* 10, 403.
54. **Bullen, C. L. and Willis, A. T.,** Resistance of the breast-fed infant to gastroenteritis, *Br. Med. J.,* 3, 338, 1971.
55. **Bullen, C. L. and Tearle, P. V.,** Bifidobacteria in the intestinal tract of infants: an in vitro study, *J. Med. Microbiol.,* 9, 335, 1976.
56. **Bullen, C. L., Tearle, P. V., and Willis, A. T.,** Bifidobacteria in the intestinal tract of infants: an in vivo study, *J. Med. Microbiol.,* 9, 325, 1976.
57. **Tomkins, A. M., Bradley, A. K., Oswald, S., and Drasar, B. S.,** Diet and the faecal microflora of infants, children and adults in rural Nigeria and urban U.K., *J. Hyg. Camb.,* 86, 285, 1981.
58. **Mata, L. J. and Urrutia, J. J.,** Intestinal colonization of breast-fed children in a rural area of low socio-economic level, *Ann. N.Y. Acad. Sci.,* 176, 93, 1971.
59. **Jacomina, A. A., Hoogkamp-Korstanje, J. G. E., Linder, M., Marcelis, J. H., Daas-Slagt, H. D., and De Vos, N. M.,** Composition and ecology of the human intestinal flora, *Antonie Van Leeuwenhoek; J. Microbiol. Serol.,* 45, 35, 1979.
60. **Stark, P. L. and Lee, A.,** The microbial ecology of the large bowel of breast-fed and formula-fed infants during the first year of life, *J. Med. Microbiol.,* 15, 189, 1982.
61. **Mitsuoka, T. and Kaneuchi, C.,** Ecology of the bifidobacteria, *Am. J. Clin. Nutr.,* 30, 1799, 1977.
62. **Mitsuoka, T., Hayakawa, K., and Kimura, N.,** The faecal flora of man. II. Communication: the composition of bifidobacterium flora of different age groups, *Zentralbl. Bakteriol. Parasitenkd. Infektionskr. Hyg. Abt. 1: Orig. Reihe A,* 226, 469, 1974.
63. **Mitsuoka, T. and Hayakawa, K.,** The faecal flora of man. I. Communication, the composition of the faecal flora of different age groups, *Zentralbl. Bakteriol. Parasitenkd. Infektionskr. Hyg. Abt. 1: Orig. Reihe A,* 223, 333, 1972.
64. **Simhon, A., Douglas, J. R., Drasar, B. S., and Soothill, J. F.,** Effects of feeding on infants' faecal flora, *Arch. Dis. Child.,* 57, 54, 1982.
65. **Beerens, H., Romand, C., and Neut, C.,** Influence of breast feeding on the bifid flora of the newborn intestine, *Am. J. Clin. Nutr.,* 33, 2434, 1980.

Chapter 13

# BACTERIAL METABOLISM AND THE DIAGNOSIS OF SMALL BOWEL AND GASTRIC OVERGROWTH

**M. J. Hill**

## TABLE OF CONTENTS

# I. INTRODUCTION

Bacterial culture of jejunal or gastric juice to detect overgrowth is time consuming and expensive. Consequently, although it is necessary to culture samples when searching for specific pathogens, there is a need for rapid and cheap methods of detecting the presence of bacteria and this can best be achieved by assaying the products of bacterial metabolism.

Bacterial metabolites have varying levels of specificity and so the choice of the metabolite may depend not only on the equipment available but also on the requirements of the clinician. If the search is for the presence of bacteria of any genus, the best tests are the most nonspecific tests and these usually involve the detection of $H_2$ or $CO_2$ following the administration of a suitable substrate such as glucose or lactose. Other metabolites used are the fatty acid end products of fermentation, metabolites of tryptophan and phenylalanine or the products of hydrolysis of specific substrates. These may be assayed either in jejunal or gastric juice, breath, urine, or in blood.

To date, no test for small bowel colonization has gained general acceptance and there is widespread disagreement about the value of any of the tests to be described. This chapter therefore attempts to indicate the various possible approaches to the problem, their advantages, and disadvantages.

# II. TYPES OF TEST

## A. Assay of Jejunal Juice

### *1. Assay of Volatile Fatty Acids*

Normal jejunal or gastric juice contains low levels of acetic acid and barely detectable levels of propionic and butyric acids. These latter acids are major end products of fermentation of carbohydrates produced by many genera of anaerobic bacteria and their presence in readily detectable amounts is diagnostic of the presence of such organisms.[1]

The usual assay technique is gas-liquid chromatography using head space or conventional sample injection after first strongly acidifying the sample (to pH $<2$) and adding $Na_2SO_4$ to aid volatilization of the fatty acids. For conventional injection the acids must be extracted into ether or another suitable solvent and the total assay time is 15 min; newer HPLC techniques appear to offer no advantages in assay speed or simplicity.

### *2. Assay of Free Bile Acids*

The biliary bile acids are secreted as conjugates of glycine or taurine and the normal jejunal juice contains no detectable free bile acid. A wide range of genera of anaerobic bacteria produce cholyglycine hydrolase, releasing the free bile acids which can then be detected by a variety of techniques.

In early studies the jejunal juice sample was deproteinized by the addition of two volumes of alcohol and the bile acids assayed by thin layer chromatography.[2] More recently, sample preparation has been simplified by the use of Lipidex® columns which in a single step removes protein and pigment from the sample.[3]

## B. Assay of Urine

### *1. Indican*

Tryptophan is metabolized by the bacterial enzyme tryptophanase to indole, ammonia, and pyruvic acid. The indole released in the gut is absorbed, hydroxylated in the third position and the metabolite is excreted in urine as the sulfate ester indican (Figure 1). In normal healthy persons the amount of indican excreted per day in urine is 0 to 100 mg; but this amount is exceeded (Table 1) in patients with protein malabsorption (due to small bowel mucosal damage), or in patients with small bowel bacterial infection.[4] Urinary indican is assayed spectrophotometrically using dimethylbenzaldehyde.[5]

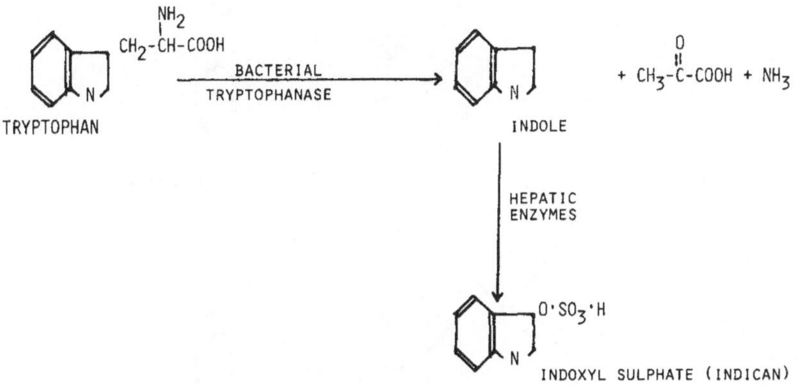


FIGURE 1. Tryptophan metabolism to indican.

**Table 1**
**INDICAN EXCRETION IN HEALTHY PERSONS AND IN PATIENT GROUPS[18]**

| | Indican excretion (mg/day) | |
|---|---|---|
| **Patient group** | **Range** | **Mean** |
| Healthy controls | 0—100 | 41.5 |
| Small bowel overgrowth | >100 | 369 |
| Healthy persons — high protein diet | | Increased over control |
| Protein malabsorption | | Increased over control |
| Antibiotic treatment (Lincomycin) | | Decreased over control |

### *2. Volatile Phenols*

The phenolic amino acids are metabolized by the gut bacterial flora to the volatile phenolic compounds phenol and *p*-cresol (Figure 2) which are excreted in urine as the sulfate or glucuronide esters. In normal healthy persons 50 to 100 mg of urinary volatile phenols (UVP) are excreted daily, with 70 to 95% usually in the form of *p*-cresol and the remainder as phenol (Table 2) but this amount is exceeded in patients with protein malabsorption (when the ratio *p*-cresol:phenol is normal) or in patients with small bowel overgrowth (when the ratio *p*-cresol:phenol is very low). The UVP are assayed by gas liquid chromatography after deconjugation using β-glucuronidase and sulfatase, or using 3 N sodium hydroxide at 120°C in an autoclave.[3]

## C. Assay of Breath

Normal products of the metabolism of amino acids and sugars include $CO_2$ and $H_2$; in addition certain organisms may produce methane either by methanogenesis from $CO_2$ and $H_2$ (as a strictly anaerobic reaction) or by other pathways yet to be determined but less restricting with regard to anaerobiosis.

### *1. Methane*

Methane is present in the breath of 20 to 40% of normal healthy persons[7-9] and a high proportion of persons with large bowel cancer.[9] Although it was thought that methanogenesis

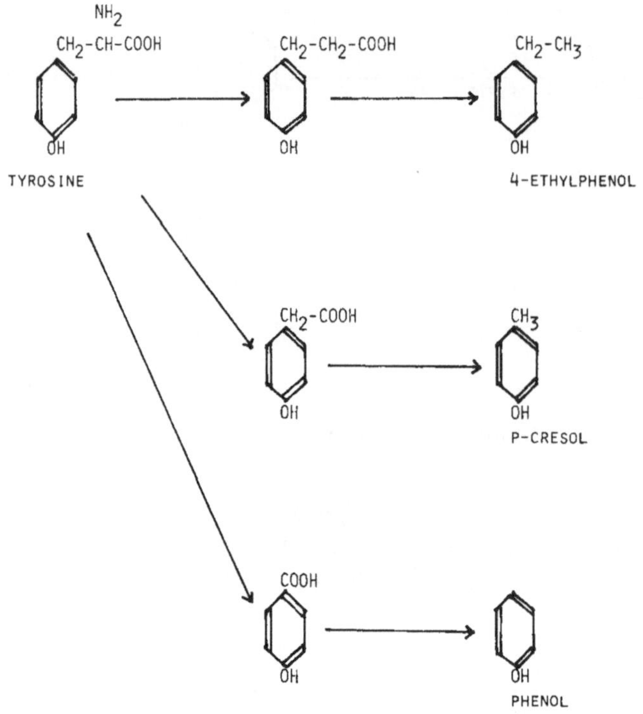


FIGURE 2. Tyrosine metabolism to the urinary volatile phenols 4-ethylphenol, 4-methylphenol (*p*-cresol) and phenol.

**Table 2**
**THE DAILY EXCRETION OF URINARY VOLATILE PHENOLS (UVP) IN VARIOUS POPULATIONS AND PATIENT GROUPS**

| Patient group or population group | UVP (mg/day) Phenol | *p*-cresol | Total | Ref. |
|---|---|---|---|---|
| Healthy persons | | | | |
| London | 11.3 | 56.8 | 68.1 | 20 |
| Copenhagen | 11.9 | 82.0 | 93.9 | 6 |
| Rural Denmark | 17.1 | 46.3 | 63.4 | 6 |
| Helsinki | 17.3 | 53.6 | 70.9 | 6 |
| Rural Finland | 10.5 | 57.9 | 68.4 | 6 |
| Norway[18] | 2.3 | 39.5 | 41.8 | 18 |
| Small bowel overgrowth[7] | 77 | 164 | 241 | 18 |
| Healthy persons | | | | |
| Low protein diet | | | 74.1 | 19 |
| High protein diet | | | 108.1 | 19 |

might be a measure of the degree of anaerobiosis and therefore of stagnation of the colonic contents[9] this cannot be so since there appears to be no correlation with colonic transit time.[10,11] It is present in the breath of a high proportion of rural Nigerians,[23] a population known to have a rapid bowel transit time. Methane is assayed in end expired breath by gas-liquid chromatography using a katharometer detector;[8] end expired air may be collected using a variety of techniques, an example of which is that described by Haines et al.[9] The amount in breath is related to diet,[12] particularly to dietary fiber components.

*2. Hydrogen*

Hydrogen is produced by the microbial catabolism of sugars, amino acids, fatty acids etc. It is assayed in breath using the same gas-liquid chromatographic system as is used for methane.

*3. Carbon Dioxide*

Carbon dioxide is produced by the microbial catabolism of sugars or amino acids. In general $^{14}C$-labeled substrates are used and the $^{14}CO_2$ trapped in alkali from the expired air and assayed radiochemically (for example, as described in Reference 13 by Fromm and Hofman, using glycocholic acid labeled in the amino acid moiety). In the $^{14}C$-glycocholic acid breath test carried out by James et al.[14] the labeled bile acid is administered by mouth as an alcoholic solution to fasting subjects; this is followed 10 min later by a standard Lundh test meal and 3 hr later by the normal meal. Samples of air were then collected at hourly intervals for 8 hr (the $^{14}CO_2$ concentration reached a maximum 5 to 6 hr after intake of the bile salt); the air samples were passed over anhydrous calcium chloride then into a tube containing 2 m$\ell$ of 0.5 *M* hyamine hydroxide until the alkali had been neutralized (as detected by thymolphthalein indicator). Scintillation fluid was added and the $^{14}C$ activity measured. In normal persons it was very low for the first 3 hr after dosage then increased to reach a maximum value 5 to 6 hr after dosage (representing the time taken for the bile salt to reach the bacterially colonized regions of the terminal ileum). In patients with small bowel overgrowth the $^{14}C$ activity should reach a maximum 1 to 3 hr after dosage.

## III. APPLICATIONS

Almost all of the types of test described have been used in attempts to detect small bowel overgrowth. Assay of jejunal juice has the major disadvantage of requiring intubation of the patient but has the advantage of allowing identification of the sample point. Although preliminary studies by Chernov et al.[1] suggested that jejunal volatile fatty acids (VFA) were of diagnostic value, a comparison of various diagnostic tests carried out at Central Middlesex Hospital[24] indicated a high percentage of false positive and false negative results and suggested that the test was of no clinical value. Similar poor results were obtained using jejunal free bile acids; only colonization by bacteria able to deconjugate bile acids were positive on this test and many cases with small bowel overgrowth did not contain jejunal free bile acids. The false-negative results may not be important since many suggest that the only small bowel infections of importance are those due to the anaerobic organisms (which are also those able to deconjugate). A recent study reported by Worsley et al.[15] gave results much more favorable to the assay of jejunal VFA (which gave a correct diagnosis in 41/44 patients).

Breath tests for bile salt deconjugation were pioneered by Fromm and Hofman;[13] the value of this test in detecting bile salt deconjugation has been confirmed by many groups.[16] In an assessment of the breath test James et al.[14] tested 34 controls of which 1 had an abnormal breath test; they also tested a range of patient groups and found a high proportion of positive breath tests in patients with ileal resection (5/6) only 1 of which had bacterial overgrowth of the small bowel, and cholangitis patients (5/6), 3 of which had bacterial overgrowth. In addition they confirmed that the breath test was abnormal in patients with small bowel overgrowth (e.g., Polya partial gastrectomy, small bowel diverticulae, gastrocolic fistula). Similar poor results were obtained by Worsley et al.[15] who obtained a correct diagnosis in only 47% of patients using this test.

In addition to the false positive results described by James et al.[14] many groups have noted false negative results when the organisms colonizing the small bowel are unable to deconjugate bile salts and this has resulted in a search for other substrates such as monosaccharides[8] which are fermented by almost all gut bacterial strains. Further, by measuring the release of hydrogen instead of $CO_2$ the need to use expensive scintillation counting equipment is removed. Bourne et al.[17] used breath hydrogen measurements to detect the presence of a gastric bacterial flora in studies of the effect of low dose cimetidine. Tadesse and Eastwood[12] studied the metabolism of dietary fiber in man using breath hydrogen and methane. They observed that the time taken for the breath $H_2$ to reach a peak value corresponded to the small bowel transit time and so the time of the peak breath $H_2$ might indicate the location of the flora; James et al.[14] had found no correlation between the time of maximum $^{14}CO_2$ and the site of small bowel colonization. In their series, Worsley et al.[15] found that the glucose hydrogen breath test gave a correct diagnosis in 88% of patients tested for small bowel bacterial overgrowth, but correctly identified only 69% of the positive samples (and 100% of the negative samples).

The major alternative to breath gas measurements has been the assay of urinary metabolites. Aarbakke and Schjonsby[18] suggested that a combination of urinary indican and UVP assay might provide a simple method for the selection of patients with small bowel overgrowth. They studied 7 patients with stagnant loop syndrome, 26 with coeliac disease, chronic pancreatitis or partial gastrectomy, and 18 control persons. In comparing the blind loop syndrome and control persons, the urinary phenol was 77 mg and 2 to 3 mg/day respectively, *p*-cresol was 164 and 39.5 mg/day respectively and indican was 369 mg and 41.5 mg/day respectively. Many factors affect the UVP production (Table 2); the amount is decreased by decreased dietary protein and decreased intestinal transit time[19] and by any manipulation which decreases the gut bacterial flora, for example antibiotic treatment, colonic wash out, total colectomy, etc.[20] It is increased in patients with protein malabsorption and after administration of tyrosine.[18] Urinary indican excretion is also affected by a wide range of factors (Table 1); some studies show an inverse correlation with fecal fat[21] and a correlation with a dietary protein.[22] When Aarbakke and Schjonsby[18] applied the UVP and urinary indican assays as diagnostic tests for stagnant loop syndrome patients in the total patient group phenol excretion showed two false negative results, *p*-cresol showed three false negative and two false positive results, while indican showed six false positive results. However, when phenol and indican assays were used in combination the false positive results were eliminated and on this basis they proposed the use of this combination in screening for small bowel overgrowth. In the study by Taylor and Taylor[24] the combination proved to be of little value having many false negatives and there have been no reports of its use in the literature.

## IV. CONCLUSIONS

The assay of bacterial metabolites as markers of bacterial overgrowth of the small bowel is extremely attractive since it offers the opportunity of diagnosis of the syndrome without intubation. Unfortunately the studies to date have not provided reliable diagnostic tools; the various types of assay and the problems involved are summarized in Table 3. Breath tests for $H_2$ following the administration of a simple sugar such as xylose or glucose offer the best prospects and there have been many reports in the literature of its successful use, despite the (varying) proportion of false positive and false negative results.

**Table 3**
**THE VARIOUS TYPES OF ASSAY OF BACTERIAL METABOLITES IN THE DIAGNOSIS OF SMALL BOWEL OVERGROWTH**

| Type of test | Major problems |
|---|---|
| Jejunal bile acid deconjugation· | Not a tubeless test<br>Does not detect nondeconjugating orgs |
| Jejunal volatile fatty acids | Not a tubeless test<br>High proportion of false negative and false positive tests |
| Urinary indican | Many false positive (due to other disease) and false negative (because organisms do not metabolize tryptophan/tyrosine)<br>Need a standardized diet |
| Breath test for $^{14}CO_2$ following cholyglycine administration | False negatives with overgrowth by nondeconjugating organisms<br>False positive results with ileal resection or cholangitis patients |
| Breath test for $H_2$ or $CH_4$ following administration of simple sugars | Nonspecificity |

## REFERENCES

1. **Chernov, A. J., Doe, W. F., and Gompertz, D.,** Intrajejunal volatile fatty acids in the stagnant loop syndrome, *Gut,* 13, 103, 1982.
2. **Northfield, T. C., Drasar, B. S., and Wright, J. T.,** Value of small intestinal bile acids analysis in the diagnosis of stagnant loop syndrome, *Gut,* 14, 341, 1973.
3. **Demark, B. R., Everson, G. T., Klein, P. D., Showater, R. B., and Kern, F.,** A method for the accurate measurement of isotope ratios of chenodeoxycholic and cholic acids in serum, *J. Lipid Res.,* 23, 204, 1982.
4. **Tabaqchali, S., Okubadejo, O., Neale, G., and Booth, C.,** Influence of abnormal bacterial flora on small intestinal function, *Proc. R. Soc. Med.,* 59, 1244, 1966.
5. **Curzon, G. and Walsh, J.,** A method for the determination of urinary indoxyl sulphate (indican), *Clin. Chim. Acta,* 7, 657, 1962.
6. **Hill, M. J., Taylor, A. J., Thompson, M. H., and Wait, R.,** Fecal steroids and urinary volatile phenols in four Scandinavian populations, *Nutr. Cancer,* 4, 67, 1982.
7. **Bond, J. H., Engel, R., and Levitt, M. D.,** Factors influencing pulmonary methane excretion in man, an indirect method of studying *in situ* metabolism of the methane producing colonic bacteria, *J. Exp. Med.,* 133, 572, 1971.
8. **Metz, G., Jenkins, D. J., Newman, A., and Blendis, L. M.,** Breath hydrogen in hyposucrasia, *Lancet,* i, 119, 1976.
9. **Haines, A., Metz, G., Diliwari, J., Blendis, L., and Wiggins, H.,** Breath methane in patients with cancer of the large bowel, *Lancet,* ii, 481, 1977.
10. **Read, N. W.,** Effect of dieting fiber on bowel transit, in *2nd George Washington Univ. Inter. Symp. Dietary Fiber,* Vahouny, G. and Kritchevsky, D., Eds., Plenum Press, New York, in press.
11. **Eastwood, M.,** Gastrointestinal responses to dietary fiber, in *2nd George Washington Univ. Inter. Symp. Dietary Fiber,* Vahouny, G. and Kritchevsky, D., Eds., Plenum Press, New York, in press.
12. **Tadesse, K. and Eastwood, M. A.,** Metabolism of dietary fibre components in man assessed by breath hydrogen and methane, *Br. J. Nutr.,* 40, 393, 1978.
13. **Fromm, H. and Hofman, A. F.,** Breath test for altered bile acid metabolism, *Lancet,* ii, 621, 1971.
14. **James, O. F., Agnew, J. E., and Bouchier, I. A.,** Assessment of the $^{14}C$-glycocholic acid breath test, *Br. Med. J.,* 3, 191, 1973.

15. **Worsley, B. W., Hamilton, I., Shoesmith, J. G., Cooke, E. M, and Axon, A. T. R.,** Evaluation of three indirect tests used in diagnosis of small bowel bacterial overgrowth, *J. Med. Microbiol.*, 18, vi, 1984.
16. **James, O. F., Agnew, J., Lydford, R., and Bouchier, I. A.,** An evaluation of a breath test to detect altered bile acid metabolism, *Gut*, 13, 845, 1972.
17. **Bourne, J. T., Mountford, R. A., Barry, R. E., and Read, A. E.,** Low dose cimetidine does not increase gastric bacterial flora, *Clin. Sci.*, 66, 69P, 1984.
18. **Aarbakke, J. and Schjonsby, H.,** Value of urinary sample phenol and indican determinations in the diagnosis of the stagnant loop syndrome, *Scand. J. Gastroenterol.*, 11, 409, 1976.
19. **Cummings, J. H., Hill, M. J., Bone, E. S., Branch, W. J., and Jenkins, D. J.,** The effect of meat protein with and without dietary fibre on colonic function and metabolism. II. Bacterial metabolites in faeces and urine, *Am. J. Clin. Nutr.*, 32, 2094, 1979.
20. **Bone, E. S., Tamm, A., and Hill, M. J.,** The production of urinary phenols by gut bacteria and their possible role in the causation of large bowel cancer, *Am. J. Clin. Nutr.*, 29, 1448, 1976.
21. **Neale, G., Gompertz, D., Schjonsby, H., Tabaqchali, S., and Booth, C.,** The metabolic and nutritional consequences of bacterial overgrowth in the small intestine, *Am. J. Clin. Nutr.*, 25, 1409, 1972.
22. **Tomkin, G. H. and Weir, D. G.,** Indicanuria after gastric surgery: an evaluation in the diagnosis of the blind loop syndrome, *Q. J. Med.*, 41, 191, 1972.
23. **Hudson, M.,** personal communiation.
24. **Taylor, A. J. and Taylor, R.,** unpublished results.

Chapter 14

# INTERRELATIONSHIP BETWEEN BACTERIA AND MUCOSA OF THE GASTROINTESTINAL TRACT

**W. E. W. Roediger**

## TABLE OF CONTENTS

## I. INTRODUCTION

Most intestinal bacteria which are associated with disease states readily secure attention of clinicians and microbiologists. In contrast microorganisms which are useful and favorable to mammalian life have not gained much interest nor been subjected to extensive scientific work in man. Without doubt the mammalian intestinal tract can exist in a germfree state when within a protected environment but total absence of bacteria in a wider sphere would probably be detrimental to mammalian life — this is difficult to test. Dependence on bacteria is consequently ill-defined and several recent reviews[1-3] for example discuss the barrier that mucosal flora present to infection, the bacterial overgrowth syndromes and the place of germfree animals in the study of colonic disease. The present review confines itself to some beneficial physiological interactions between the whole bacterial population and the epithelium of the small and large bowel.

## II. ENZYMATIC AND BACTERIAL DIGESTION IN THE INTESTINAL TRACT

Ingested food in man is foremostly degraded by acid hydrolysis or enzymatic cleavage by proteases, by lipases and by saccharidases found in small bowel which is sparsely populated by bacteria.[4] The opposite prevails in the pre- and post-gastric fermentative chambers and organs (Table 1) in which not enzymes but bacteria degrade oligosaccharides, oligopeptides, and fiber. To group pre- and post-gastric fermentative organs together is perhaps scientifically unjustifiable but such grouping has nevertheless promoted study of physiological mechanisms common to the rumen and colon in mammals[5] and in man.[6] Differences between the rumen and colon lie in the lining epithelium (squamo-cuboidal in the rumen and columnar, mucus-producing in the colon) and in the bacterial populations (cellulose digesters and numerous protozoa in the rumen, hemi-cellulose digesters, and fewer protozoa in the colon). Despite these differences there are striking analogies: both organs absorb sodium at a high rate, secrete bicarbonate, and absorb short-chain fatty acids produced by bacteria. In the process of absorption short-chain fatty acids provide sustenance to the lining epithelium in both the rumen and colon, a point requiring further elucidation.

## III. NUTRITION OF THE MUCOSA IN THE SMALL AND LARGE BOWEL

Much new information on the intestinal mucosa has appeared due to use of newer biochemical techniques. By means of a vascularly perfused small bowel[9-11] and isolated surface epithelial cells of the mucosa,[12,13] amino acids have been shown to play a prominent part in the nutrition of the mucosa of the small bowel.[7,8] Windmueller and Spaeth[11] simultaneously with Hanson and Parsons[10] showed that the amino acids glutamine and aspartate accounted for well over 70% of the $CO_2$ of respiration and that during states of fasting these amino acids and ketone bodies were the preferred metabolic fuels in the small bowel mucosa. Oxidation of glucose and unesterified fatty acids constitute less than 10% of the total oxidative fuels in the mucosa of fed animals. Further experiments revealed that in a contest between luminal and vascular metabolic substrates 45% of the total mucosal metabolism was derived from luminal nutrients such as glutamine, glutamate, and aspartate.[14] Thus, products of enzymic digestion, in the main amino acids, support the metabolic welfare of the small bowel mucosa and of the total support, half is derived from luminal contents.

In the fermentative organs products of bacterial fermentation nourish the colonic mucosa. Evidence in support comes from experiments performed in vitro with mucosal sheets and colonocytes. Henning and Hird[15] and McNeil et al.[16] have shown that bacterial short-chain fatty acids are absorbed in the colon along a concentration gradient from lumen to serosa

**Table 1**
**SITES OF DIGESTION IN THE INTESTINAL TRACT**

| | | |
|---|---|---|
| Enzymatic digestion | Duodenum | |
| | Jejunum | |
| | Ileum | |
| Bacterial digestion | Rumen | Pre-gastric |
| | Omasum | Pre-gastric |
| | Cecum | Post-gastric |
| | Colon | Post-gastric |
| | Rectum | Post-gastric |

**Table 2**
**DEGREE OF DEPENDENCE (OXIDATIVE METABOLISM) BY MUCOSAL CELLS OF THE INTESTINAL TRACT UPON LUMINAL AND VASCULAR SUBSTRATES**

| Intestinal tract | % $CO_2$ derived from luminal substrate | Vascular substrate (%) |
|---|---|---|
| Small bowel (jejunum ileum) | 45 | 38 |
| Large bowel (colon) | 70 | 21—30 |

*Note:* Percentages are rounded values from various studies[9-18] of fed animals.

without evidence of active transport. When these same bacterial short-chain fatty acids are used in conjunction with those nutrients of value to the small bowel mucosa, several points emerge. Amino acids, while of some importance in the proximal colon, become less important as a metabolic fuel in the distal colon.[17,18] In their place fatty acids ascend in utility and contribute mostly to production of $CO_2$ by the mucosa. Disregarding regional differences along the length of the colon, epithelial cells prepared from the entire length of colon reveal a preference of metabolic fuels in the order of short-chain fatty acids first and then followed by ketone bodies, amino acids, and glucose.[17] These conclusions emanate from fed animals where luminal nutrients were continually available. It appears that 70% of the energy supply is derived from bacterial fermentation products (Table 2). Bacteria in the colon play a far greater part in nutrition of the mucosa from the lumen than do amino acids from the lumen in the small bowel. Thus, dependence on luminal contents for nourishment of intestinal mucosa is greatest in the distal gastrointestinal tract. The utility of bacterial fatty acids for nutrition of the small bowel mucosa has been studied. Short-chain fatty acids do not stimulate oxygen consumption[12,13] of enterocytes as does glutamine but short-chain fatty acids are nevertheless oxidized. The extent to which bacterial fatty acids can sustain the small bowel mucosa metabolically has yet to be determined accurately.

## IV. BACTERIA AND MUCOSAL CELL GROWTH OF THE LARGE BOWEL

The view has frequently been supported that bacteria stimulate cell growth of the large bowel to provide better protection against bacterial invasion. Whether bacteria through their

Table 3
INTERACTION BETWEEN BACTERIA AND MUCOSA OF THE COLON

| Mucosal contribution to bacteria | Bacterial contribution to mucosal cells |
|---|---|
| $O_2$ | Acetate, propionate, butyrate |
| Urea | $NH_3$ |
| $NO_3^-$ | $NO_2^-$ |
| $Lactate^-$ | Amines and other nitrogenous substances |
| $K^+$ | |
| $HCO_3^-$ | |

physical presence or as a result of their fermentative products alone stimulate mucosal cell growth has recently become clear and should be assessed in the light of nutrition of the colonic mucosa.

Control of mucosal growth and control of intestinal adaptation are mediated by various mechanisms most of which were recently reviewed.[19,20] Nutrition is one of these mechanisms. As early as 1959[21] it was shown that infusion of bacterial fatty acid in the rumen caused marked development of the rumenal mucosa while infusions of glucose and saline did not produce such an effect. These observations on mucosal growth have been confirmed in other ruminants such as sheep.[22]

In the colon cell renewal is markedly depressed by fasting,[23] elemental diets,[24] and interruption of the fecal stream either by colostomies[25] or ligation.[26] The mechanisms involved are unclear but Stragand and Hagemann[26] showed that physical exposure to luminal contents is needed to initiate colonic cell hyperplasia following induction of mucosal atrophy by deviation of the fecal stream. Many factors, for instance short-chain fatty acids, ammonia, and amines, are removed in a colon defunctioned by ligation or colostomy. Only fatty acids on their own have been tested for their effect on mucosal cell turnover. A mixture of bacterial fatty acids markedly stimulate the mitotic index in the defunctioned colon,[27,28] whereas control infusions with saline do not. While confirmation is needed the results indicate that the metabolic products of bacteria, rather than bacteria themselves, moderate cell proliferation. Hyperplasia of the colonic mucosa seen after surgical closure of colostomies[25] is very likely related to the re-exposure to bacterial fatty acids. *n*-Butyrate has a pronounced effect on nucleic acid metabolism in general[29] and hyperplasia may possibly be linked to this effect on colonic mucosal cells.

## V. FUNCTIONAL RELATION BETWEEN BACTERIA AND COLONIC MUCOSA

In order to maintain adequate production of short-chain fatty acids for mucosal nutrition the pH and $pO_2$ or degree of anaerobiosis has to remain within a narrow physiological range. Control of each is ill understood but both conditions involve interaction between the colonic mucosa and bacterial population. Some interacting factors are listed in Table 3 and reviewed further.

### A. Maintenance of Luminal pH

Luminal pH depends upon the production rate of short-chain fatty acids by bacteria and the luminal appearance of bicarbonate which maintains the luminal pH at 7.4 with little variation.[30] Bicarbonate is secreted at a high rate into the colon and controlled by carbonic anhydrase found in large quantities in the nonmucus surface epithelial cells of the colonic

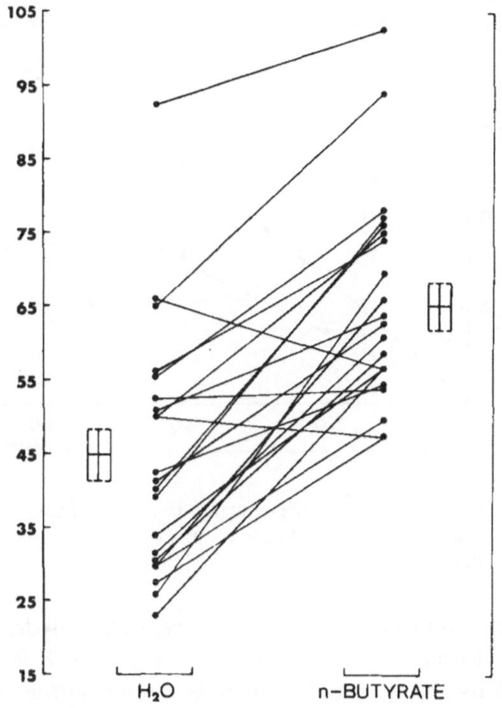


FIGURE 1. Bicarbonate output of the human rectum after 60 min measured with water and *n*-butyrate contained in dialysis tubing. Mean ± SEM of 21 cases. Paired rank sum test $p = < 0.01$.

mucosa.[31] Apart from maintaining luminal pH bicarbonate also controls cellular pH through a process of bicarbonate exchange for luminal chloride anions and anions of bacterial fatty acids (butyrate and acetate). We have shown that luminal appearance of bicarbonate was raised when the rectal mucosa was exposed to sodium butyrate in dialysis bags[32] (Figure 1). Bicarbonate apppearance elevated the juxto-mucosal pH from 7.40 to 7.64 reflecting an interaction between bacterial metabolites and mucosa in health. In disease such as *E. coli,* colitis exotoxins acting upon the colonic mucosa markedly stimulate bicarbonate secretion.[33] In this case another bacterial product rather than the organism itself acts to induce metabolic changes in the colonic mucosa.

### B. Regulation of Anaerobiosis

The $pO_2$ on the luminal surface of the colonic mucosa ranges between 85 to 100 mm of Mercury when measured by equilibration across dialysis membranes closely applied to the mucosa. The $pO_2$ in the center of bacteria is less than 5 mm of Mercury and correspondingly the $pCO_2$ much higher.[43] Several ill understood factors contribute to this extreme anaerobiosis. The mucosal interactions involved in this control are twofold:

1. Lactate produced in large quantities by the colonic mucosa[34] diffuses into the lumen where it becomes available for metabolism by facultative anaerobes especially to short-chain fatty acids. It is possible that mucosally associated organisms are more likely to be lactate fermentors which assist in lactate removal and in turn production of fatty acids for mucosal nutrition. This metabolic shuttle aids in the maintenance of an anaerobic environment as some oxygen is removed in the process.

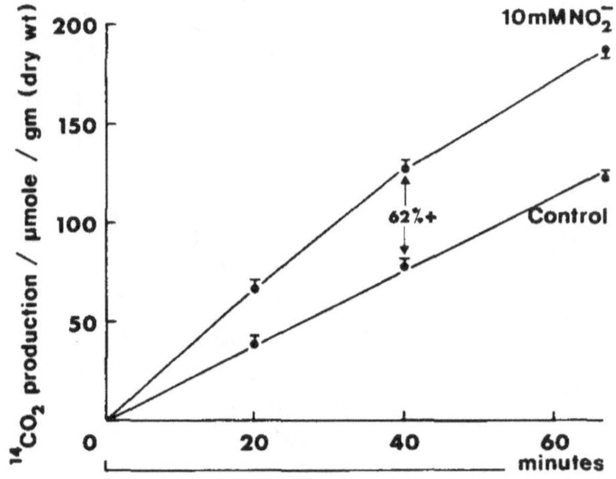


FIGURE 2. Rate of $^{14}CO_2$ production from [1-$^{14}$C]-butyrate by isolated colonic epithelial cells of the rat with and without $NO_2^-$. Paired $t$-test $p = < 0.01$.

2. Availability of nitrate and nitrite. In aerobic respiration hydrogen is passed via the cytochrome system to the terminal electron acceptor which is oxygen. In anaerobic respiration bacteria use inorganic ions such as nitrate, nitrite, sulfate, and sulfite as an alternative to oxygen as a terminal electron acceptor (Reaction 1).

$$NO_3^- \xrightarrow[\text{nitrate reductase}]{2e^-} NO_2^- \xrightarrow[\text{nitrite reductase}]{6e^-} NH_3 \text{ (Reaction 1)} \qquad (1)$$

Microorganisms which have the enzymes to reduce nitrate are mainly *Escherichia coli* and *Pseudomonadaceae*.

The interdependence between nitrate, mucosa, and bacteria is demonstrated by the fact that the colon extracts ingested nitrite. By using isotopically labeled nitrite, ingested or inhaled nitrite appears in the colonic lumen either as labeled ammonia or as labeled glutamate no doubt due to nitrite which enters the colonic lumen directly from the circulation through the colonic mucosa.[35,36] Furthermore large quantities of nitrite may enter the colon from the ileum. Measurement from ileostomies of ulcerative colitis patients indicate that levels of nitrite can be as high as 9.0 m*M*[37] yet the appearance of nitrite in stool is very low. If nitrite is incubated in stool under anaerobic conditions almost all nitrite is reduced and appears as ammonia.[38] Thus, circulating and ingested nitrite contribute towards respiration of anaerobic bacteria thereby assisting in the maintenance of bacterial flora.

The action of nitrite anions traversing the colonic mucosa has not been studied before but we have recently shown that colonocytes are not endangered by nitrite (10 m*M*). Quite the opposite, nitrites up to 20 m*M* concentration stimulate oxidation of bacterial fatty acids, (Figure 2) the major oxidative fuel for the colonic mucosa. Elsewhere such concentrations of nitrite are generally held to be detrimental to tissues. Other nitrogenous substances such as ammonia and methylamine derived from nitrate and produced by bacteria do not have a stimulatory effect on mucosal oxidation of fat.[43] These observations suggest that anaerobic respiration of the colonic bacteria and aerobic respiration of the colonic mucosa are interrelated and that nitrites are mutually of benefit and not harmful.

**Table 4**
**KNOWN OR CONJECTURED EFFECTS OF BACTERIAL FATTY ACIDS ON BIOCHEMICAL FUNCTIONS IN THE COLONIC EPITHELIUM**

| | |
|---|---|
| Mitochondrial respiration | Endogenous repiration[17] replaced by fatty acid oxidation |
| Mitochondrial enzyme activity (HMG-coA synthetase) | Probably stimulated |
| DNA content | Probably stimulated[25] |
| Nucleic acid acetylation | Enhanced[29] |
| Cell mitosis rate | Increased[27,28] |
| Membrane enzyme activity (alkaline phosphatase) | Probably increased |
| Microsomal enzyme activity | Probably increased |
| Cytosol acetylation | ? |

## VI. EFFICIENCY OF SODIUM ABSORPTION IN THE COLON DEPENDS ON BACTERIA

Short-chain fatty acids stimulate sodium absorption in the colon of animals and man[6,34] by mechanisms which are not dependent on mere diffusion of fatty acids through the mucosa along a concentration gradient. Oxidative respiration has been invoked as an explanation for stimulation of sodium absorption by short-chain fatty acids in the colonic mucosa but the proximal colon absorbs more sodium than the distal colon,[39] when in fact bacterial fatty acids make a greater contribution to mucosal metabolism in the distal colon.[18] This suggests that mechanisms other than oxidative respiration, for example enzyme induction, may play a part in stimulation of sodium absorption. Irrespective of mechanisms, absence of bacterial fatty acids leads to net luminal appearance of sodium.[40] Reintroduction of short-chain fatty acids in the colon redresses the imbalance and restores the colonic mucosa to a resorptive state. Thus, fatty acids appear to have a trophic effect on the colonic mucosa not unlike in other fermentative organs such as the rumen where withdrawal of short-chain fatty acids severely restricts the functional capacity of the ruminal mucosa.[41] A trophic change due to short-chain fatty acids may possibly explain the diarrhea that occurs after closure of colostomy where the atrophied and defunctioned colon is incapable of absorbing a substantial water load.[42] The above observations clearly point to a strong interdependence between luminal bacteria of the fermentative organs and the lining mucosa.

Throughout the review attention was focussed on short-chain fatty acids and their effects on the mucosa. Their action on the colonic mucosa has been summarized in Table 4 wherein is indicated the wide range of effects that these simple chemical substances exert on the epithelium of the colon.

## VII. CONCLUSIONS

The lining cells of the mucosa in the fermentative part of the intestinal tract have metabolic pathways which are so adjusted to bacterial metabolites that mucosal function is impaired by the absence of bacteria. In numerous aspects — cell proliferation, sodium absorption, maintenance of anaerobiosis — bacterial short-chain fatty acids play an intermediary role between the microbiological and mucosal cell systems, a further study of which may help to unravel diseases such as rumenitis and ulcerative colitis which are confined to the mucosa of the fermentative organs.

## REFERENCES


1. **Mackowiak, P. A.,** The normal microbial flora, *N. Engl. J. Med.*, 307, 83, 1982.
2. **Simon, G. L. and Gorbach, S. L.,** Intestinal microflora, *Med. Clin. N. Am.*, 66, 557, 1982.
3. **Gustafsson, B. E.,** The physiological importance of the colonic microflora, *Scand. J. Gastroenterol., Suppl.*, 77, 117, 1982.
4. **Bentley, D. W., Nichols, R. L., Condon, R. E., and Gorbach, S. L.,** The microflora of the human ileum and intra abdominal colon: results of direct needle aspiration at surgery and evaluation of the technique, *J. Lab. Clin. Med.*, 79, 421, 1972.
5. **Wolin, M. J.,** Fermentation in the rumen and human large intestine, *Science*, 213, 1463, 1981.
6. **Cummings, J. H.,** Short chain fatty acids in the human colon, *Gut*, 22, 763, 1981.
7. **Evered, D. F.,** Advances in amino acid metabolism in mammals, *Biochem. Soc. Trans.*, 9, 159, 1981.
8. **Windmueller, H. G.,** Glutamine utilisation by the small intestine, in *Advances in Enzymology*, Vol. 53, Meisler, A., Ed., John Wiley & Sons, New York, 1982, 201.
9. **Hanson, P. J. and Parsons, D. S.,** The utilization of glucose and production of lactate by in vitro preparations of rat small intestine: effects of vascular perfusion, *J. Physiol.*, 255, 775, 1976.
10. **Hanson, P. J. and Parsons, D. S.,** Factors affecting the utilisation of ketone bodies and other substrates by rat jejunum: effects of fasting and of diabetes, *J. Physiol.*, 278, 55, 1978.
11. **Windmueller, H. G. and Spaeth, A. E.,** Identification of ketone bodies and glutamine as the major respiratory fuels in vivo for post absorptive rat small intestine, *J. Biol. Chem.*, 253, 69, 1978.
12. **Watford, M., Lund, P., and Krebs, H. A.,** Isolation and metabolic characteristics of rat and chicken enterocytes, *Biochem. J.*, 178, 589, 1979.
13. **Porteous, J. W.,** Glutamate, glutamine, aspartate, asparagine, glucose and ketone body metabolism in chick intestinal brush-border cells, *Biochem. J.*, 188, 619, 1980.
14. **Windmueller, H. G. and Spaeth, A. E.,** Respiratory fuels and nitrogen metabolism in vivo in small intestine of fed rats, *J. Biol. Chem.*, 255, 107, 1980.
15. **Henning, S. J. and Hird, F. J. R.,** Transport of acetate and butyrate in the hind gut of rabbits, *Biochem. J.*, 130, 791, 1972.
16. **McNeil, N. I., Cummings, J. H., and James, W. P. T.,** Short chain fatty acid absorption by the human large intestine, *Gut*, 19, 819, 1978.
17. **Roediger, W. E. W.,** Utilisation of nutrients by isolated epithelial cells of the rat colon, *Gastroenterology*, 83, 424, 1982.
18. **Roediger, W. E. W.,** Role of anaerobic bacteria in the metabolic welfare of the colonic mucosa in man, *Gut*, 21, 793, 1980.
19. **Williamson, R. C. N.,** Intestinal adaptation, *N. Engl. J. Med.*, 298, 1393, 1978.
20. **Johnson, L. R.,** Regulation of gastrointestinal mucosal growth, *World J. Surg.*, 3, 477, 1979.
21. **Sander, E. G., Warner, R. G., Harrison, H. N., and Loosli, J. K.,** The stimulatory effect of sodium butyrate and sodium propionate on the development of rumen mucosa in the young calf, *J. Dairy Sci.*, 42, 1600, 1959.
22. **Sakata, T. and Tamate, H.,** Rumen epithelial cell proliferation accelerated by rapid increase in intra ruminal butyrate, *J. Dairy Sci.*, 61, 1109, 1978.
23. **Hagemann, R. F. and Stragand, J. J.,** Fasting and refeeding: cell kinetic response of the jejunum, ileum and colon, *Cell Tissue Kinet.*, 10, 3, 1977.
24. **Janne, P., Carpentier, Y., and Willems, G.,** Colonic mucosal atrophy induced by a liquid elemental diet in rats, *Am. J. Dig. Dis.*, 22, 808, 1977.
25. **Terpstra, O. T., Dahl, E. P., Williamson, R. C. N., Ross, J. S., and Malt, R. A.,** Colostomy closure promotes cell proliferation and dimethylhydrazine-induced carcinogenesis in rat distal colon, *Gastroenterology*, 81, 475, 1981.
26. **Stragand, J. J. and Hagemann, R. F.,** Effect of luminal contents on colonic cell replacement, *Am. J. Physiol.*, 233, E208, 1977.
27. **Sakata, T., Hikosaka, K., Shiomura, Y., and Tamate, H.,** The stimulatory effect of butyrate on epithelial cell proliferation in the rumen of the sheep and its mediation by insulin; differences between in vivo and in vitro studies, in *Cell Proliferation in the Gastrointestinal Tract*, Appleton, D. R., Sunter, J. P., and Watson, A. J., Eds., Pitman, Tunbridge Wells, England, 1980.
28. **Sakata, T. and Von Engelhardt, W.,** Stimulatory effect of short chain fatty acids on the epithelial cell proliferation in rat large intestine, *Comp. Biochem. Physiol.*, 74A, 459, 1983.
29. **Kruh, J.,** Effects of sodium butyrate, a new pharmacological agent, on cells in culture, *Mol. Cell Biochem.*, 42, 65, 1982.
30. **Wrong, O. M., Edmonds, C. J., and Chadwick, V. S.,** *The Large Intestine: Its Role in Mammalian Nutrition and Homeostasis*, MTP Press, Lancaster, England, 1981, 95.
31. **Spicer, S. S., Sens, M. A., and Tashian, R. E.,** Immunocytochemical demonstration of carbonic anhydrase in human epithelial cells, *J. Histochem. Cytochem.*, 30, 864, 1982.

32. **Lawson, M., Kwok, K., Kerr Grant, A., and Roediger, W. E. W.,** A functional test to assess disease activity and severity in ulcerative colitis, *Gastroenterology*, 84, 1226, 1983.
33. **Argenzio, R. A. and Whipp, S. C.,** Effect of Escherichia coli heat-stable enterotoxin, cholera toxin and theophylline on ion transport in porcine colon, *J. Physiol.*, 320, 469, 1981.
34. **Roediger, W. E. W. and Moore, A.,** Effect of short chain fatty acid on sodium absorption in isolated human colon perfused through the vascular bed, *Dig. Dis. Sci.*, 26, 100, 1981.
35. **Witter, J. P., Gatley, S. J., and Balish, E.,** Distribution of nitrogen - 13 from labelled nitrate ($^{13}NO_3^-$) in humans and rats, *Science*, 204, 411, 1979.
36. **Thayer, J. R., Chasko, J. H., Swartz, L. A., and Parks, N. J.,** Gut reactions of radioactive nitrite after intratracheal administration in mice, *Science*, 217, 151, 1982.
37. **Tannenbaum, S. R., Fett, D., Young, V. R., Land, P. D., and Bruce, W. R.,** Nitrite and nitrate are formed by endogenous synthesis in the human intestine, *Science*, 200, 1487, 1978.
38. **Archer, M. C., Saul, R. L., Lee, L.-J., Bruce, W. R.,** Analysis of nitrate, nitrite and nitrosamines in human feces, in *Gastrointestinal Cancer: Endogenous Factors*, Cold Spring Harbor Laboratory, Cold Spring Harbor, New York, 1981, 321.
39. **Rechkemmer, G. and Von Engelhardt, W.,** Absorptive processes in different colonic segments of the guinea-pig and the effects of short chain fatty acids, in *Colon and Nutrition*, Kasper, H. and Goebell, H., Eds., MTP Press, Lancaster, England, 1983, 61.
40. **Roediger, W. E. W. and Rae, D.,** Trophic effect of short chain fatty acids on mucosal handling of ions by the defunctioned colon, *Br. J. Surg.*, 69, 23, 1982.
41. **Hecker, J. F.,** *Experimental Surgery on Small Ruminants*, Butterworths, London, 1974, 116.
42. **Tilson, M. D., Fellner, B. J., and Wright, H. K.,** A possible explanation for postoperative diarrhea after colostomy closure, *Am. J. Surg.*, 131, 94, 1976.
43. **Roediger, W. E.,** unpublished observation.

Chapter 15

# BACTERIA AND INFLAMMATORY BOWEL DISEASE

**M. J. Hill**

## TABLE OF CONTENTS

## I. INTRODUCTION

A role for the gut bacterial flora in the etiology of chronic inflammatory bowel disease (IBD) has been thought likely since the early isolation of etiological agents from patients with acute colitis (e.g., Reference 1). In the early stages the symptoms of acute and chronic IBD are shared and only when the disease has been established for some time can a diagnosis of chronic IBD be considered. It is plausible to believe, therefore, that chronic IBD may originate as one of the common enteric infections.[2,3] It has been suggested that the acute infection only becomes chronic when the gut flora has a characteristic composition (Figure 1); the overall composition of the gut flora may have a key role, therefore, in maintaining the disease state.

The study of the role of bacteria in the causation of IBD is complicated by the fact that chronic IBD is not a single entity but has three major and a host of minor forms. The major forms are Crohn's disease (CD), ulcerative colitis (UC), and irritable bowel syndrome (IBS). UC is an inflammatory disease of the mucosal layer of the large intestine and, although it usually affects the distal colon and rectum, it may involve the whole of the large bowel. CD affects all layers of the mucosal wall and may involve the whole of the intestine, although it is usually most severe in the ileum and proximal colon. IBS is usually diagnosed by exclusion of a wide range of other bowel diseases. The differential diagnosis of these and the minor forms of IBD have been reviewed recently by Price[4] and by Lennard-Jones.[5] A further complication is that bacteria may be involved either in the initiation of IBD, or in the maintenance of the disease state, or in both phases of the disease by the same or by different mechanisms; in addition, the role of bacteria may differ between the various forms of IBD.

The original suggestion that chronic IBD might have a bacterial etiology was based on the similarity beween the symptoms of the diseases, particularly UC, and those of acute bowel infections, and was supported by the limited success of antibiotics (particularly of salazopyrine and, more recently, of metronidazole) in the treatment of UC. Similarly, the symptoms of intestinal tuberculosis can be difficult to distinguish from those of CD and the latter is often confused with the former in those countries where tuberculosis is still common. Studies of such a role for gut bacteria have been facilitated by the recent improvements in the techniques of isolation, enumeration, and identification of anaerobic bacteria.

Investigations of the role of bacteria in chronic IBD can be divided into two types. One line of research has been the search for a specific pathogen or group of pathogens, based on the assumption that each bacterial disease has a clearly identifiable causal organism with an identifiable mode of action. The alternative has been to study the profile of organisms on the assumption that IBD is a complex disease with no simple cause. Both types of approach have been applied to studies of the initiation and maintenance of the disease state.

The demonstration of the role of a specific pathogen in the causation of IBS would normally indicate misdiagnosis (since the diagnosis is usually by exclusion). No studies of the profile of bacteria in the gut flora of IBS patients have been made although if the current popular hypothesis that IBS is due to a food allergy is correct then the fecal flora during attacks of IBS might be abnormal as a result rather than as a cause of the disease. Consequently IBS will not be considered further, and in this review only the role of bacteria in CD and UC will be discussed.

## II. INITIATION OF IBD

The search for a single causative organism has a long history. For example, *Shigella* was implicated in UC by Hirst[2] and in both UC and CD by Felsen,[3] *Diplococci* were implicated by Bargen,[6] and *Bacteroides* by Dack et al.[7] The early work was reviewed by Kirsner and

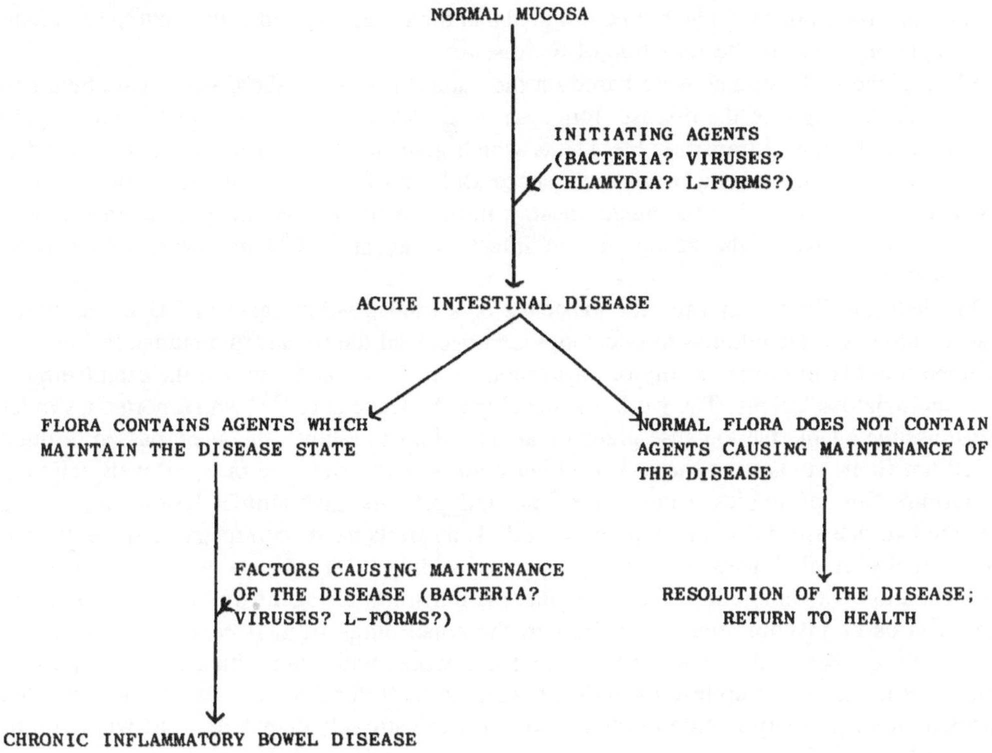


FIGURE 1. Suggested etiology of IBD in which the initial event is acute infection which either becomes chronic or is self-limiting.

## Table 1
## TRANSMISSIBLE AGENTS SUSPECTED OF BEING CAUSAL OF IBD

| Nature of evidence of transmission | Characteristics of agent | Ref. |
|---|---|---|
| Causes granulomas in mice | Obtained from CD tissue | 11 |
| Causes granulomas in rabbits | Obtained from CD or UC tissue, filterable through 220 nm filter | 12—15 |
| Causes granulomas in rabbits | Obtained from CD tissues, passes through 220 nm filter, destroyed by ampicillin | 16 |
| Virus seen in infected tissue but not in controls | RNA virus, 50—60 nm diameter (reovirus); only in CD tissue | 20, 21 |
| Agent isolated from CD tissue but not from controls | Cell wall deficient *Pseudomonas* present in CD but not UC tissue | 26 |
| Agent isolated from IBD but not from controls | Cell wall deficient organism found in both UC and CD tissue, reverted to *Mycobacterium kansasii* | 29 |
| Antibodies by *Chlamydia* found in IBD cases but not in controls | *Chlamydia* are filterable agents, found in CD but not UC | 31 |

Palmer[8] and by Weinstein[9] and both concluded that there was no good evidence to incriminate any single organism in the initiation of the disease.

Most of the early studies were based on the natural history of the disease and attempted to identify the origin of the disease from case notes. More recently, groups have attempted to isolate and identify transmissible agents which give rise to disease in an animal model, and these studies have been much more successful with CD than with UC (Table 1). This approach presupposes that the agent causing initiation of IBD is still present in the later stages of the disease; the search for an infectious agent in CD has been reviewed by Whorwell.[10]

Mitchell and Rees[11] reported the existence of a transmissible agent in CD tissue which was able to cause granulomas in mice; they homogenized the tissue from human lesions and injected it either intraperitoneally or intravenously into mice and reported the establishment of granulomatous lesions. The work was developed by Cave et al.[12-14] who reported a similar transmissible agent able to cause granulomatous lesions in rabbits; the agent passed through a 220 nm filter, suggesting that it was either a virus, a mycoplasma or a cell wall deficient bacterium. Similar studies using tissue from UC patients gave similar lesions suggesting that the two diseases have a similar etiology.[15] The results have been repeated and extended by Donnelly et al.[16] who showed that the tissue from CD patients was inactivated by preincubation with ampicillin suggesting that the transmissible agent was a *Bacterium* rather than a virus or mycoplasma, in contrast to the conclusions from filtration studies. There has, however, been much controversy about this work, with many other groups repeating consistent failure to set up lesions in the animals (e.g., References 17, 18). In addition, the histopathological interpretation of the lesions has been strongly disputed.[19] The whole problem has been confused by variations in the choice of species and strains of test animals and in the preparation of the material for injection. The high proportion of negative results might simply reflect the failure of the causal organisms to persist into the chronic phase of the disease reproducibly. Alternatively, it might indicate that the problem is not a simple one and that although an agent is present, transmission of the disease involves procedures that are not presently understood.

In addition to the high proportion of negative results there is a wide variation in the observations of those able to transmit the disease. Cave et al. reported that their agent was filterable[12-15] and the various possible candidates all have their protagonists. Gitnick et al.[20,21] reported the isolation of an RNA virus of particle size 55 to 60 nm in diameter from CD tissue but not from controls; this observation has been repeated by Whorwell et al.[22] but not by many other groups[23-25] (e.g., Kapikian et al., 1979; Philpotts et al., 1979, 1980). The virus was thought to be a reovirus, but serological tests indicated that antibody levels in CD patients were similar to those in controls, suggesting that such viruses are unlikely to be other than coincidental observations. Alternative filterable forms are cell-wall deficient bacteria. These have been isolated by a number of groups and the organisms could only be identified by obtaining reversion to the complete bacterial form in vitro, a notoriously hazardous procedure the results of which are readily misinterpreted. Parent and Mitchell[26,27] reported the isolation of cell-wall deficient bacteria from CD but not from UC tissues; in vitro the organisms reverted to *Pseudomonas* spp. but such organisms have not been isolated by others and antibodies to them have not been visualized by immunofluorescent staining of CD tissues.[28]

Burnham et al.[29] and Stanford et al.[30] have isolated cell-wall deficient organisms from both UC and from CD tissues, a single strain of which reverted in vitro to *Mycobacterium kansasii*. This is an extremely attractive observation; the histopathology of CD is readily confused with that due to *Mycobacterium tuberculosis*, mycobacterial cell walls can cause granulomatous lesions in experimental animals, and a disease in cattle — Johne's disease — with symptoms and lesions very similar to those in CD is known to be caused by a

**Table 2**
**REPORTED RECURRENCE OF IBD FOLLOWING ACUTE INTESTINAL INFECTION**

| Agent | Evidence | Ref. |
|---|---|---|
| Any acute intestinal infection (e.g., foreign travel) | Provokes relapse | 32 |
| *Campylobacter jejunum* | Provokes relapse | 33 |
| *Clostrium difficile* | Provokes relapse | 34 |
| *Streptococcus mobilis* | Associated with IBD | 38 |
| Some component of the non-sporing anaerobes flora | Mucosal antibodies | 45 |
| *Bacteroides vulgatus* | Serology | 46—48 |
| *Peptostreptococcus* spp. | | 50—51 |
| *Eubacterium* spp. | Serology | |
| *Coprococcus* spp. | Fecal culture | |
| *B. fragilis* | Immunological evidence in animals | 53 |

*Mycobacterium* spp. Skin testing studies with *M. kansasii* tended to support an etiological role for the organism. However others have failed to isolate the organism, *M. kansasii* does not produce granulomatous lesions in animals but tended to cause septicemia or abscesses[29] and the organism was not detectable by immunofluorescence in either CD or UC tissues.[28]

*Chlamydia* spp. (which are filterable agents) have been implicated in CD serologically[31] but other studies have failed to repeat this observation.

In summary, there have been many reports of transmissible agents (Table 1) able to initiate granulomatous lesions in test animals, but in no case has the phenomenon been defined so clearly that it can be repeated readily in other laboratories. The evidence for a transmissible agent is very weak indeed for UC, but is somewhat stronger for CD; identification of the agent (if such an agent exists) is even less certain.

## III. MAINTENANCE OF THE DISEASE STATE

In comparison with initiation, the evidence for a role for bacteria in the maintenance of the disease state is much stronger, especially for UC. The evidence comes from the effect of antibiotics in controlling the disease, from the effect of acute intestinal infection on recurrence of chronic IBD, from immunological and serological studies, and from animal studies. The nature of that role is extremely obscure, however, with different types of study indicating different causal organisms.

### A. Individual Organisms

There have been many reports of recurrence of IBD following acute intestinal infection (Table 2). In one report[32] diarrhea of any origin resulted in a 50% relapse rate in UC patients. As a result of foreign travel, approximately the same proportion of UC patients (13%) as controls (9%) developed diarrhea but 59% of the UC patients who developed diarrhea suffered a relapse in their UC; patients who developed gastroenteritis had a 42% relapse rate, those with infections of the gut requiring antibiotic treatment had a 47% relapse rate, and there was even a 10% relapse rate following a barium enema (in which senna was used in the bowel preparation). Newman and Lambert[33] showed that in a high proportion of patients *Campylobacter jejuneum* infection caused a flare-up in their IBD. Since these studies incriminated acute infection of any etiology, the conclusion must be that recurrence may be caused by any nonspecific disturbance of the gut flora.

Perhaps the single organism receiving most attention recently has been *Clostridium difficile*. Bolton et al.[34] showed that infection with toxigenic strains of *C. difficile* was associated with relapse of IBD; treatment of the infection with metronidazole caused remission of the symptoms of IBD. Similar results were obtained by Lamont and Trinka,[35] but Meyers et al.[36] found no relation between the presence of toxigenic strains and the relapse of IBD. In what appeared to be a definitive study Dorman et al.[37] studied a group of 50 CD patients longitudinally compared with 24 healthy controls; two fecal samples were taken with an interval of 3 months and although *C. difficile* was isolated from 4 patients none had active disease at that time. In the 6 patients with active disease on both occasions and the 12 who had active disease on one occasion there was no evidence of infection with *C. difficile* or any other pathogen. A conclusion from these studies could be that *C. difficile,* like any other enteric pathogen, can cause relapse of IBD but that it is not the only cause of relapse and other causes remain to be identified.

In an interesting observation Wiel-Korstanje and Winkler[38] compared UC patients and healthy controls and observed high numbers of *Enterococci,* and in particular *S. mobilis,* in colitics. These organisms were capable of producing hyaluronidase, chondroitinase, and mucinase and so were good candidates as etiological agents. However the observation has yet to be confirmed, and our own attempts to repeat the work have, so far, failed.[59]

**B. The Effect of Antibiotics**

The first antibiotic to be shown to be effective against IBD was sulfasalazine, which includes an anti-inflammatory agent and a sulfonamide antibiotic linked by a diazo bond. The diazo bond is readily reduced in the gut releasing the two agents, either of which could be the more important. Azad Khan and Truelove[39] showed that the anti-inflammatory agent was the more important, but more recent studies showing that metronidazole is as effective as sulfasalazine suggest that antibiotics are also important. There have been many studies of the fecal flora of patients in relapse or in remission, or on sulfasalazine or metronidazole[40-44] and no clear pattern has emerged incriminating any specific organism or group of organisms.

The observation of the effectiveness of antibiotic therapy is clinically very useful and is evidence of a role for bacteria, but gives no evidence as yet regarding the causation of relapse in the disease state in IBD.

**C. Immunological Studies**

Reports of a role for autoimmunity (e.g., Reference 8) and for hypersensitivity in the maintenance of the disease state of IBD provoked more detailed studies of the immunology of IBD. Monteiro et al.[45] reported higher mucosal antibody levels to fecal nonsporing anaerobic bacteria in UC patients than in controls, and suggested that maintenance of the disease state might be due to an antibody mediated hypersensitivity reaction (Arthus reaction) between the gut flora and the colonic mucosa. This would explain the value of sulfasalazine in the maintenance of remission of the disease (by suppressing the antigen load and decreasing the inflammation) and also the fact that it is of less value in treating acute relapses (since this would be due to an inflammatory reaction that had already been stimulated).

Other groups have studied serum antibodies (Table 2) and elevated titres have been reported to *Bacteroides vulgatus* (e.g., References 46 to 48), especially in CD patients. Wensinck and van der Merwe[49] reported high levels of serum agglutinins to a *Peptostreptococcus* spp., a *Eubacterium* spp. and a *Coprococcus* spp. in the serum of 50 to 60% of CD patients compared with 20 to 30% of UC patients and less than 10% of control persons. An international study involving 19 centers in 16 countries, in which sera from 937 patients were studied blind, confirmed these results;[50] 59% of CD patients had antibodies (67% of patients with CD of the colon) compared with 29% of UC patients, 8% of diseased controls and 8% of healthy controls. In a previous study Wensinck et al.[51] had shown that these organisms were commonly present in the feces of CD patients but were normally absent from feces of control persons. Clearly there is an association between fecal carriage of these organisms, serum antibodies to the same organisms, and CD but the association is not clear cut. Further, the organisms are absent from the flora of ileostomy fluid from IBD patients even though the site affected most severely by CD is the terminal ileum and any inflammatory reaction to the causal organisms must presumably take place there. The conclusion must be that the association between the organisms and the disease is not directly causal but must be indirect.

**D. Animal Studies**

There are few good models of IBD. A model of UC can be set up using degraded carrageenan in guinea pigs. No colitis occurs in germfree animals but an ulcerative lesion is readily established in animals with a normal gut flora; no lesions are produced in animals with a normal flora but treated with metronidazole.[52] Kitano and Kobayashi[53] established an ulcerative lesion in the colon of guinea pigs by injecting oily suspensions of lipopolysaccharide from *B. fragilis* to immunize the animals and then challenging them with the same organisms.

## IV. RECENT BASIC STUDIES

All of the studies described to date concern the luminal bacterial flora, but the mucosal flora is in a unique position to be involved in IBD. Very little is known about the human mucosal flora, particularly of the normal healthy mucosa, but there has been a recent upsurge of interest in the subject. It is already clear that a wide range of organisms incriminated by some as possible causal agents are not present in the mucosa in large numbers regardless of the severity of the disease; these include *S. mobilis, B. vulgatus, Peptococcus* spp., *Coprococcus* spp., *Eubacterium* spp., *Campylobacter* spp., *Chlamydia* spp. etc. Peach et al.[54] studied the colonic mucosal flora in normal persons and the same group[55] studied the mucosal flora of the jejunum, ileum, colon, and rectum in healthy persons and in persons with Crohn's disease. They noted that the ratio of anaerobic:aerobic organisms in the mucosal flora of the colon and rectum was about 1:1 to 10:1 (compared with 100 to 1000 for the luminal flora at these sites). Whereas the flora of the lumen was dominated by anaerobic nonsporing rods (*Bacteroides* spp., *Bifidobacterium* spp., *Eubacterium* spp.) with Gram-positive and Gram-negative organisms being present in similar numbers, in the mucosa the flora was much more Gram-negative and much less dominated by rod-shaped organisms (most of the rods observed by Peach et al.[54] were *Bacteroides* spp.). Throughout the intestinal tract the mucosal flora remained qualitatively constant with *Bacteroides* spp. being the dominant anaerobes and enterobacteria the dominant aerobes. A similar study by Hudson[59] produced essentially similar results except that the dominant facultative organisms were *Enterococci*.

Peach et al.[55] and Peach and Tabaqchali[56] compared the mucosal flora of patients with CD and control persons; the flora in CD patients was richer and was more dominated by Gram-negative rods although the differences were not statistically significant. Hill and Hudson[57] studied mucosal tissue from the distal ileum, transverse colon and rectum of UC patients, and control persons but no statistically significant differences were seen; neither were there any consistent differences in the mucosal flora between UC patients in relapse or in remission.

There have been a few studies in which the effects of antibiotics on the mucosal flora of CD and UC patients was compared with their clinical effects. Marks et al.[58] studied a range of antibiotics and found little effect on the mucosal flora even when the luminal flora was greatly altered. Hudson et al.[44] however showed that in CD patients treated with metronidazole and cotrimaxazole (a combination that has proved to be clinically beneficial) there was a large decrease in the numbers of mucosal bacteroides and coliforms.

## V. CONCLUSIONS

Despite all of the efforts and the large number of interesting preliminary studies, it has to be admitted that there is still at present no clear evidence implicating bacteria either in the initiation or in the maintenance of IBD. Nevertheless there is sufficient inconclusive evidence to persuade many groups of researchers to maintain the search. Evidence continues to accumulate indicating a major role for bacteria in the maintenance of the disease state in IBD but the possible mechanism remains unclear and it is quite likely that none of the proposed mechanisms is of major importance.

It is unlikely that the disease is caused by a single organism and it is unlikely that much headway will be made until we know much more about the interactions between bacterial species and about the activities of the consortia of organisms in the human gut.

## REFERENCES


1. **Flexner, S. and Sweet, J. E.,** The pathogenesis of experimental colitis and the relation of colitis in animals and man, *J. Exp. Med.*, 8, 514, 1906.
2. **Hurst, A. F.,** Ulcerative colitis, *Guy's Hosp. Rep.*, 71, 26, 1921.
3. **Felsen, J.,** The relationship of bacillary dysentery to distalileitis, chronic ulcerative colitis and non-specific intestinal granuloma, *Ann. Med.*, 10, 645, 1936.
4. **Price, A. B.,** Difficulties in the differential diagnosis of ulcerative colitis and Crohn's disease, in *The Gastrointestinal Tract,* Yardley, J., Morson, B. C., and Abell, M. R., Eds., Williams & Wilkins, Baltimore, 1977, 1.
5. **Lennard-Jones, J. E.,** Crohn's disease: definition, pathogenesis and astiology, in *Recent Advances in Gastrointestinal Pathology,* Wright, R., Ed., W. B. Saunders, Philadelphia, 1980, 173.
6. **Bargen, J. A.,** Experimental studies on the actiology of chronic ulcerative colitis, *JAMA*, 83, 332, 1924.
7. **Dack, G. M., Heinz, T. E., and Dragstedt, L. R.,** Ulcerative colitis: studies of bacteria in the isolated colons of three patients by cultures and by inoculation of monkeys, *Arch. Surg.*, 31, 225, 1935.
8. **Kirsner, J. B. and Palmer, W. L.,** Ulcerative colitis: considerations of its actiology and treatment, *JAMA*, 155, 341, 1954.
9. **Weinstein, L.,** Bacteriologic aspects of ulcerative colitis, *Gastroenterology*, 40, 323, 1961.
10. **Whorwell, P. J.,** Infectious agents in Crohn's disease — fact or artefect?, *Scand. J. Gastroenterol.*, 16, 161, 1981.
11. **Mitchell, D. M. and Reese, R. J. W.,** Agent transmissible from Crohn's disease tissue, *Lancet*, 2, 168, 1970.
12. **Cave, D. R., Mitchell, D. N., Kane, S. P., and Brooke, B. N.,** Further animal evidence of a transmissible agent in Crohn's disease, *Lancet*, 2, 1120, 1973.
13. **Cave, D. R., Mitchell, D. N., and Brooke,** Experimental animal studies on the actiology and pathogenesis of Crohn's disease, *Gastroenterology*, 69, 618, 1975.
14. **Cave, D. R., Mitchell, D. N., and Brooke, B. N.,** Introduction of granulomas in mice by Crohn's disease tissue, *Gastroenterology*, 75, 632, 1978.
15. **Cave, D. R., Mitchell, D. N., and Brooke, B. N.,** Evidence of an agent transmissible from ulcerative colitis tissue, *Lancet*, 1, 1311, 1976.
16. **Donnelly, B. J., Delany, P. V., and Healy, T. M.,** Evidence for a transmissible agent in Crohn's disease, *Gut*, 18, 360, 1977.
17. **Bolton, P. M., Owen, E., Heatley, R. V., Williams, W. J., and Hughes, L. E.,** Negative findings in laboratory animals for a transmissible agent in Crohn's disease, *Lancet*, 2, 1122, 1973.
18. **Ahlberg, J., Bergstrand, O., Gillstrom, P., Hollstrom, B., Kronev, T., and Reiland, S.,** Negative findings in search for a transmissible agent in Crohn's disease, *Acta Chir. Scand.*, 482, 45, 1978.
19. **Sachar, D. B. and Auslander, M. D.,** Missing pieces in the puzzle of Crohn's disease, *Gastroenterology*, 75, 745, 1978.
20. **Gitnick, G. L. and Rosen, V. J.,** Electron microscopic studies of viral agents in Crohn's disease, *Lancet*, 2, 217, 1976.
21. **Gitnick, G. L., Rosen, V. J., Arthur, M. H., and Hertweck, S. A.,** Evidence for the isolation of a new virus from ulcerative colitis patients. Comparison with virus derived from Crohn's disease, *Am. J. Dig. Dis.*, 24, 609, 1979.

22. **Whorwell, P. J., Phillips, C. A., Beeken, W. L., et al.**, Isolation of reovirus-like agents from patients with Crohn's disease, *Lancet*, 1, 169, 1977.
23. **Kapikian, A. Z., Barile, M. F., Wyatt, R. G., Yolken, R. H., Tully, D. G., Greenberg, H. G., Karlica, A. R., and Channock, R. M.**, Mycoplasma contamination in all culture of Crohn's disease material, *Lancet*, 2, 466—467, 1979.
24. **Phillpotts, R. J., Hermon-Taylor, J., and Brooke, B. N.**, Virus isolation studies in Crohn's disease: a negative report, *Gut*, 20, 1057, 1979.
25. **Phillpotts, R. J., Hermon-Taylor, J., Teich, N. M., and Brooke, B. N.**, A search for persistent virus infection in Crohn's disease, *Gut*, 21, 202, 1980.
26. **Parent, K. and Mitchell, P. D.**, Bacterial variants: etiologic agent in Crohn's disease?, *Gastroenterology*, 71, 368, 1976.
27. **Parent, K. and Mitchell, P. D.**, Cell-wall defective variants of Pseudomonas-like (Group Va) bacteria in Crohn's disease, *Gastroenterology*, 75, 368, 1978.
28. **Whorwell, P. J., Davidson, I. W., Beeken, W. L., and Wright, R.**, Search by immunofluorescence for antigens to rotavirus. *Pseudomonas maltophilia* and *Mycobacteria kansasii* in Crohn's disease, *Lancet*, 2, 697, 1978.
29. **Burnham, W. R., Lennard-Jones, J. E., Stanford, J. L., and Bird, R. G.**, Mycobacteria as a possible cause of inflammatory bowel disease, *Lancet*, 2, 693, 1978.
30. **Stanford, J. L., White, S. A., Burnham, W. R., Lennard-Jones, J. E., and Bird, R. G.**, Mycobacteria and inflammatory bowel disease, *Lancet*, 1, 444, 1979.
31. **Schiller, J. L., Piket-Van Ulsen, J., Veeken, I. V. D., Michel, M. F., and Stolz, E.**, Antibodies against chlamydia of lymphogranuloma-venereum type in Crohn's disease, *Lancet*, 1, 19, 1979.
32. **Isgar, B., Harman, M., and Whorwell, P. J.**, Factors preceding relapse of ulcerative colitis, *Digestion*, 26, 263, 1983.
33. **Newman, A. and Lambert, J. R.**, *Campylobacter jejunum* causing flare-up in inflammatory bowel disease, *Lancet*, 2, 919, 1980.
34. **Bolton, R. P., Sherriff, R. J., and Read, A. E.**, *Clostridium difficile* associated with diarrhoea: a role in inflammatory bowel disease?, *Lancet*, 1, 383, 1980.
35. **Lamont, J. T. and Trnka, Y. M.**, Therapeutic implications of *Clostridium difficile* toxin during relapse of chronic inflammatory bowel disease, *Lancet*, 1, 381, 1980.
36. **Meyer, S. S., Mayer, L., Buttone, E., et al.**, Occurrence of *Clostridium difficile* toxin during the course of inflammatory bowel disease, *Gastroenterology*, 80, 697, 1981.
37. **Dorman, S. A., Liggoria, E., Winn, W. C., and Beeken, W. L.**, Isolation of *Clostridium difficile* from patients with inactive Crohn's disease, *Gastroenterology*, 82, 1348, 1982.
38. **Wiel-Korstanje, J. A. and Winkler, K. C.**, The faecal flora in ulcerative colitis, *J. Med. Microbiol.*, 8, 491, 1975.
39. **Azad Khan, A. K. and Truelove, S. C.**, Circulating levels of sulphasglazine and its metabolites and their relation to the clinical efficiency of the drug in ulcerative colitis, *Gut*, 21, 706, 1980.
40. **Gorbach, S. L., Nahas, L., Plaut, A. G., Weinstein, L., Patterson, J. F., and Levitan, R.**, Studies on intestinal microflora. Faecal microbiol ecology in ulcerative colitis and regional enteritis: relationship to severity of disease and chemotherapy, *Gastroenterology*, 54, 575, 1968.
41. **Cooke, E. M.**, Faecal flora of patients with ulcerative colitis during treatment with salicylazosulphapyridine, *Gut*, 10, 565, 1969.
42. **West, B., Lendrum, R., Hill, M. J., and Walker, G.**, Effects of sulphasalazine (salazapyrin) on faecal flora in patients with inflammatory bowel disease, *Gut*, 15, 960, 1974.
43. **Krook, A., Danielsson, D., Khellander, J., and Jarnerot, G.**, The effect of metronidazole and sulfasalazine on the faecal flora in patients with Crohn's disease, *Scand. J. Gastroenterol.*, 16, 183, 1981.
44. **Hudson, M. J., Hill, M. J., Elliot, P. R., Berghouse, L., Burnham, W. R., and Lennard-Jones, J. E.**, The microbiol flora of the rectal mucosa and faeces of patients with Crohn's disease before and during antimicrobial chemotherapy, *J. Med. Microbiol.*, 25, 1071, 1984.
45. **Monteiro, E., Fossey, J., Shiner, M., Drasar, B. S., and Allison, A. C.**, Antibacterial antibodies in rectal and colonic mucosa in ulcerative colitis, *Lancet*, 1, 247, 1971.
46. **Helphingstine, C. J., Hentges, D. J., Campbell, B. J., Butt, J., and Barrett, T.**, Antibodies detectable by counter-immunoelectrophoresis against bacteroides antigens in serum of patients with inflammatory bowel disease, *J. Clin. Microbiol.*, 9, 373, 1979.
47. **Lindberg, A. A., Weintraub, A., and Nord, C. E.**, The humoral antibody response to *Bacteroides fragilis* infection in humans, *Scand. J. Infect. Dis.*, 19, 46, 1979.
48. **Persson, S. and Danielsson, D.**, On the occurrence of serum antibodies to *Bacteroides fragilis* and serogroups of *E. coli* in patients with Crohn's disease, *Scand. J. Infect. Dis.*, 19, 61, 1979.
49. **Wensinck, F. and Van der Merwe, J. P.**, Serum agglutinin to *Eubacterium* and *Peptostreptococcus* species in Crohn's and other diseases, *J. Hyg. (Cambridge)*, 87, 13, 1981.

50. **Wensinck, F., Van der Merwe, J. P., and Mayberry, J. F.,** An international study of agglutinins to *Eubacterium, Peptostreptococcus* and *Coprococcus* species in Crohn's disease, ulcerative colitis and control subjects, *Digestion,* 27, 63, 1983.
51. **Wensinck, F., Custers-Van Bieshout, L. M. C., Poppelaars-Kustermans, P. A., and Schroder, A. M.,** The faecal flora of patients with Crohn's disease, *J. Hyg. (Cambridge),* 87, 1, 1981.
52. **Nakaya, R.,** Bacterial species from mesenteric lymph nodes in IBD, in *Etiopathogenesis and Treatment of Inflammatory Bowel Disease,* Japan Medical Research Foundation, Tokyo, 1984, 43.
53. **Kitano, A. and Kibayashi, K.,** Experimental ulcerogenicity of *Bacteroides,* in *Etiopathogenesis and Treatment of Inflammatory Bowel Disease,* Japan Medical Research Foundation, Tokyo, 1984, 65.
54. **Peach, S. L., Drasar, B. S., Hawley, P. R., Hill, M. J., and Marks, C. G.,** Mucosal flora of the human colon, *Gut,* 16, 824, 1975.
55. **Peach, S. L., Lock, M. R., Katz, D., Todd, I. P., and Tabaqchali, S.,** Mucosal-associated bacterial flora of the intestine in patients with Crohn's disease and in a control group, *Gut,* 19, 1034, 1978.
56. **Peach, S. L. and Tabaqchali, S.,** Mucosa-associated flora of the human gastrointestinal tract in health and disease, *Eur. Chemother. Antibiol.,* 2, 41, 1982.
57. **Hill, M. J. and Hudson, M. J.,** Intestinal microflora in inflammatory bowel disease, in *Etiopathogenesis and Treatment of Inflammatory Bowel Disease,* Japan Medical Research Foundation, Tokyo, 1984, 3.
58. **Marks, C. G., Hawley, P. R., Peach, S. L., Drasar, B. S., and Hill, M. J.,** The effects of phthalylsulphathiazole on the bacteria of the colonic mucosa and intestinal contents as revealed by the examination of surgical samples, *Scand. J. Gastroenterol.,* 14, 891, 1979.
59. **Hudson, M. J.,** unpublished results.

Chapter 16

# BACTERIAL METABOLISM AND HUMAN CANCER

**M. J. Hill**

## TABLE OF CONTENTS

# I. INTRODUCTION

Bacteria are able to form carcinogens, mutagens, or promoters of carcinogenesis or mutagenesis either by releasing them from conjugates or by forming them as accidental byproducts of their catabolic activity. The function of such compounds in the ecology of the producer organism is unclear.

In this review some of the mechanisms by which carcinogens, mutagens, or promoter substances are formed will be discussed. The questions that must then be answered are (1) *can* bacteria produce carcinogens or mutagens in vitro by the proposed mechanism, (2) if so, *do* they in fact do so in vivo, (3) if so, are the amounts produced of importance in human carcinogenesis. Those questions will be discussed later in the chapter.

# II. PRODUCTION OF CARCINOGENS/MUTAGENS

Carcinogens/mutagens may be released from conjugates by hydrolysis, or may be formed as a byproduct of amino acid, fatty acid, or steroid metabolism.

### A. Deconjugation Reactions

The conjugates that may be hydrolyzed to release carcinogens may be naturally occurring or may be the result of hepatic detoxification reactions. An example of a naturally occurring conjugate is cycasin — methylazoymethanol-β-D-glucoside — which is hydrolyzed by β-glucosidase as discussed in Chapter 3. When cycasin is fed to germfree rodents or when it is administered intraperitoneally (i.p.) to conventional rodents it is nontoxic even at very high doses. In contrast, when it is fed to conventional rodents in high doses it is hepatotoxic and when given in small doses is carcinogenic causing tumors of the intestine, liver, and earduct. When the aglycone methylazoxymethanol is administered orally or i.p. to conventional or germfree rodents it is both hepatotoxic and carcinogenic. The toxicity of cycasin in conventional rodents is caused by its hydrolysis by the bacterial β-glucosidase; the enzyme present in the intestinal mucosa has a high degree of substrate specificity and is inactive on cycasin and consequently the glucoside is nontoxic in germfree rodents.

Many lipid soluble carcinogens (for example, the polycyclic aromatic hydrocarbons — PAH's) are absorbed from the small intestine by passive diffusion, transported to the liver via the portal blood and detoxified by the liver by conjugation as glucuronides (after prior hydroxylation if necessary). These water soluble conjugates are retained in the lumen of the intestine until they reach the bacterially colonized region of the gut where they are hydrolyzed by bacterial β-glucuronidase. The aglycone released, if it had been hydroxylated, is usually noncarcinogenic but Kinoshita and Gelboin[1] showed that an active intermediate is formed during hydrolysis which is able to bind to DNA and is therefore potentially mutagenic or carcinogenic.

### B. Amino Acid Metabolism

Carcinogens or tumor promoters are formed by bacterial action on a number of amino acids including methionine, tyrosine, and tryptophan.

Methionine can be converted to its *s*-ethyl analogue ethionine by the action of a number of bacterial species including *Escherichia coli;* the organisms carry out the reaction in a mineral salts medium containing added glucose and sulfate, and the product is extracellular. It is a potent carcinogen in rodents and a mutagen in tissue culture cell lines; although its mode of action is unclear it is known to cause ethylation instead of methylation of RNA bases and of proteins, and to inhibit *s*-methyltransferase. The carcinogenicity of ethionine has been reviewed recently by Farber.[2]

Tyrosine is metabolized by the gut bacterial flora to a range of phenolic acids and volatile phenols, principally phenol and *p*-cresol. Normally 50 to 100 mg of volatile phenols are

**Table 1**
**THE URINARY VOLATILE PHENOLS EXCRETED BY VARIOUS POPULATIONS AND PATIENT GROUPS**

| Subject group | Number of persons | Urinary volatile phenols Phenol | *p*-Cresol | Total | Ref. |
|---|---|---|---|---|---|
| Normal British | 10 | 9.8 | 51.8 | 61.6 | 55 |
| Finland | | | | | |
| Rural | 30 | 10.7 | 57.7 | 68.4 | 56 |
| Helsinki | 30 | 17.3 | 53.6 | 70.9 | 56 |
| Denmark | | | | | |
| Rural | 30 | 17.1 | 46.3 | 63.4 | 56 |
| Copenhagen | 30 | 11.9 | 82.0 | 93.9 | 56 |
| Normal German | | 10.5 | 86.4 | 96.9 | 57 |
| Patients with colostomy | 3 | 4.8 | 56.7 | 61.5 | 55 |
| Patients with ileostomy | 2 | 5.1 | 1.2 | 6.3 | 55 |
| After preoperative bowel preparation | 3 | 3.0 | 17.9 | 20.9 | 55 |
| Patients with colorectal cancer | 6 | 4.3 | 54.1 | 58.4 | 55 |

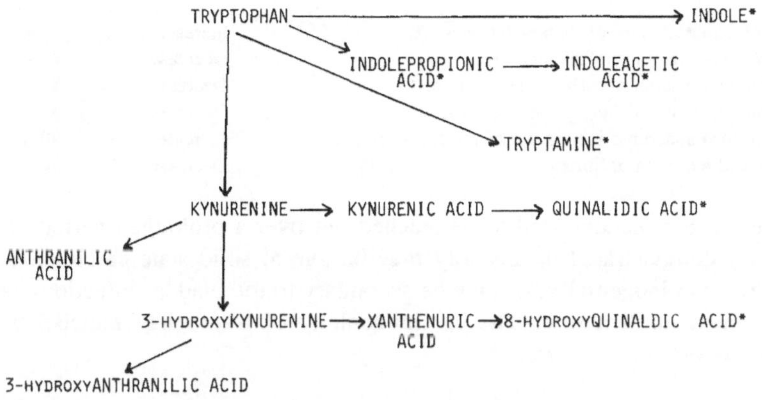


FIGURE 1. The metabolism of tryptophan in the human gut and liver.

excreted in the urine per day; normally more than 80% as *p*-cresol and the remainder as phenol; the amount is related to the amount of dietary protein and the intestinal transit time[3] but since Africans excrete high levels of urinary volatile phenols (UVP) there must be other additional factors of importance; the amount of UVP is greatly decreased by pre-operative bowel preparation, by colectomy, or by any other procedure which grossly interferes with the gut bacterial flora (Table 1). A range of volatile phenols (including *p*-cresol and phenol) has been shown to promote the skin carcinogenicity of dimethylbenzanthracene.[4]

Tryptophan is metabolized to a wide range of metabolites, (Figure 1) all of which have been demonstrated in normal human urine either as a result of bacterial, or of hepatic metabolism. A number of these (principally those on the quinoline pathway) have been shown to be mutagenic or carcinogenic (Table 2); much of this work was done using the bladder implantation method in which the test compound is impregnated into a waxy vehicle

**Table 2**
**EVIDENCE OF TUMORIGENICITY IN TRYPTOPHAN METABOLITES**

| Test system | Metabolite | Indication[5] |
|---|---|---|
| AAF treated rats | Indole | Tumor promotion? |
| | Indoleacetic acid | Cocarcinogenicity? |
| Bladder implantation | 3,Hydroxykynurenine | |
| | 3,Hydroxyanthranilic acid | Carcinogenicity? |
| | 8,Hydroxquinaldic acid | Cocarcinogenicity? |
| | Xanthenuric acid | Tumor promotion |
| | Quinaldic acid | |
| Cultured mammalian tissue cells | 3,Hydroxykynurenine | Mutagenicity |
| | 3,Hydroxyanthranilic acid | |

**Table 3**
**FACTORS AFFECTING THE FORMATION OF THE FECAL MUTAGEN OF BRUCE ET AL.[7]**

| | Effect on mutagen activity | Ref. |
|---|---|---|
| Increased amount of dietary fat or meat | Increase | 7, 8 |
| Increased amount of dietary fiber | Decrease | 7, 8 |
| Supplements of ascorbic acid in the diet | Decrease | 8 |
| Supplements of dietary α-tocopherol | Decrease | 8 |
| In vitro anaerobic incubation of the fecal sample | Increase | 58 |
| Treatment with antibiotics | Decrease | 58 |

and implanted in the bladder wall to be leached out over a prolonged period. The bladder carcinogenicity demonstrated in this way may be due to solid state carcinogenesis, tumor promotion, true carcinogenicity, or may be secondary to the bladder infection that undoubtedly follows such implantation. The carcinogenicity of tryptophan metabolites has been reviewed by Bryan[5] and by Price.[6]

### C. Fatty Acid Metabolism

The fecal mutagen described by Bruce[7,8] has now been isolated and identified as a gylcerol ether of a pentenol[9] and now a family of such mutagens has been identified.[10] Early experiments showed that the amount of mutagen was increased by a high meat/fat diet, decreased by dietary supplements of fiber or ascorbic acid, and increased by in vitro anaerobic incubation of the fecal sample containing the mutagen (Table 3). There is a wide range of activity temporally and between persons in fecal mutagen activity;[11] the strain specificity of the mutagen in the microbial mutagenesis assay suggests that the mutagen causes frame shift mutations.

### D. Bile Acid Metabolism

The bile acids produced by the liver and secreted in bile are the glycine and taurine conjugates of the primary bile acids cholic and chenodeoxycholic acids; these are metabolized by the gut bacteria by

1. Deconjugation to release the free bile acids
2. Dehydroxylation to produce the major secondary bile acids deoxycholic and lithocholic acids

**Table 4**
**STRUCTURE ACTIVITY RELATIONSHIPS IN THE C-MUTAGENICITY OF BILE ACIDS[14]**

| Structural change | Effect on co-mutagenicity |
|---|---|
| Removal of the 7α-hydroxyl group | Increased activity |
| Inversion of the 3α hydroxyl to 3β | Decreased activity |
| Inversion of the 5β carbon to 5α (allo-form) | Increased activity |

**Table 5**
**EVIDENCE THAT BILE ACIDS ARE CO-CARCINOGENIC**

| Type of study | Bile acid tested | Observation |
|---|---|---|
| Skin painting on rodents using an oily vehicle | Deoxycholic acid | Co-carcinogenic |
| | Bisnor-5-cholenic acid | Co-carcinogenic |
| | Primary bile acids | Inactive |
| Rectal instillation in rats treated with a colon carcinogen | Deoxycholic acid | Co-carcinogenic |
| | Lithocholic acid | Co-carcinogenic |
| | Cholic acid | Inactive |
| | Chenodeoxycholic acid | Inactive |
| *Salmonella* mutagenesis | Deoxycholic acid | Co-mutagenic |
| | Lithocholic acid | Co-mutagenic |
| | Cholic acid | Inactive |
| | Chenodeoxycholic acid | Inactive |
| Cell transformation assay | Lithocholic acid | Mutagenic |
| | Other bile acids | Inactive |
| *Drosophila* mutagenesis | Deoxycholic acid | Mutagenic |

3. Hydroxyl oxidoreduction to give the keto bile acids and inversion of the hydroxyl groups to the β-configuration
4. 4-5 Dehydrogenation to give bile acids with the 4-en-3-one configuration, which can then undergo reduction to the 5α (allo) bile acids

These have been discussed in detail in Chapter 5. Detailed studies of the mutagenicity of the bile acids showed that they were not direct acting mutagens and were inactive after incubation with rat liver microsomes. However, dimethylhydrazine was found to be mutagenic and its mutagenicity was increased in the presence of certain bile acids.[12-14] Structure activity studies[14] showed that the co-mutagenic activity was decreased greatly by the presence of a 7-oxygen function and by conjugation (illustrating the importance of deconjugation and of 7-dehydroxylation), was decreased somewhat by inversion of the 3α hydroxyl group but was increased by inversion of the 5 β carbon atom to give the allo bile acids (Table 4).

These results have been tested to a limited extent in animals and those studies have confirmed the mutagenicity results. Thus, the secondary bile acids deoxycholic and lithocholic acids have been shown to be co-carcinogenic (Table 5) and the primary bile acids have proved to be inactive.

### E. *N*-Nitroso Compounds

The *N*-nitroso compounds have been found to be highly carcinogenic in all of the animal species in which they have been tested and there is no good reason to believe that humans are uniquely resistant to their action. *N*-nitrosation can occur via an acid-catalyzed reaction, the pH-optimum of which depends in part on the substrate but also on the catalysts present.

**Table 6**
**EVIDENCE OF *N*-NITROSATION BY BACTERIA**

| Observation | Ref. |
|---|---|
| 1968 —*N*-nitrosation is catalyzed by *E. coli* at pH 7 | 59 |
| 1971 —*N*-nitrosation is catalyzed by a range of organisms requires growing cultures | 60 |
| 1972 —*N*-nitrosamines demonstrated in the urine of two patients with urinary tract infections | 61 |
| 1974 —*N*-nitrosation by bacterial action demonstrated in an animal model of urinary tract infection | 62 |
| 1976 —Kinetics of *N*-nitrosation studied in *E. coli* | 63 |
| 1979 —Kinetics studied further. Concluded that the reaction is enzymic | 64 |
| 1984 —*N*-nitrosation by bacterial action confirmed. pH optimum varies with the base used | 15 |

**Table 7**
**SITES OF THE BODY IN WHICH THE FORMATION OF *N*-NITROSO COMPOUNDS HAS BEEN DEMONSTRATED**

| Site | Subject group | Ref. |
|---|---|---|
| Stomach | Gastric achlorhydria | 18, 19, 44, 45, 47 |
| | Normal healthy persons | 20, 21 |
| Colon | Ureterocolic anastomosis | 50 |
| Urinary bladder | Urinary tract infection | 61, 65 |
| | Bilharzia | 48 |
| Vagina | *Trichomonas vaginalis* infection | 52, 53 |

The reaction can also be catalyzed at neutral pH values by bacteria; this reaction has been studied by many groups (Table 6) but its mechanism is still far from clear (see Chapter 9). Whereas the acid catalyzed reaction is bimolecular for nitrite the bacterial reaction is unimolecular and whereas the acid-catalyzed *N*-nitrosation of piperidine has a pH optimum close to 2 that of the bacterially catalyzed reaction is 7.5.[15]

The in vivo production of *N*-nitroso compounds has been the subject of much dispute. It is clear that they are formed in the normal and achlorhydric stomach, the urine of persons with urinary tract infection and the urine of persons with a ureterocolic anastomosis (Table 7) but the relative amounts produced under different conditions is unclear. Thus, it is generally agreed that the concentration of nitrite in the resting gastric juice is very low if the pH is below 4 and increases greatly with increasing pH;[16-22] if the most important factor in determining the rate of *N*-nitrosation is the nitrite concentration it would be expected that the concentration of *N*-nitroso compound would increase with pH (as shown by Reed et al.[18]) while if the hydrogen ion concentration is the most important factor then an inverse correlation between *N*-nitroso compound concentration and pH should be observed (as reported by Hall et al.[20]) The reason for these two groups getting such different results is not clear. The production of *N*-nitroso compounds in vivo is certain, their significance will be discussed later in this review.

## III. RELEVANCE TO HUMAN CANCER

A number of the carcinogens/tumor promoters discussed in the previous section have yet to be implicated in human cancer. Thus, the release of carcinogens from conjugates by

**Table 8**
**EVIDENCE RELATING FECAL MUTAGEN ACTIVITY TO RISK OF COLORECTAL CARCINOGENESIS**

| Study | Observation | Ref. |
|---|---|---|
| Effect of diet | Dietary changes which decrease colorectal cancer incidence decrease mutagen activity | 8 |
| Study of 3 populations in South Africa | Fecal mutagen activity highest in the population with highest colon cancer risk | 66 |
| Comparison of 2 diet groups in New York | Group at higher risk of colorectal cancer had the higher fecal mutagen concentration | 67 |
| Comparison of vegetarians and nonvegetarians | Vegetarians with the lower risk of colon cancer had lower fecal mutagen activity | 68 |

bacterial β-glucuronidase has been implicated in human colorectal cancer by Renwick and Drasar[23] but no evidence to support this has been reported to date and indeed, it is difficult to see how such evidence might be obtained. Similarly, a role for ethionine in human cancer has still to be hypothesized and tested. In this section only those aspects that have been seriously studied will be discussed.

### A. Cycasin

Cycasin is present in cycads and these form a major starch source in South and East Asia after the cycasin has been removed by repeated aqueous extraction. Following climatic and natural disasters (typhoons, earthquakes, etc.) when food supplies are interrupted it is common for people to try to eat raw cycads; some die as a result of the hepatotoxicity and some suffer hepatic symptoms (and so have clearly been exposed to cycasin) and survive. These latter are the subject of continuing epidemiological study to determine whether they will have an increased risk of colon cancer (or of cancer of any other site); to date no strongly positive results have been reported, suggesting that cycasin may not be a potent carcinogen in humans.

### B. Fecal Mutagens

Early studies suggested a possible role for the fecal mutagen of Bruce et al.[8] in human colorectal cancer (Table 8), but later results failed to support the relationship and it is now concluded that the mutagen plays no role in human colorectal cancer either in the formation of the premalignant adenoma or in the progression from adenoma to carcinoma.

Most of the studies have related the colorectal cancer risk to the mutagen activity in single samples of feces; Vennitt et al.[11] showed that the temporal variation in fecal mutagen activity is so wide that such results are virtually useless and perhaps it would be wise to suspend judgment on the role of the mutagen — the evidence is too weak either to confirm or to refute such a role.

### C. Tryptophan Metabolites

Tryptophan metabolites on the quinoline pathway have been implicated by Bryan and others (reviewed by Bryan[5]) in the causation of those bladder cancers not related to industrial exposure to chemicals, to bilharzia, or to other known causes of bladder cancer. The evidence was based on case control studies in conjunction with the implantation tests which suggested activity as bladder carcinogens. The relationship has been questioned more recently by Teulings et al.[24] who demonstrated that the amount of quinoline metabolites was also increased in patients with urethral strictures and was therefore probably a secondary relationship.

**Table 9**
**EVIDENCE IMPLICATING FECAL BILE ACIDS (FBA) IN COLORECTAL CANCER (CRC) CAUSATION**

| Study | Observation | Ref. |
|---|---|---|
| Comparison of populations | FBA concentration correlated with CRC incidence | 56, 69, 70 |
| Animal studies | Dietary changes — those which increase CRC incidence increase FBA | 69, 70 |
| | Surgical changes — those which increase FBA increase CRC | 69, 70 |
| Studies with pure bile acids | Animals — secondary bile acids are tumor promoters | 32 |
| | Animals — secondary bile acids cause dysplasia in the colonic mucosa | 26, 27 |
| | In vitro tests — secondary bile acids are co-mutagens | 12—14 |
| Bile acid receptors | Bile acid receptors are found in the mucosa of CRC patients, not of controls | 29 |
| | Patients with colorectal adenomas take up bile acids from the colon faster than healthy controls | 28 |
| Precancerous lesions | FBA concentration and CRC risk correlated with adenoma size | 38 |
| | FBA concentration higher in colitics who develop CRC than in those who do not | 70 |

## D. Bile Acids

Bile acids have been implicated in a number of cancer sites, principally those of the digestive tract (colon, rectum, stomach, gallbladder, pancreas, and kidney) by Lowenfels.[25] The evidence is strongest for cancer of the large bowel but there is growing evidence for a role in gastric cancer.

### *1. Large Bowel*

The evidence for a role for bile acids in colorectal carcinogenesis is summarized in Table 9. In humans there have been many studies of populations showing that the risk of large bowel cancer is correlated with the mean fecal bile acid (FBA) concentration; bile acids have been shown to be toxic to the colonic mucosa causing dysplastic changes in adenomatous tissues;[26,27] bile acids are taken up much faster from the colonic lumen by adenomatous tissue than by normal tissue;[28] the colonic mucosa of bowel cancer cases contains bile acid receptor sites which are rare in the mucosa of normal healthy persons.[29] However, the failure to show a higher FBA concentration in colorectal cancer cases than in controls (Table 10) suggested a need to refine the hypothesis. In parallel with the work of Wilpart et al.,[13,14] Owen et al.[30,31] have shown that in a comparison of colorectal cancer cases and controls there was no difference in the mean concentration of deoxycholic acid (DA) or lithocholic acid (LA) or the sum of DA + LA (which might have been expected to provide a good discriminant from the animal studies of Narisawa et al.[32]) but the ratio LA/DA was very much higher in the cases than in the controls (which is explained by the observation of Wilpart et al.[13]). A number of studies have implicated the *Clostridia* which produce 4 to 5 dehydrogenase (the NDC) in colorectal carcinogenesis; the observation of Wilpart et al.[14] that allo bile acids were more potent co-mutagens than the 5 β isomers suggests that the NDC may be acting to produce more allo bile acids. Wait et al.[33,34] have demonstrated the presence of a number of allo bile acids in human feces and the relation between their amount and the concentration of NDC will be of great interest.

## Table 10
## CASE-CONTROL STUDIES OF THE RELATION BETWEEN FECAL BILE ACID (FBA) CONCENTRATION AND LARGE BOWEL CANCER (LBC)

| Cancer cases | Controls | FBA in cases compared with controls | Ref. |
|---|---|---|---|
| Colorectal | Nonmalignant bowel disease | FBA higher in cases | 71 |
| Colorectal | Healthy persons | FBA higher in Dukes A and B cases of the left colon and rectum. Lower in advanced cases | 70 |
| Colon | Healthy persons | FBA higher in cases than in controls | 72 |
| Colorectal | Healthy persons | FBA lower in cases than in controls | 73 |
| Colorectal | Healthy persons | FBA similar in cases than controls | 74 |
| Colorectal | Healthy persons | FBA similar in cases than controls | 75 |

## Table 11
## RELATIONSHIP BETWEEN FECAL BILE ACID (FBA) AND NEUTRAL STEROID (FNS) CONCENTRATION AND ADENOMA CARRIAGE

| | Number of patients | Fecal steroid concentration (mg/g dry weight) | |
|---|---|---|---|
| | | FBA | FNS |
| Patients with no adenomas | 100 | 8.7 ± 2.5 | 16.2 ± 6.4 |
| Patients with adenomas | 133 | 8.8 ± 2.2 | 14.1 ± 4.4 |
| 1 adenoma | 57 | 8.3 ± 2.2 | 12.0 ± 4.8 |
| 2 adenomas | 32 | 9.4 ± 2.5 | 16.4 ± 7.4 |
| 3 or 4 adenomas | 21 | 8.8 ± 2.0 | 10.6 ± 2.7 |
| More than 4 adenomas | 20 | 9.4 ± 3.4 | 15.5 ± 5.5 |

The other major interest in refining the bile acid concept of colorectal carcinogenesis has been to determine the stage at which it acts. Colorectal carcinogenesis is a multistage phenomenon,[35-37] in which the first stage is adenoma formation; the second stage is adenoma growth from a small size (with a low malignant potential) to a large size (with a high risk of malignancy); the third stage is the increase in the severity of the dysplasia (an adenoma is, by definition, an area of dysplasia) from the initial mild grade to severe and then finally to carcinoma. Studies by Hill et al.[38] of 133 patients with adenomas compared with healthy controls showed that

1. There was no relation between FBA concentration and the number of adenomas formed (Table 11) suggesting that bile acids play no role in adenogenesis
2. There was a close correlation between FBA concentration and adenoma size (Table 12) ($p < 0.002$)
3. There was a relation between FBA concentration and the severity of epithelial dysplasia ($p < 0.05$) (Table 13)

**Table 12**
**THE RELATION BETWEEN FECAL BILE ACID (FBA) CONCENTRATION AND MEAN DIAMETER OF THE LARGEST ADENOMA IN 133 PATIENTS**

| Mean diameter of the largest adenoma (mm) | Number of patients | Fecal steroid concentration (mg/g dry weight) | |
|---|---|---|---|
| | | FBA | FNS |
| 0—5 | 26 | 6.7 ± 1.2 | 14.8 ± 4.3 |
| 5—10 | 30 | 7.9 ± 1.5 | 9.5 ± 3.6 |
| 10—20 | 45 | 8.0 ± 2.6 | 13.0 ± 9.4 |
| More than 20 | 32 | 11.4 ± 2.4 | 15.0 ± 2.2 |

**Table 13**
**THE RELATION BETWEEN FECAL BILE ACID (FBA) CONCENTRATION AND THE SEVERITY OF EPITHELIAL DYSPLASIA IN THE MOST SEVERELY DYSPLASTIC ADENOMA IN 123 PATIENTS**

| | Severity of epithelial dysplasia | |
|---|---|---|
| | Mild | Moderate or severe |
| Number of patients | 84 | 39 |
| Mean FBA concentration (mg/g dry weight) | 8.3 ± 2.4 | 10.0 ± 2.1 |
| Mean FNS concentration | 12.9 ± 4.1 | 17.7 ± 7.5 |

In a further study of 110 patients with ulcerative colitis of more than 10 years duration at the beginning of the investigation for 10 years, the mean FBA concentration in the 43 patients with epithelial dysplasia was similar to that of the 67 who did not develop dysplasia (Table 14) but there was a correlation between FBA concentration and the severity of epithelial dysplasia. These two studies suggest that the FBA concentration is unimportant in colorectal carcinogenesis until dysplasia has been initiated but is very important in the subsequent stages.

*2. Gastric Cancer*

There is a histopathological sequence in the causation of gastric cancer in which the first stage is gastric atrophy and the consequent loss of the gastric acid and the ability to sterilize the gastric lumen; the second stage is atrophic gastritis then intestinal metaplasia followed by increasingly severe epithelial dysplasia and finally carcinoma. The reflux of bile into the atrophied stomach has been well known for many years and Domellof[39] suggested that the refluxed bile acids might be the cause of atrophic gastritis and there have been many studies showing that free bile acids, particularly deoxycholic acid, are toxic to the gastric mucosa.[40,41] More recent studies have concentrated on the relation between refluxed bile acids and epithelial dysplasia particularly in patients with Polya partial gastrectomy. Mortensen et al.[42] showed a good correlation between refluxed bile acids and the severity of epithelial dysplasia in these patients (a result which has been confirmed by Hall et al.[20]) but not with atrophic

**Table 14**
**FECAL BILE ACID (FBA) CONCENTRATION IN PATIENTS WHO HAVE HAD ULCERATIVE COLITIS WITH TOTAL COLONIC INVOLVEMENT FOR MORE THAN 10 YEARS**

| Patients | Number | FBA concentration (mg/g dry weight) |
|---|---|---|
| Controls | 100 | 8.7 ± 2.5 |
| Ulcerative colitis patients | 110 | 7.8 |
| Those with no dysplasia | 67 | 7.2 |
| Those with dysplasia | 43 | 8.7 |
| Mild | 22 | 7.4 |
| Moderate | 8 | 9.5 |
| Severe | 13 | 10.5 |

gastritis. Keighley et al.[21] studied patients with idiopathic achlorhydria as a result of various types of vagotomy and found only small amounts of refluxed bile which did not correlate with histopathological abnormalities. Similarly Hall et al.[20] also reported little bile reflux in pernicious anemia patients and no correlation with gastritis, intestinalization, or dysplasia. In summary, although refluxed bile acids may cause gastritis and dysplasia, the amount of refluxed bile is significant only in patients with partial gastric resection.

## IV. *N*-NITROSO COMPOUNDS

The in vivo formation of *N*-nitroso compounds has been claimed in gastric luminal contents, the infected urinary bladder, the urine of patients with urine diversion, and in the cervix of women with *Trichomonas vaginalis* infection, and has been incriminated in the local cancer associated with those situations.

### A. Gastric Cancer

The histopathological sequence in gastric carcinogenesis postulated by Correa et al.[43] has gastric atrophy (associated gastric bacterial colonization) as its first stage. On this hypothesis, all causes of gastric atrophy (e.g., pernicious anemia, simple aging, partial gastrectomy, vagotomy, etc.) should be associated with a high risk of gastric cancer and this is so. There have been many studies investigating the hypothesis that the progression from gastric atrophy to gastric cancer is mediated by the formation of *N*-nitroso compounds from nitrite (produced by bacterial reduction of nitrate). In most studies the presence of a bacterial flora and the nitrite concentration were assayed (Table 15). *N*-nitrosation in gastric juice was demonstrated in vitro at neutral pH in the presence of bacteria and nitrate.[44] All of these studies have shown an association between the gastric bacterial flora, nitrite concentration, and the histopathology of the stomach. Only three groups have attempted to assay *N*-nitroso compounds in gastric juice (Walters et al.,[18,47] in association with various clinical groups, Bavin et al. in association with various clinical groups and Schlag et al.[45]); Schlag et al.[45] showed a strong association between the *N*-nitroso compound concentration and the histopathological stage of the gastric mucosa, but the results have been strongly criticized.[46] Walters et al.[18,47] have shown a strong positive correlation between *N*-nitroso compound concentration and gastric pH strongly supporting a role for *N*-nitroso compounds in gastric carcinogenesis. In direct contrast, Bavin et al.[20,21] suggested no role for *N*-nitroso compounds in gastric carcinogenesis. This conflict has no obvious explanation and, until it has been resolved, judgment must be suspended on this matter.

### Table 15
### NITRITE AND *N*-NITROSO COMPOUNDS IN THE PROGRESSION FROM GASTRIC ATROPHY TO GASTRIC CANCER

| Study group | Observation on gastric juice | Ref. |
|---|---|---|
| PA and PG patients and controls | Nitrate-reducing activity of the gastric flora increased with pH | 19—21, 22, 44, 45 |
| PA and HGG | As above | 76 |
| PA, PG, GC and controls | Nitrite concentration is high in patients at high risk of GC and low in controls | 16, 17, 19, 20, 21, 45, 76, 77 |
| PA, PG patients and controls | *N*-nitroso compounds concentration is high in patients at high risk of GC and low in controls | 18, 19, 45 |

*Note:* PA = pernicious anemia; PG = Polya partial gastrectomy; GC = gastric carcinoma; and HGG = hypogammaglobulinemia.

## B. Bladder Cancer

Hicks et al.[48] studied the possible role of *N*-nitroso compounds in the causation of the bladder cancer associated with bilharzia infection. Bilharzial infection of the urinary bladder was shown to be associated with bacterial overgrowth; the urinary flora was mixed containing a large number of species of both aerobic and anaerobic bacteria. In such patients the urinary nitrite concentration was not extraordinarily high but there was a very high urinary *N*-nitroso compound concentration.

It was suggested that the *N*-nitroso compounds formed at the result of the metabolic action of the secondary bacterial flora acted as initiators of bladder carcinogenesis and that the primary bilharzial infection acted as a promoter of tumorogenesis.

## C. Cancer in Urine Diversion

The three major types of operation for diverting the urine following major bladder surgery are ureterocolic anastomosis, ileal conduit, and colonic conduit. In ureterocolic anastomosis the urine (rich in nitrosatable amines and in nitrate) is conducted into the colon (rich in bacteria) providing conditions which might be considered ideal for bacterial *N*-nitrosation. The flora of the urine of such patients[49] contains a wide range of bacterial genera and species, both anaerobic and aerobic (Table 16); the flora in patients with an ileal conduit was also found to be complex in a high proportion of patients while that in patients with a colonic conduit was comparatively simple.

The risk of colon cancer in patients with ureterocolic anastomosis is extremely high (the estimated excess over the normal population being estimated as 200-fold) there was a latency of more than 20 years; the subject has been summarized by Stewart et al.[50] The first ileal conduit was performed in 1951 and the operation rapidly gained popularity; the first report of a carcinoma in the isolated ileal segment was in 1977[51] and a number of further cases have been reported. There have also been a few isolated reports of cancers in isolated colonic loops. A role for *N*-nitroso compounds in the colonic cancer associated with ureterocolic anastomosis was postulated by Stewart et al.[50]

## D. Cervical Cancer

There is a high incidence of cervical cancer in the black populations of southern Africa and this has been associated with infection with *Trichomonas vaginalis*.[52] The presence of *N*-nitroso compounds in cervical exudate of cervical cancer patients was detected by

### Table 16
### THE BACTERIAL FLORA OF PATIENTS WITH URINE DIVERSION[49]

| Organisms isolated | Proportion of patients carrying the given bacterial genus in the various urinary diversions: Ureterocolic anastomosis | Ileal conduit | Colonic conduit |
|---|---|---|---|
| Anaerobes | | | |
| *Bacteroides* spp. | 21/26 | 4/15 | 0/16 |
| *Bifidobacterium* spp. | 3/26 | 0/15 | 0/16 |
| *Clostridium* spp. | 19/26 | 2/15 | 0/16 |
| *Veillonella* spp. | 7/26 | 1/15 | 0/16 |
| Aerotolerant | | | |
| *Enterobacteria* | 20/26 | 7/15 | 4/16 |
| *Streptococci* | 25/26 | 9/15 | 2/16 |
| *Lactobacilli* | 10/26 | 2/15 | 0/16 |
| Proportion of urines infected | 100% | 73% | 38% |
| 1 Genus only | 0% | 20% | 38% |
| More than 3 genera | 88% | 27% | 0% |

### Table 17
### EXCESS RISK OF CANCERS OF THE COLON AND BILIARY TRACT IN PATIENTS WITH POLYA PARTIAL GASTRECTOMY AND PERNICIOUS ANEMIA

| | Pernicious anemia | | | Polya partial gastrectomy | | |
|---|---|---|---|---|---|---|
| | Expected deaths | Observed deaths | SMR | Expected deaths | Observed deaths | SMR |
| Gastric cancer | 12 | 3.2 | 375 | 13 | 5.2 | 251 |
| Colorectum | 7 | 3.6 | 194 | 10 | 7.4 | 135 |
| Biliary tract | 0 | 0.1 | | 3 | 0.3 | 937 |

Harrington et al.[53] but this suggestion has received little attention and the observation has still to be confirmed.

## E. Other Cancers

If *N*-nitroso compounds formed in, for example, the achlorhydric stomach are responsible for the gastric cancer associated with that condition, it is likely that a number of *N*-nitroso compounds will be formed and that not all of them will be locally acting carcinogens. Cancers at distant sites associated with gastric achlorhydria would also be expected therefore. In a study of patients with Polya partial gastrectomy and with pernicious anemia[54] preliminary results indicate that the disease is associated with an excess risk of colorectal cancer and of biliary tract cancer (Table 17). The latency of the cancers in both sites was very similar to that of gastric cancer in those patients suggesting a common etiology. This observation has to be confirmed, but offers support for the hypothesis that *N*-nitroso compounds formed in the achlorhydric stomach are responsible for the causation of local tumors and of tumors in the large bowel, biliary tract, and possibly in other sites.

## REFERENCES


1. **Kinoshita, N. and Gelboin, H. V.**, β-Glucuronidase catalysed hydrolipsis of benzo[a]pyrene-3-glucuronide and binding to DNA, *Science*, 199, 307, 1978.
2. **Farber, E.**, Ethiomine carcinogenesis, *Adv. Cancer Res.*, 7, 383, 1963.
3. **Cummings, J. H., Hill, M. J., Bone, E. S., Branch, W. J., and Jenkins, D. J.**, The effect of meat protein with and without dietary fibre on colonic function and metabolism, *Am. J. Clin. Nutr.*, 32, 2094, 1979.
4. **Boutwell, R. K. and Bosch, D. K.**, The tumour promoting action of phenol and related compounds on the mouse skin, *Cancer Res.*, 19, 413, 1959.
5. **Bryan, G. T.**, The role of urinary tryptophan metabolites in the etiology of bladder cancer, *Am. J. Clin. Nutr.*, 24, 841, 1971.
6. **Price, J. M.**, Bladder cancer, *Can. Cancer Conf.*, 6, 224, 1966.
7. **Bruce, W. R., Varghese, A. J., Furrer, R., and Land, P. C.**, A mutagen in the faeces of normal humans, in *Origins of Human Cancer*, Hiatt, H., Watson, J., and Winsten, J., Eds., Cold Spring Harbor Lab Press, Cold Spring Harbor, New York, 1977, 1641.
8. **Bruce, W. R., Varghese, A. J., Wang, S., and Dion, P.**, The endogenous production of nitroso compounds in the colon and cancer at that site, in *Naturally Occurring Carcinogens — Mutagens and Modulators of Carcinogenesis*, Miller, E. C. et al., Eds., Japanese Science Society Press, Tokyo, 1979, 221.
9. **Hirai, N., Kingston, D. G., Tassell, R. L. V., and Wilkins, T. D.**, Structure elucidation of a potent mutagen from human faeces, *J. Am. Chem. Soc.*, 104, 6149, 1982.
10. **Gupta, I., Baptista, J., Bruce, W. R., et al.**, Structures of faecal pentaenes, the mutagens of bacterial origin isolated from human faeces, *Biochemistry*, 22, 241, 1983.
11. **Venitt, S.**, Faecal mutagens in the acticology of colon cancer, in *Colonic Carcinogenesis*, Malt, R. and Williamson, R., Eds., MTP Press, Lancaster, England, 1982, 59.
12. **Silverman, S. J. and Andrews, A. W.**, Bile acids: co-mutagenic activity in the Salmonella-mammalian-microsome mutagenicity test, *J. Natl. Cancer Inst.*, 59, 1557, 1977.
13. **Wilpart, M., Mainguet, P., Maskens, A., and Roberfroid, M.**, Mutagenicity of 1.2-dimethylhydrazine towards *Salmonella typhimurium*, co-mutagenic effects of secondary bile acids, *Carcinogenesis*, 4, 45, 1983.
14. **Wilpart, M., Mainguet, P., Maskens, A., and Roberfroid, M.**, Structure-activity relationship amongst biliary bile acids showing co-mutagenic activity towards dimethylhydrazine, *Carcinogenesis*, 4, 1239, 1983.
15. **Leach, S. A., Challis, B., Cook. A. R., Hill, M. J., and Thompson, M. H.**, The bacterial catalysis of the N-nitrosation of secondary amines, *Trans. Biochem. Soc.*, 13, 380, 1985.
16. **Ruddell, W. S. J., Bone, E. S., Hill, M. J., Blendis, L. M., and Walters, C. L.**, Gastric juice nitrite: a risk factor for cancer of the hypsochlorhydric stomach, *Lancet*, ii, 1037, 1976.
17. **Ruddell, W. S. J., Bone, E. S., Hill, M. J., and Walters, C. L.**, Pathogenesis of gastric cancer in pernicious anaemia, *Lancet*, i, 521, 1978.
18. **Reed, P., Haines, R., Smith, P. L., House, F. R., and Walters, C. L.**, Gastric juice N-nitrosaminos in health and gastroduodenal disease, *Lancet*, ii, 550, 1981.
19. **Bartholomew, B. A., Hill, M. J., Hudson, M. J., Ruddell, W. S. J., and Walters, C. L.**, Gastric bacteria, nitrate, nitrite, and nitrosamines in patients with pernicious anemia and in patients treated with cimetidine, in *N-Nitroso Compounds: Analysis, Formation and Occurrence*, Walker, E., Griciute, L., Castagnero, M., and Barszonyi, M., Eds., International Agency Research on Cancer, Lyon, 1979, 595.
20. **Hall, C. N., Cook, A., Darkin, D., Viney, N., et al.**, Evaluation of the nitrosamine hypothesis of gastric carcinogenesis in man, *Clin. Sci.*, 66, 34, 1984.
21. **Keighley, M. R. B., Youngs, D., Poxon, V., et al.**, Intragastric N-nitrosation is unlikely to be responsible for gastric carcinoma developing after operations for duodenal ulcer, *Gut*, 25, 238, 1984.
22. **Watt, P. C. H., Sloan, J. M., Donaldson, J. D., Patterson, C. C., and Kennedy, T. L.**, Relationships between histology and gastric juice pH and nitrite in the stomach after operation for duodenal ulcer, *Gut*, 25, 246, 1984.
23. **Renwick, A. G. and Drasar, B. S.**, Environmental carcinogens and large bowel cancer, *Nature (London)*, 263, 234, 1976.
24. **Teulings, F. A. G., Peters, H. A., Hop, W. C., et al.**, A new aspect of the urinary excretion of tryptophan metabolites in patients with cancer of the bladder, *Int. J. Cancer*, 21, 140, 1978.
25. **Lowenfels, A.**, Does bile promote extracolonic cancer?, *Lancet*, ii, 239, 1978.
26. **Cassidy, M. M., Lightfoot, F. G., and Vahouny, G. V.**, Dietary fibre, bile acids and intestinal morphology, in *Dietary Fiber in Health and Disease*, Vahouny, G. V. and Kmtchevsky, D., Eds., Plenum Press, New York, 1982, 239.
27. **Wargovich, M. J., Eng, V. W., Newmark, H. L., and Bruce, W. R.**, Calcium ameliorates, the toxic effect of deoxycholic acid on colonic epithelium, *Carcinogenesis*, 4, 12, 1983.

28. **Van der Werf, S. D. J., Nagengast, F. M., Berge-Henegouwen, G. P., et al.,** Intracolonic environment and the presence of colonic adenomas in man, *Gut,* 24, 876, 1983.
29. **Summerton, J., Flynn, M., Cooke, T., and Taylor, I.,** Bile acid receptors in colorectal cancer, *Br. J. Surg.,* 70, 549, 1983.
30. **Owen, R. W., Dodo, M., Thompson, M. H., and Hill, M. J.,** The faecal ratio of lithocholic to deoxycholic acid may be an important actiological factor in colorectal cancer, *Eur. J. Cancer Clin. Oncol.,* 19, 307, 1983.
31. **Owen, R. W., Dodo, M., Thompson, M. H., and Hill, M. J.,** The faecal ratio of lithocholic to deoxycholic acid and large bowel cancer, *Trans. Biochem. Soc.,* 12, 861, 1984.
32. **Narisawa, T., Magadia, N. E., and Wynder, E. L.,** Promoting effect of bile acids on colon carcinogenesis after intrarectal instillation of N-methyl-N-nitroso-N-nitroguanidine in rats, *Cancer Res.,* 53, 1093, 1974.
33. **Wait, R. and Thompson, M. H.,** The role of minor faecal bile acids in colorectal cancer, *Trans. Biochem. Soc.,* 12, 113, 1984.
34. **Wait, R., Hill, M. J., and Thompson, M. H.,** The role of minor faecal bile acids in the actiology of colon cancer, *Eur. J. Cancer Clin. Oncol.,* 19, 1323, 1983.
35. **Morson, B. C.,** The polyp-cancer sequence in the large bowel, *Proc. R. Soc. Med.,* 67, 451, 1974.
36. **Hill, M. J., Morson, B., and Bussey, H. J.,** Actiology of the adenoma-carcinoma sequence in large bowel, *Lancet,* ii, 535, 1978.
37. **Morson, B. C., Bussey, H. J., Day, D. W., and Hill, M. J.,** Adenomas of large bowel, *Cancer Surveys,* 2, 451, 1983.
38. **Hill, M. J., Morson, B. C., and Thompson, M. H.,** The role of faecal bile acids in large bowel carcinogenesis, *Br. J. Cancer,* 48, 143, 1983.
39. **Domellof, L.,** Gastric carcinoma promoted by alkaline reflex gastritis with special reference to bile and other surfactants as promoters of post-operative gastric cancer, *Med. Hypothesis,* 5, 463, 1979.
40. **Black, R. B., Roberts, G., and Rhodes, J.,** The effect of healing on bile reflux in gastric ulcer, *Gut,* 12, 552, 1971.
41. **Rhodes, R. B., Barnard, D. E., Phillips, S. F., Ravelstad, R. A., and Hofmann, A. F.,** Increased reflux of bile into the stomach in patients with gastric ulcer, *Gastroenterology,* 57, 241, 1969.
42. **Mortensen, N. J., Thomas, W. E., Cooper, M. J., et al.,** Relationship of duodenogastric reflux, gastric pH and nitrite levels to histological changes in the gastric mucosa, *Br. J. Surg.,* 1984, in press.
43. **Correa, P., Haenszel, W., Cuello, C., Archer, M., and Tannenbaum, S.,** A model for gastric cancer epidemiology, *Lancet,* ii, 58, 1975.
44. **Walters, C. L., Hill, M. J., and Ruddell, W. S. J.,** Gastric juice nitrite: its source and relationship to hydrogen ion concentration, in *Environmental Aspects of N-Nitroso Compounds,* Walker, E., Castognero, M., Griciute, L., and Lyle, R., Eds., International Agency Research on Cancer, Lyon, 1978, 279.
45. **Schlag, P., Bockler, R., Ulrich, H., et al.,** Are nitrate and N-nitroso compounds in gastric juice risk factors for carcinomas in the operated stomach?, *Lancet,* i, 727, 1980.
46. **Ruddell, W. S. J. and Walters, C. L.,** Nitrate and N-nitroso compounds in gastric juice, *Lancet,* i, 1187, 1980.
47. **Reed, P. I., Smith, P. L. R., Haines, K., House, F. R., and Walters, C. L.,** Effect of cimetidine on gastric juice N-nitrosamine concentration, *Lancet,* ii, 553, 1981.
48. **Hicks, R. M., Walters, C. L., Elsebai, I., et al.,** Demonstration of nitrosamines in human urine: preliminary observations on a possible etiology for bladder cancer in association with chronic urinary tract infections, *Proc. R. Soc. Med.,* 70, 413, 1977.
49. **Hill, M. J., Hudson, M. J., and Stewart, M.,** The urinary bacterial flora in patients with three types of urinary tract diversion, *J. Med. Microbiol.,* 16, 221, 1983.
50. **Stewart, M., Hill, M. J., Pugh, R. C. B., and Williams, J. P.,** The role of N-nitrosamine in carcinogenesis at the uretero-colic anastomosis, *Br. J. Urol.,* 53, 115, 1981.
51. **Shousha, S., Scott, J., and Polak, J.,** Heal loop carcinoma after cystectomy for bladder exstophy, *Br. Med. J.,* 2, 397, 1978.
52. **Allsobrook, A. J., Du Plessis, L. S., Harrington, J. S., Nunn, A. J., and Nunn, J. R.,** Nitrosamines in the human vaginal vault, in *N-Nitroso Compounds in the Environment,* Bogovski, P. and Walker, E., Eds., International Agency Research for Cancer, Lyon, 1974, 197.
53. **Harrington, J., Nunn, J. R., and Irwig, L.,** Dimethylnitrosamine in the human vaginal vault, *Nature (London),* 241, 49, 1973.
54. **Caygill, C., Hall, R., and Hill, M. J.,** Association between biliary tract cancer and gastric cancer, *Lancet,* ii, 1204, 1983.
55. **Bone, E. S., Tamm, A., and Hill, M.,** The production of urinary phenols by gut bacteria and their possible role in the causation of large bowel cancer, *Am. J. Clin. Nutr.,* 29, 1448, 1976.
56. **Hill, M. J., Taylor, A. J., Thompson, M. H., and Wait, R.,** Fecal steroids and urinary volatile phenols in four Scandinavian populations, *Nutr. Cancer,* 4, 67, 1982.

57. **Schmidt, E. G.,** Urinary phenols. IV. The simultaneous determination of phenol and p-cresol in urine, *J. Biol. Chem.,*, 179, 211, 1949.
58. **Ederman, M., Tressell, R. L., West, S. E., Ehrich, M. F., and Wilkins, T. D.,** *In vitro* production of human faecal mutagen, *Mutat. Res.*, 79, 115, 1980.
59. **Sander, J.,** Nitrosaminsynthese durch Bakterien, *Z. Physiol. Chem.*, 349, 429, 1968.
60. **Hawksworth, G. M. and Hill, M. J.,** Bacteria of N-nitrosation of secondary amines, *Br. J. Cancer,* 25, 520, 1971.
61. **Brooks, J. B., Cherry, W. B., Thacker, L., and Alley, C. C.,** Analysis by gas chromatography of amines and nitrosamines produced *in vivo* and *in vitro* by *Proteus mirabilis, J. Infect. Dis.*, 126, 143, 1972.
62. **Hawksworth, G. M. and Hill, M. J.,** *In vivo* formation of N-nitrosamines in the rat bladder and their subsequent absorption, *Br. J. Cancer,* 29, 353, 1974.
63. **Coloe, P. S. and Heywood, N. J.,** The importance of prolonged incubation for the synthesis of dimethylnitrosamine by enterobacteria, *J. Med. Microbiol.*, 9, 211, 1976.
64. **Kunisaki, N. and Hayashi, M.,** Formation of N-nitrosamines from secondary amines and nitrite by resting cells of *Esch. coli* B, *Appl. Env. Microbiol.*, 37, 279, 1979.
65. **Radomski, J. L., Greenwald, D., Hearn, W. L., Block, N. L., and Woods, F. M.,** Nitrosamine formation in bladder infections and its role in the etiology of bladder cancer, *J. Urol.*, 120, 48, 1978.
66. **Ehrich, M., Aswell, J., Tassell, R. V., et al.,** Mutagens in the faeces of 3 South African populations at different levels of risk for colon cancer, *Mutat. Res.*, 64, 231, 1979.
67. **Reddy, B. S., Sharma, C., Darby, L., et al.,** Metabolic epidemiology of large bowel cancer. Fecal mutagens in high and low risk populations for colon cancer, *Mutat. Res.*, 72, 511, 1980.
68. **Kuhnlein, U., Bergstrom, D., and Kuhnlein, H.,** Mutagens in faeces from vegetarians and non-vegetarians, *Mutat. Res.*, 85, 1, 1981.
69. **Hill, M. J.,** Metabolic epidemiology of large bowel cancer, in *Gastrointestinal Cancer,* De Cosse, J. and Sherlock, P., Eds., Martinus Nijhoff, The Hague, 1981, 187.
70. **Hill, M. J.,** The etiology of colorectal cancer, in *Recent Advances in Gastrointestinal Pathology,* Wright, R., Ed., W. B. Saunders, Philadelphia, 1980, 297.
71. **Hill, M. J., Drasar, B. S., Williams, R. E. D., et al.,** Faecal bile acids and clostridia in patients with cancer of the large bowel, *Lancet,* i, 535, 1975.
72. **Reddy, B. S. and Wynder, E. L.,** Metabolic epidemiology of colon cancer. Faecal bile acids and neutral sterils in colon cancer patients and patients with adenomatous polyps, *Cancer,* 39, 2533, 1977.
73. **Blackwood, A., Murray, W. R., Mackay, C., and Carman, K.,** Faecal bile acids and clostridia in the etiology of colorectal cancer and breast cancer, *Br. J. Cancer,* 38, 175, 1978.
74. **Mudd, D. G., McKelvey, S. T., Sloan, J. M., and Elmore, D. T.,** Faecal bile acid concentrations in patients at increased risk of large bowel cancer, *Acton. Gastro. Belg.*, 41, 241, 1978.
75. **Kaibara, N., Sasaki, T., Ikeguchi, M., Koga, S., and Ikawa, S.,** Fecal bile acids and neutral sterols in Japanese with large bowel carcinoma, *Oncology,* 40, 255, 1983.
76. **Dolby, J., Webster, A. D., Borriello, S. P., et al.,** Bacterial colonisation and nitrite concentration in the achlorhydric stomachs of patients with primary hypogammaglobulinemia or classical pernicious anemia, *Scand. J. Gastroenterol.*, 19, 105, 1984.
77. **Caygill, C., Hill, M., Craven, J., Hall, R., and Miller, C.,** Relevance of gastric achlorhydria to human carcinogenesis, in N-Nitroso Compounds in the Environment, IARC Publ. 57, 1984, 895.

# INDEX

## A

## C

## D

## E

## G



## H

## I



## J

## K



## L

## M



## N

## O



## P

## Q



## R

# S

## T

U



V



W



X



Y